AF572419

Crude Oil Chemistry

Crude Oil Chemistry

Vasily Simanzhenkov

University Duisburg-Essen
Duisburg, Germany

Raphael Idem

University of Regina
Regina, Saskatchewan, Canada

MARCEL DEKKER, INC. NEW YORK • BASEL

Library of Congress Cataloging-in-Publication Data
A catalog record for this book is available from the Library of Congress.

ISBN: 0-8247-4098-X

This book is printed on acid-free paper.

Headquarters
Marcel Dekker, Inc., 270 Madison Avenue, New York, NY 10016, U.S.A.
tel: 212-696-9000; fax: 212-685-4540

Distribution and Customer Service
Marcel Dekker, Inc., Cimarron Road, Monticello, New York 12701, U.S.A.
tel: 800-228-1160; fax: 845-796-1772

Eastern Hemisphere Distribution
Marcel Dekker AG, Hutgasse 4, Postfach 812, CH-4001 Basel, Switzerland
tel: 41-61-260-6300; fax: 41-61-260-6333

World Wide Web
http://www.dekker.com

The publisher offers discounts on this book when ordered in bulk quantities. For more information, write to Special Sales/Professional Marketing at the headquarters address above.

Current printing (last digit):

10 9 8 7 6 5 4 3 2 1

PRINTED IN THE UNITED STATES OF AMERICA

Preface

This book is devoted to students as well as scientists and process engineers involved in petroleum science, refining and engineering. Part I of the book gives a historical overview of the origin of petroleum. The first chapter shows how crude oil is linked with human civilization. In this chapter it is also shown that the energy used to run most of today's machinery derives from petroleum. It also provokes our imagination on how the various moving parts of machinery would operate without mineral oil or lubricating oil, both of which are also produced from crude oil. The first chapter also provides insight into the problems that have arisen as a result of applying different standards by different countries to similar crude oil products and how these problems are tackled.

The second chapter of Part I deals with modern analytical methods used in crude oil chemistry. Modern and classical methods of petroleum and petroleum product characterization are explained. This chapter is an essential chapter for present and potential crude oil chemists since analytical chemistry constitutes an important part of crude oil chemistry. Besides, crude oil products have so many special properties that are important for the industry. The need to determine these properties gives rise to the use of very many analytical methods in the petroleum industry.

Part II (i.e. chapters 3 and 4) shows the geopolitical and economic nature of petroleum chemistry. In this chapter, the initial stages of development of various petroleum companies are shown. It gives a historical run-down on how small companies of the past have blossomed into giant multinational companies of worldwide repute. For readers, it is especially interesting to learn the history of the development of the Eastern Bloc companies, especially the Russian companies. In this chapter, it is shown how the political situation in this country has had a great influence on the economic development of all Russian industries in general and the petroleum industry in particular. It is also shown how it has been possible for the big Russian petroleum concerns to be formed in less than twenty years during the difficult times of political and economic reforms.

Chapters 5 and 6 of Part III of the book introduce the reader to the science of crude oil refining. An illustration of the complete process scheme that starts from crude oil all the way to final products is given. In particular, the process route of crude oil from the well to the gas station and indeed the car tank is provided. These chapters also present the techniques and technologies involved in most of the important processes used in modern petroleum refineries for processing light and heavy distillate fractions. Chapter 7 in Part III looks at ecological problems that arise in the crude oil industry. The chapter also shows that, in order to develop an appropriate technology for crude oil treatment, not only are economically rational decisions required, but also, ecologically acceptably decisions are needed. This chapter looks at our environment as a very sensitive system that must be protected with respect to the many processes that take place in the petroleum refining industry.

Good quality crude oil is often defined in terms of high API gravity and low sulfur content. However, reserves of this type of oil are disappearing, giving rise to increasing contributions from oil of lower API gravity and higher sulfur content. Is the chemistry of processing of the two types of oil different? This question is answered in Part IV of this book. The chapter also introduces the presence of asphaltenes in crude oil. A list of problems that occur during oil production, transportation, and processing that could be attributed to asphaltene presence in crude oil is presented. New concepts and approaches that aid in the processing of crude oils with significant amounts of asphaltenes are introduced. Part IV also looks at future processes that may be introduced in the petroleum refinery. These are hybrid fuel production processes that involve combining the well-known modern cracking process with the recycling of waste plastics or bio-fuels. The advantages than can be derived from co-processing of asphaltene-containing crude oils with plastics or biomass materials are given. Part IV also presents some analytical techniques that could be used by petroleum engineers and chemists to better understand the nature of heavy crude oil and residues, and possible ways to positively influence their processing. In all, Part IV presents critical material that can contribute towards further development of the petroleum industry. This is especially true for the non-conventional petroleum industry, and it can be particularly important for North America, since Canada has the largest reserves of non-conventional petroleum in the world.

Both authors have worked extensively in the areas of petroleum science and engineering. We hope that this book will go a long way in introducing the reader to the fascinating world and concepts of the black gold of our planet.

Vasily Simanzhenkov
Raphael Idem

Contents

Preface		*iii*
Part I	Classification and Characterization of Crude Oil	1
1	Nature and Classification of Crude Oil	3
1.1	History and nature of petroleum	3
1.1.1	Origin of petroleum	13
1.1.2	Oil formation in the world's oceans	16
1.1.3	Modern concept of formation of petroleum	18
1.1.4	Consequences of intensive extraction and processing of oil	26
1.1.4.1	Dangerous fogs	28
1.1.4.2	Black oceans	31
1.2	General properties and classification of petroleum: comparison of petroleum from different countries	33
1.2.1	Fractions and chemical composition of petroleum	33
1.2.2	Chemical classification of petroleum	35
1.2.3	Classification by density	37
1.2.4	Characterization by viscosity-gravity constant (vgc)	38
1.2.5	Technological classification of petroleum	39
1.3	Products from crude oil	39
1.3.1	Gasoline	40
1.3.2	Jet fuel (kerosene)	48
1.3.3	Diesel	49
1.3.4	Residual fuel	55
1.3.5	International standards for fuels	58
1.4	Lubricating oils and lubricants	59
1.4.1	International standards for lubricating oils	66
1.4.1.1	Industrial oils	66
1.4.1.2	Motor oils	67
1.4.1.3	Transmission oils	70
1.4.1.4	Hydraulic oils	72
2	Modern Characterization and Analysis Techniques for Crude Oil	73
2.1	Chromatographic methods	73
2.1.1	Gas chromatography	74
2.1.2	High performance liquid chromatography	97
2.1.3	Thin layer chromatography	107
2.2	Spectroscopic methods	112
2.2.1	Infrared spectroscopy	112

2.2.2	Raman spectroscopy	127
2.2.3	Colorimetry and photometry	131
2.2.4	Fluorescence and phosphorescence spectroscopy	136
2.2.5	Atomic absorption spectroscopy and atomic emission spectroscopy	139
2.2.6	X-ray fluorescence spectroscopy	144
2.3	Other methods for elucidating the structure of crude oil	146
2.3.1	Separation methods	146
2.3.2	Chemical analysis methods of crude oil products: determination of unsaturated compounds	149
2.3.3	Structural bulk analysis of heavy crude oil fractions: n-d-M method	150
2.4	Methods of characterization of colloidal properties of crude oil and its products	153
2.4.1	Direct methods	153
2.4.2	Indirect methods	156
2.5	Determination of the physical properties of crude oil	157
2.5.1	Density determination	157
2.5.2	Viscosity determination	159
2.5.3	Refractive index determination	161
	Bibliography	163
Part II	Regional Petroleum Industry	171
3	Petroleum Producing Countries: OPEC and Non-OPEC	173
3.1	Introduction	173
3.1.1	Short background on OPEC	174
3.2	North America	175
3.2.1	United States	175
3.2.2	Canada	178
3.3	Russia	178
3.3.1	The role of the petroleum industry for Russia	179
3.3.2	Reforms in the Russian oil industry	180
3.3.3	Russian petroleum and gas in the world market	182
3.3.4	Structure of the petroleum sector in Russia	183
3.4	Arabian East	185
3.4.1	Oman	187
3.4.2	Iraq	187
3.4.3	Iran	188
3.4.4	Qatar	190
3.4.5	Kuwait	191
3.4.6	United Arab Emirates	194
3.4.7	Saudi Arabia	194

4	International Petroleum Companies	197
4.1	British Petroleum	197
4.2	Castrol	200
4.3	ExxonMobil	201
4.4	Neste/Fortum	205
4.5	Shell	206
4.6	Total / Fina / Elf	208
4.7	LUKOil	214
4.8	Yukos	215
4.9	TNK	216
	Bibliography	217
Part III	Main Processes in the Petroleum Refining Industry	219
5	Crude Oil Distillation	221
5.1	Petroleum and gas preparation	221
5.1.1	Formation of petroleum emulsions and their basic properties	222
5.1.2	Separation of water–oil emulsions	224
5.1.3	Mechanical petroleum drying	227
5.1.4	Thermal petroleum drying	227
5.1.5	Chemical methods of petroleum drying	228
5.1.6	Thermal chemical petroleum drying	229
5.1.7	Stabilization of petroleum	230
5.1.8	Technological schemes for petroleum preparation	231
5.1.9	Pressure extraction system	231
5.2	Desalting	234
5.3	Atmospheric rectification	235
5.4	Vacuum rectification	251
5.5	Heat exchangers and separators	256
6	Processing of Light and Heavy Distillates	261
6.1	Thermal cracking	261
6.2	Catalytic cracking	275
6.3	Visbreaking	285
6.4	Coking	285
6.5	Hydroprocessing	287
6.6	Reforming	292
6.7	Isomerization	296
6.8	Alkylation	298
6.9	Blending	301
7	Environmental Issues Facing the Refining Industry	303
7.1	Introduction	303
7.2	Methods of cleaning crude oil contaminated water and soil	304

7.3	Methods of air and gas cleaning used in the crude oil industry	309
7.4	Conclusion to Part III	318
	Bibliography	320
Part IV	Heavy Oil Processing - Chemistry of Asphaltenes	325
8	Chemistry of Crude Oil Asphaltenes	327
8.1	Introduction	327
8.2	Problems of crude oil residue treatment with respect to asphaltenes	328
8.2.1	Coke formation and reduction of heavy metals	329
8.2.2	Treatment possibilities for crude oil residues	331
8.2.2.1	Physical treatment - deasphalting	331
8.2.2.2	Chemical treatments	333
8.2.2.3	Visbreaking	334
8.2.2.4	Coking	335
8.2.3	Coke forming reactions during residue treatment	337
8.2.3.1	Catalytic treatment	337
8.2.3.2	Thermal treatment	340
8.3	Methods of analysis of crude oil residue	342
8.3.1	Methods and main definitions for the determination of coke formation tendency	342
8.3.2	Analytical characterization of heavy oil residues and asphaltenes	343
8.3.2.1	Solution analysis	343
8.3.2.2	Coagulation analysis	346
8.3.2.3	Distillation method	349
8.3.2.4	Chromatography	350
8.3.2.5	^{13}C-NMR analysis	351
8.3.2.6	Ultimate analysis	353
8.3.2.7	Molecular weight determination	354
8.3.3	Temperature influence on molecular weight determination	357
9	Processing of Heavy Crude Oils and Crude Oil Residues	359
9.1	Introduction	359
9.2	Chemistry and reaction of asphaltenes during co-processing of crude oil residue and plastics	360
9.2.1	Change of asphaltene structure during thermal processing	360
9.2.2	Evaluation of possibilities of various asphaltene reactions based on thermodynamics	368
9.2.3	Hydrogen transfer	375
9.3	Co-processing with cracked products	377
9.3.1	Co-processing with cracked products from aromatics containing plastics (e.g. polystyrene)	380
9.3.2	Co-processing with cracked products from plastics containing	

	paraffin groups	383
9.3.3	Possibilities that exist for carrying out co-processing of heavy crude oils and various co-feeds	385
9.3.4	Behavior of heavy metals during co-processing	387
9.3.5	Conclusions of co-processing of crude oil residue and co-feed	387
9.4	Industrial methods of crude oil residue treatment	388
9.4.1	Fluid catalytic cracking	388
9.4.2	Hydrocracking	390
9.4.3	Coking	392
	Bibliography	394
Appendix A: Conversion Factors Important for Crude Oil Chemists		397
Appendix B: Glossary		399
Index		407

Part I

CLASSIFICATION AND CHARACTERIZATION OF CRUDE OIL

OVERVIEW

In this part of the book, fundamental chemical information of interest to petroleum chemists is given. The first chapter deals with the main definitions as well as the important properties of crude oil and petroleum products used by petroleum specialists. A short history is given as to how crude oil became the most important power source for our civilization for over thousands of years.

A short discussion is also given on the ecological consequences of crude oil production and treatment as well as problems that generally arise in the petroleum industry which crude oil chemists have to confront.

A detailed discussion on the properties of crude oil and crude oil products and some methods for their improvement prepares the reader for the problems the crude oil chemist faces daily. Some early traditional solutions for these problems, which never became popular at the industrial scale, will show the reader that there are many yet-to-be-researched ways to improve the methods for crude oil treatment. This chapter also highlights the general chemistry of crude oil and crude oil products.

The last chapter (i.e. chapter 2) of Part I deals with modern analytical methods used in crude oil chemistry. Modern and classical methods of petroleum and petroleum products characterization are explained. This chapter is an essential chapter for present and potential crude oil chemists since analytical chemistry constitutes an important part of crude oil chemistry. Besides, crude oil products have so many special properties that are important for the industry. The need to determine these properties gives rise to the very many analytical methods used in petroleum chemistry.

1

Nature and Classification of Crude Oil

1.1 HISTORY AND NATURE OF PETROLEUM

Petroleum or crude oil has been known for a long time. Archeologists have shown that it had already been extracted and used for about 5-6 thousand years before Christ. The most ancient known oil wells are those at Ephrata and the Kerch coast in the Chinese province of Sychuan. The mention of petroleum has been found in many ancient manuscripts and books. For example, the Bible writes about "pitch wells in the vicinities of the Dead Sea".

In ancient times, petroleum had some applications in medicine as well as civil works. For example, the ancient Greek scientist Hippocrates (IV-V century B.C.) has described many recipes of medicines which included petroleum. In one ancient manuscript is written: "we shall rub the patients with petroleum in such a way that the illness is taken away. White petroleum takes away the illness (cough in this case). Black petroleum takes away a reasoning of the cough". The Egyptians used petroleum oils to manufacture preservation mixtures.

Petroleum was also widely applied during construction work. Petroleum bitumen was added to cement and the resulting product used during the construction of the tower of Babylon. In the Bible, there is a narration that goes: "Also each other has told to each other: "we shall do bricks and heat it by fire". And they used the bricks instead of stones, and earthen pitch instead of cement". Modern chemical analyses show that "earthen pitch" is "asphalt", the viscous resinous substance remaining after the natural evaporation of the light fractions from petroleum. Asphalt was applied in the construction of the Great Chinese Wall as well as the trailing gardens of Semiramida. It was used as a water-resistant medium for the construction of most of the ancient dams on Ephrata

River. In the ruins of the ancient Indian city Mohengo-Daro was found a huge pool constructed five thousand years ago. The walls of the pool were covered with a layer of asphalt.

However, the greatest glory petroleum got was not for its use for construction. For more than two thousand years, petroleum was applied in military actions and served as a source of military power. This was found in the discovery of the antiquity based on the invention of "Greek fire". This new kind of weapon considerably strengthened the military power of the countries that knew how to make and operate them. It is still not proven scientifically who first invented the napalm. Some people attribute the invention to the Byzantium alchemists, while others think that the secret of its preparation was already known in ancient Greece. The Greeks used to bind a vessel with a mysterious mixture to the end of a stick, and threw it with the huge fire. Historians indicate that the fire flew with the speed of light and with sound of thunder. When this vessel impinged on the wall, an explosion occurred that gave rise to a huge cloud of smoke. The flame was distributed in all directions. Water could not extinguish this fire.

Byzantium won a lot of fights using "Greek fire". The antic napalm was especially of great service to Byzantium in the VII century during the attack by the Arabs on Constantinople. The Arabian fleet had besieged the capital of Byzantium. Besieged inhabitants of Constantinople had lost any hope of rescue when the great idea came. During one of the attacks, they allowed most of the Arabian fleet to come very close and unexpectedly unleashed a huge quantity of "Greek fire" on the sea and burned it. The flame burned all the Arabian ships. It seemed as if the sea was burning.

The composition of "Greek fire" was kept as a top secret. However Arabian alchemists solved the secret of the "Greek fire" after almost four hundred years after the fight at Constantinople. The main component of "Greek fire" was petroleum with the addition of sulfur and saltpeter.

Up till now, petroleum has been used in many branches of construction work or military service. It is thus hard to imagine what our life today would be without crude oil. It brings power to all our machines and our houses. It is used as a lubricant for various parts of machines. Hardly any modern device would work without relying on various products derived from crude oil.

Even though the history of crude oil could be traced back by more than two thousand years, real production of crude oil perhaps began in August 27, 1859, when the first industrial-scale crude oil well with a depth of 22 meters was opened in Oil Creek, Pennsylvania. After this first industrial crude oil well was opened, there was the commencement of a rapid development of crude oil production and treatment. Probably, this day could be said to mark the birth of modern crude oil chemistry. In 1878, the Swedish businessman Alfred B. Nobel together with his brothers formed the Naphtha Company Brothers Nobel. The company extracted the crude oil in Baku, Russia and transported it to the first crude oil refineries via the pipelines built by Naphtha Co., which still exists now.

It may sound strange but petroleum refers to a mountain mineral. It usually exists together with sand, clay, stone, salt, etc. We normally think of a mineral as a firm substance. However, there also exist minerals in the liquid form and even in the gaseous form. One important property of petroleum is its ability to burn. Other minerals that have this property are peat, brown and stone coal, and anthracite. These combustible minerals form the special family of minerals named "caustobolites" (derived from the Greek words *causthos*, combustible; *bios*, life; *cast*, stone) meaning combustible organic stone [1]. There is a distinction between coal caustobolites and petroleum caustobolites.

All caustobolites, however, contain carbon, hydrogen and oxygen even though in different proportions for different caustobolites. Specifically, petroleum is a complex mixture of hydrocarbons and other carbon compounds. At the elemental level, it consists of elements such as carbon (84-87%) and hydrogen (12-14%) as well as oxygen, nitrogen and sulfur (1-2%). The sulfur content can sometimes be up to 3-5%. Overall, petroleum consists of hydrocarbons, asphaltenes and resins, paraffins, sulfur and ash. There are three main groups of hydrocarbons in petroleum—namely, paraffinic, naphthenic and aromatic hydrocarbons [2].

The paraffinic series of hydrocarbons have the general formula $C_nH_{(2n+2)}$ and can be either straight chains (normal) or branched chains (isomers) of carbon atoms. The lighter, straight-chain paraffins are found in gases and paraffin waxes. Examples of straight-chain paraffinic hydrocarbons are methane, ethane, propane, and butane (gases containing one to four carbon atoms, respectively), and pentane and hexane (liquids with five and six carbon atoms, respectively). The branched-chain (isomer) paraffins are usually found in the heavier fractions of crude oil. They usually have higher octane numbers than the normal paraffins. Paraffinic hydrocarbons are saturated compounds with all carbon bonds saturated (i.e., the hydrocarbon chain carries the full complement of hydrogen atoms).

The amount of paraffins in different crude oils varies from 2 to 50%. The light paraffins are mainly components of natural gas, which dissolve in the crude oil in the oil wells. Depending on the composition and conditions in the oil well, one can specify well classes such as gas wells, gas condensate wells and crude oil wells. Gas wells contain mainly such light paraffins as methane, ethane, propane and butane, all of which are gases at normal conditions (0.1 Mpa and 20 °C). Apart from these hydrocarbon gases, gas wells also contain carbon dioxide (CO_2), hydrogen sulfide (H_2S) and inert gases such as nitrogen (N_2), argon (Ar), helium (He), neon (Ne) and xenon (Xe).

Often, gas condensate wells contain compounds with higher molecular weights than compounds of gas wells. At natural conditions in the oil well (pressures ranging from 25 to 45 MPa), these high molecular weight compounds dissolve the gas. Initially during oil production from gas condensate wells, pressure will decrease thereby releasing the low molecular weight compounds and leaving the high molecular weight compounds behind. This high molecular weight fraction is called condensate.

Crude oil wells contain crude oil as well as gas. The amount of gas in the crude oil varies from very little to hundreds of cubic meters per ton of crude oil. These gases, solved in crude oil, can be released from the crude oil at normal pressures. After production, crude oil is stabilized by separating the gas from the oil (see Part III). The crude oil coming to the refinery usually contains less than 1% of dissolved gas.

All paraffins from C_1 to C_{78} can be separated from crude oil. However, it has been shown that the largest fraction of paraffins in the crude oil is composed of molecules from C_7 to C_{14}. Lighter or heavier paraffins are present in crude oil in smaller amounts or as trace compounds.

All types of paraffins (i.e. *n*-paraffins and *iso*-paraffins) are present in crude oil. The methyl-substituted paraffins were analytically proven to be present in crude oil in the 1960s. It has been shown that methyl groups in paraffins are located in positions 2, 6, 10, 14, 18 and further. Over 20 such isomers have been found. The most abundant compounds of this kind of isomer are phitane $C_{20}H_{42}$ and pristane $C_{19}H_{40}$ (each was found in different crude oils in amounts up to 1.5%).

It is known that paraffins from methane to butane are gases, from C_5 till C_{17} are liquids, and from C_{18} onwards are solid substances. The solid paraffins are present in all crude oils in different amounts, often up to 5%, but in some crude oils up to 7% or even 12% have been found. Solid fractions of crude oils do not only contain paraffins, but indeed these solids are complicated mixtures of paraffins, naphthenes, aromatics and other compounds. It has been shown that some heavy fractions from paraffinic oils can contain up to 50% paraffins, 47% naphthenes and up to 3% aromatic compounds. It is known that the higher the boiling temperatures of the crude oil fraction, the less the amount of paraffinic compounds present in the fraction. However, paraffins are present in smaller or higher amounts in all crude oils, crude oil fractions and products. The kind and how the paraffins are present in oil (i.e. gas, solved or dispersed) depend on the properties of the crude oil and the chemical conditions of paraffins.

The carbon atoms in the paraffin molecule are connected by a covalent sigma (σ) bond. The length of these bonds for the free isolated molecule in the gas phase is 0.154 nm. The covalent angle between these C-C bonds is 112°. The length and the valent angle can be different from the numbers shown for the liquids and real gas paraffins. This difference can be explained on the basis of the formation of hydrogen bonds between paraffin molecules. Through these bonds, the conditions for intermolecular equilibrium in the paraffin will be changed. However, it is well known that the power of the crystal field can strongly influence the geometrical parameters of molecules by the formation of hydrogen bonds. At the moment, there are very limited studies on the geometrical differences between free isolated molecules and condensed molecules.

Paraffins can be present in crude oil as molecular paraffins as well as associated molecules. The fraction of associated or molecular paraffins in crude oil depends on many factors. However, one of the more important factors is temperature; the higher the temperature, the less the fraction of associated paraffins in crude oil.

Usually paraffins are less prone to most known industrial reactions. The most important industrial reactions of paraffins are oxidation, catalytic isomerization and sulfurization.

Naphthenic hydrocarbons have the formula C_nH_{2n}. All bonds of carbon with hydrogen are saturated. As such, naphthenic hydrocarbons in petroleum are also relatively stable compounds.

Naphthenic hydrocarbons are the most abundant class of hydrocarbons in most crude oils. Their composition in oil can vary from 25 to 75%. Usually, the amount of naphthenes in crude oil fractions increases as the boiling point of the fraction also increases. However there is an exception: The amount of naphthenic hydrocarbons decreases with an increasing boiling temperature for heavy oils. This can be explained on the basis of the increasing amounts of aromatic compounds in heavy oils.

The distribution of monocyclic naphthenes is well investigated at the moment in comparison to polycyclic naphthenes. Monocyclic naphthenic compounds are distributed mainly in the light fractions of crude oil. So, naphthenic hydrocarbons in the gasoline fraction are mainly present as substituted cyclopentanes and cyclohexanes. The amount of these compounds in gasoline fractions varies from 10 to 85%. The polycyclic naphthenes can be found mainly in the heavy fractions of crude oil (with boiling temperatures over 350°C).

At the moment, chemical analysis has identified only 25 dicyclic, five tricyclic and four tetra- and pentacyclic naphthenic compounds in crude oil. In cases where there are over one naphthenic ring in one molecule, a part of the molecule normally consists of a polycondensed ring.

Bicyclic naphthenes (C_7 – C_9) are usually used as an indication of a naphthenic crude oil. The following bicyclic naphthenic compounds were observed in different crude oils: bicyclo[3,3,0]octane, bicyclo[3,2,1]octane, bicyclo[2,2,2]octane, bicyclo[4,3,0]nonane, bicyclo[2,2,1]heptane and their isomers or substituted compounds.

The tricyclic naphthenes are mainly present by alkylperyhydrophenantrens. The following compounds of this class, a), b) and c), have already been analytically identified.

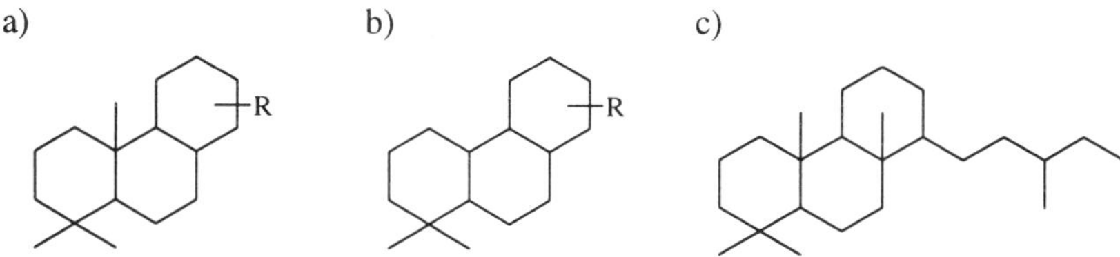

Tetracyclic naphthenic compounds are mainly isomers and substituted cyclopentanperhydrophenanthrene (C_{27} – C_{30}). Examples are presented as compounds d), e) and f).

d)

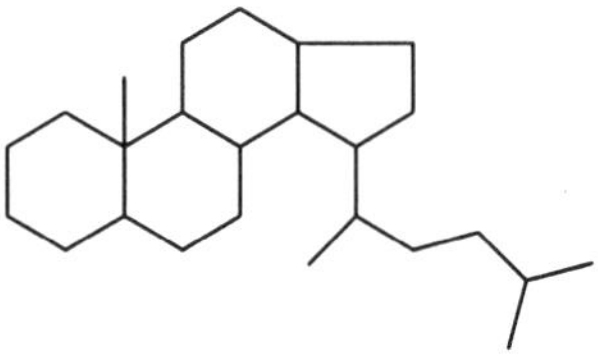

e)

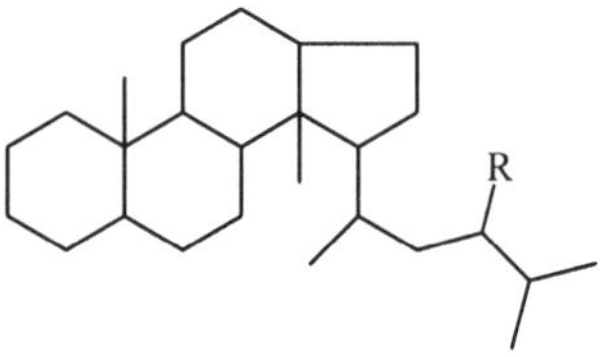

R

f)

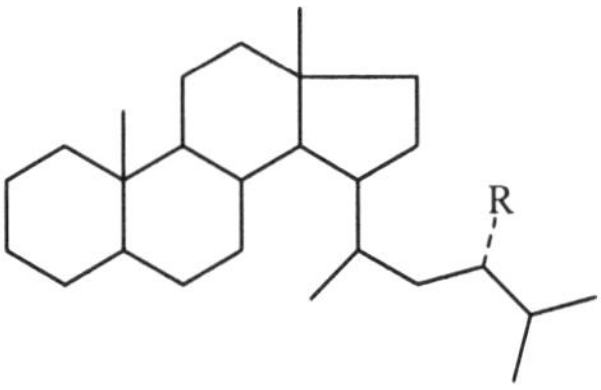

R

The most important compounds of the class of pentacyclic naphthenes are represented by gopan (g), lupan (h) and phridelan (i).

g)

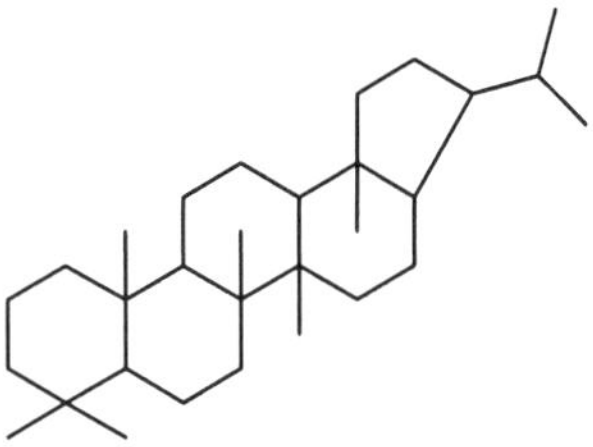

h)

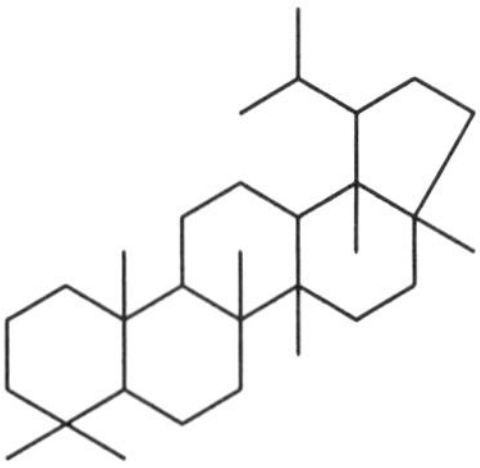

i)

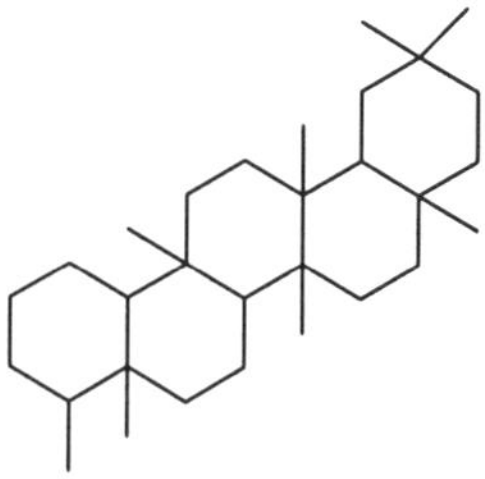

There are no analytical proofs for the exact structure of polycyclic naphthenic compounds with number of rings over five. However, based on the results of mass spectral analysis of heavy oil fractions, it can be said that there are polycyclic naphthens with seven or eight rings in their structure. At the moment, it is very difficult to be specific in analytical terms of the exact chemical structure of such molecules.

Since naphthenes are saturated hydrocarbons, the chemical activity of the naphthenic compounds is similar to the chemical activity of paraffins. During thermal treatment of the naphthenes, it takes part in reactions involving C-C bond cleavage, dehydration and, to a lesser extent, aromatization reactions.

Aromatics are unsaturated ring-type (cyclic) compounds that react readily since they have carbon atoms that are deficient in hydrogen. All aromatics have at least one benzene ring as part of their molecular structure. Aromatics may also have two or more of the ring structures fused together. An example of a fused double-ring aromatic compound is naphthalene. The most complex aromatics are polynuclear (i.e. they have three or more aromatic rings fused together). These are found in the heavier fractions of crude oil.

The amount of aromatics in different crude oils varies from 15 to 50%. The highest amounts of aromatics are typically found in naphthenic oils. The amounts of different types of aromatic compounds decrease in the following order: benzols > naphthalenes > phenanthrenes > hriezenes > pyrenes > anthracenes.

The highest amounts of aromatic compounds are concentrated in crude oil fractions with high boiling temperatures. It has been shown analytically that aromatics are usually present as substituted aromatic compounds with the length of the substituents up to thirty carbon atoms.

Asphaltenes and resins are dark substances (from dark red to brown). They are soluble in aromatic solvents but insoluble in paraffin solvents [3]. Asphaltenes have various types of heteroatoms present in crude oil in their structure. Asphaltenes are the most complicated known compounds in crude oil.

Special properties of asphaltenes include the tendency to associate, high molecular weight and paramagnetism. All these properties make asphaltenes very difficult to analyze or investigate. This is why approximately since the 1970s, asphaltenes chemistry exists as a separate science independent from crude oil chemistry. In part four of this book, some problems that arise in asphaltenes studies will be discussed.

Porphyrins, special nitrogen compounds of organic origin, are also present in petroleum. They are believed to be formed from chlorophyll of plants and hemoglobin of animals. Porphyrins can be cracked at temperatures ranging from 200 to 250°C. The basic structural unit of porphyrins is given in Fig. 1.1.

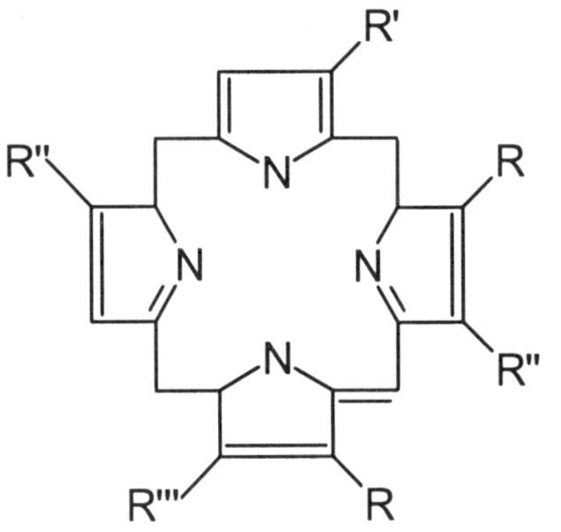

R'
R''
N
R
N
N
N
R''
R'''
R

Fig. 1.1: The structure of porphine basic structural unit of porphyrins.

The amount of nitrogen in different crude oils varies from 0.02 to 1.5%. There are many types of nitrogen compounds in crude oil. The example shown in Figure 1.1 is only one of them.

Studies of nitrogen compounds present in crude oil are made possible in two ways. First, these compounds can be analyzed directly in crude oil. The biggest advantage of such an analysis is the possibility to investigate these compounds both in their natural form and natural environment. However, the concentration of nitrogen compounds in crude oil is relatively small, and this makes the analysis not only difficult but results in a rather wide divergence of the measurements. The second method is that the nitrogen compounds can be separated before analysis. The disadvantage of this method is the possibility that during separation, the native structure could be destroyed. However, despite the difficulties in investigating nitrogen compounds, analysis has shown that, at the moment, nitrogen compounds are present in crude oil mainly as cyclic compounds. Nowadays, nitrogen compounds can be classified as alkaline (lye) nitrogen and neutral nitrogen compounds.

Pyridines (a), hinolines (b) and acredines (c) belong to the strong alkaline (lye) nitrogen, because of their free, non-compensated electron pair.

a) b) 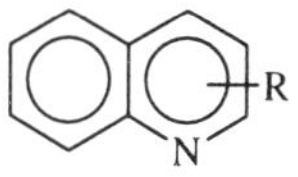 c)

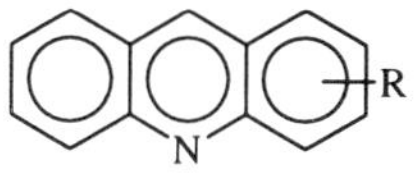

The substituted anilines (d), amides and/or imides belong to the weak alkaline (lye) nitrogen compounds.

d)

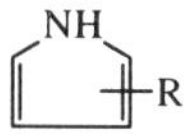

The non-substituted compounds such as indols (e) or carbozoles (f) are typical nitrogen neutral compounds.

e) f)

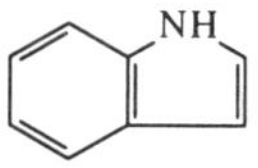

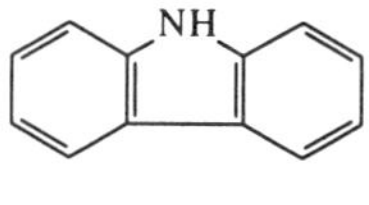

The above are some examples of nitrogen compounds in crude oil. It is however difficult to show all the possible nitrogen compounds present in crude oil. Recent investigations have shown that compounds with two nitrogen atoms or one nitrogen atom and one sulfur atom in one molecule can be found in crude oil.

The nitrogen compounds are very important in their role as natural surfactants. The concentration of these compounds in crude oil has a great influence on the chemical and physical activities of the crude oil, on metal/crude oil interface and ground/crude oil interface. This property of nitrogen compounds is used during the production of crude oil from the oil well. For example, hinoline can prevent the corrosion of metal parts; this is very important for the continuous working of many oil production plants.

The next class of heteroatom compounds in crude oil is the oxygen compounds. The amount of oxygen in crude oil can vary from 0.1 to 3% or even 4%. The amount of oxygen in crude oil fractions increases with the boiling temperature of the fraction. Over 20% of all oxygen compounds are concentrated in asphaltenes and resins.

Almost similar to nitrogen compounds, the oxygen compounds can be classified as neutral oxygen and acidic oxygen compounds. The cyclic and aromatic compounds, ethers, anhydrides, furans and so on usually belong to the neutral oxygen compound class.

The acidic oxygen compounds are usually represented by carbon acids. The presence of these compounds in crude oil has been known for a very long time. It was noticed during the production of light kerosene. In the production of high quality light kerosene, it was necessary to clean the kerosene with lye. Compounds with strong emulsifying properties were produced during this process. At the end of the nineteenth century, it was shown that these compounds were sodium salts of carbon acids.

Sulfur may be present in crude oil either as hydrogen sulfide (H_2S), or as compounds such as mercaptans (a), thiophenols (b), cycloalkanethiols (c), thiophenes (d), benzothiophenes (e), alkylbenzothiophenes (f), etc., or as elemental sulfur.

a) $C_nH_{2n+1}SH$

b) SH

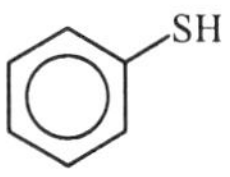

c) CH_2 S CH_2 R

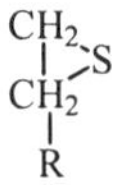

d) S

e) S

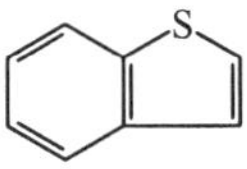

f) S R

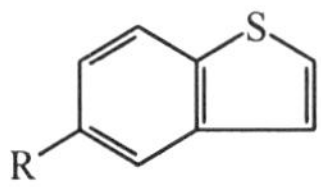

Each crude oil has its own types and proportions of sulfur compounds. As a general rule, however, the proportion, stability, and complexity of the compounds are greater in the heavier crude oil fractions. Hydrogen sulfide is a primary contributor to corrosion in refinery processing units. Other corrosive sulfur materials are elemental sulfur and mercaptans. Pyrophoric iron sulfide results from the corrosive action of sulfur compounds on the iron and steel materials used in refinery process equipment, piping and tanks. The combustion of petroleum products containing sulfur compounds results in the production of undesirable by-products such as sulfuric acid and sulfur dioxide. Catalytic hydrotreating processes such as hydrodesulfurization remove sulfur compounds from refinery product streams. Sweetening processes either remove obnoxious sulfur compounds (example, mercaptans) or convert them to odorless disulfides. The amount of sulfur in petroleum of different origins ranges from 0.1 to 5% [2].

Sulfur compounds in crude oil sharply decreases the quality of fuels and oils produced from the crude oil. They cause corrosion of equipment during treatment, reduce activity of antidetonation additives and antioxidizing stability of gasoline, raise the propensity to form hard residues in cracking gasoline fractions, and result an environment pollution.

Metals (including heavy metals) have been found in all crude oils. Their composition varies from 0.01 to 0.04% of crude oil. About thirty different metals are found in different crude oils. The most common are vanadium, nickel, iron, zinc, mercury, boron, sodium, potassium, calcium and magnesium.

Unsaturated compounds like alkenes are not presented in crude oil. However, these compounds can be produced during the thermal or/and catalytic treatment of the crude oil. These compounds differ from all crude oil compounds by their high chemical activity. Based on the high chemical activity of unsaturated compounds, it is clear why this class of compounds does not exist in crude oils.

Ash forms the balance in petroleum. It is the noncombustible portion that is left behind after petroleum is burned. Ash is composed of various metallic compounds such as compounds of iron, nickel and vanadium as well as various salts.

Petroleum is also characterized by physical properties such as density, viscosity, temperature of hardening, boiling temperature and solubility as well as electrical and optical properties [4].

1.1.1 Origin of Petroleum

The first attempt to explain the origin of petroleum dates back to antiquity. For example, the Greek scientist Strabon, who lived about 2000 years ago wrote: "At the place named Nymphey, there is a rock spiting fire, and under it are the sources of warm water and asphalts... ". Strabon united two facts: the eruption of volcanoes and the formation of asphalts (the way he named petroleum). This connection between the two facts was a mistake. In the places mentioned by his work, there were no erupting volcanos. The events which Strabon described as "eruptions" were actually "emissions", i.e. breaking out of underground waters (so-called geysers), accompanied by outputs of petroleum and gas on the surface.

M.V. Lomonosov was one of the first scientists to introduce a reasonable scientific concept of the origin of petroleum. In his mid-eighteenth century work on "terrestrial layers", this Russian scientist wrote: "It is expelled from underground with heat, prepared from stone coal and brown coal, this black oily material... And this is a birth of a different grade of combustible liquid and dry hard matter. This is the essence of stone oil, liquid pitch, petroleum, and similar materials which are different by cleanliness, but occur from the same origin" [5]. It can therefore be stated that the idea of the organic origin of petroleum from stone coal was conceived more than 200 years ago. The initial substance was an organic material transformed at first into coal and then into petroleum.

Lomonosov was not the only one who addressed the question of the origin of petroleum in the eighteenth century. However, some of the other hypotheses formed at this time were less than scientific. For example, a hypothesis credited to a Warsaw priest was that the Earth was very fertile in the paradise period. The core of the earth contained a fatty impurity. After the paradise period, this fat was partially evaporated, and the vapor partially condensed on the ground where it mixed up with a variety of materials. This was later transformed to petroleum by the world flood.

There are many other less scientific hypotheses about the origin of petroleum even by scientists. At the end of the nineteenth century, the authoritative German geologist H. Hefer reported of an American petroleum industrialist who considered petroleum to have resulted from wet whales that existed at the bottom of polar seas. This petroleum penetrated into Pennsylvania by seeping through underground channels [5].

In any case, the most widespread ideas among the scientists in the nineteenth century centered on the organic origin of petroleum. Disputes were mainly around the initial material for petroleum formation: animals or plants? German scientists H. Hefer and K. Engler carried out experiments in 1888 in which they sought to prove that petroleum formation was from animal origin. The experiments were performed by evaporation of fish fat at 400°C and 1 bar. Oil, combustible gases, water, fats and different acids were formed from the 492 kg of fat used. The largest fraction of evaporated material was oil (299 kg, or 61%) with a density of 0.8105 g/cm^3. Subsequent evaporation of the oil product yielded saturated hydrocarbons (ranging from pentane to nonane), paraffin, lubricant oils as well as olefins and aromatic hydrocarbons. Later, a Russian scientist (N.D. Zelinskiy) carried out a similar experiment in 1919. However, his initial material was organic silt of mainly vegetative origin from Lake Balhash. The evaporation products in this case were: crude pitch – 63.2%, coke – 16.0%, and gases (methane, carbon oxides, hydrogen, hydrogen sulfide) – 20.8% [5]. Subsequent processing of the pitch yielded gasoline, jet oil and heavy oil.

By the end of the nineteenth century, two different hypotheses of petroleum origin had emerged: organic and inorganic hypothesis. The main concept of inorganic petroleum origin was illustrated by the experiments of Berthelot. In 1866, Berthelot considered that acetylene was the basic material. Large quantities of acetylene were assumed to be produced by the reaction of water with carbides which, themselves, were formed by the reaction of alkali metals with carbonates. The conversion of acetylene to petroleum was accomplished at an elevated temperature and pressure according to the following:

$$CaCO_3 \rightarrow CaC_2 + H_2O \rightarrow C_2H_2 \downarrow \text{petroleum}$$

Indeed, the idea of the inorganic origin of crude oil did not initially have any success with geologists, who considered that experiments carried out in the laboratory considerably were different from processes that occur in a nature. However, the inorganic theory of crude oil formation unexpectedly received support due to new evidence from astrophysics. Research on the spectra of planets showed that, there are hydrocarbon compounds in the atmosphere of Jupiter and other large planets as well as in gas environments of comets. If hydrocarbons are widespread in space, it means natural processes of synthesis of organic substances from inorganic substances are possible.

In the 1950s, the Russian scientist N.A. Kudryavzev collected a lot of geological material involving petroleum and gas deposits in the world. First of all, Kudryavzev noticed that many gas and petroleum deposits were found in zones of deep cracks of the terrestrial core. This knowledge was not new at this time since other scientists had noticed this fact much earlier. However, Kudryavzev extended the application of such ideas to a great extent. For example, in the north of Siberia, near the area of the so-called Marhiinskij shaft, there are frequent outbursts of petroleum onto the surface. At a depth of about two kilometers, the mountain layers are literally impregnated with petroleum. At the same time, it has been shown that the amount of carbon formed simultaneously with mountain layers is extremely small (only 0.02 to 0.4%). But further from the shaft, the amount of organic compounds in the layers increases. Nevertheless, the quantity of petroleum sharply decreases. Based on these extra data, Kudryavzev suggests that crude oil formation in the Marhiinskij shaft can most likely be explained not on the basis of formation from organic substance, but by an inorganic theory of oil formation in the deep layers (or shells) of the planet. Similar oil wells have been found in other regions of the world as well. A long time ago in Wyoming (USA), the inhabitants heated their houses using pieces of asphalt, which they collected from the cracks in mountain layers in the Copper Mountains. But the minerals, of which these mountains consisted, could not accumulate petroleum and gas. This means that the asphalt (similar to oil) could only be formed according to the inorganic theory.

The space hypothesis of the origin of oil deserves mention as well. In 1892, Sokolov stated that the dust cloud from which the Earth and other planets of the solar system were formed consisted of hydrocarbons. In the process of the formation of the Earth, hydrocarbon substances were buried in the core of the earth. Further, during the cooling of the planet, the hydrocarbons were pushed out. As a result, they penetrated into cracks of friable minerals. This hypothesis is also one of the representations of petroleum synthesis from minerals.

However, the origin and formation of petroleum are very difficult questions and it is almost impossible to answer them using only one theory. A more detailed discussion concerning the origin of crude oil formation can be found in references [6 – 9].

1.1.2 Oil Formation in the World's Oceans

All seas and oceans are populated with biomass which are essentially a wide variety of animals and plants. Of all sea biomass, the ones with the most significant role in petroleum formation are microorganisms, typically plankton, 90% of which is microscopic seaweed (phytoplankton). Plankton is the basic source of organic material in the sea. Plankton is contained not only in the silts at the bottom of seas or lakes but also dispersed or dissolved in the water. Approximate quantities of organic material dissolved per m^3 of water are 2 g in the Atlantic and Pacific oceans, 5-6 g in the Baltic and Caspian sea, and 10 g in the Azov sea. It is interesting that the dissolved organic material is like greasy acids that is structurally similar to plankton fats. The concentration of organic material is highest at the bottom of the oceans. This is obvious because, for the most part, these organisms are denser than the liquid medium. As such, they fall down to the bottom by gravity. Shallow conditions are the preferable places for accumulation of organic material. Generally, the process of mineral (clay, sandy minerals, etc.) accumulation promotes fast trapping or collection of organic material as well as its protection from decomposition. On the other hand, for organic material found deep in the ocean water, there is sufficient time for it to be substantially dissolved and dispersed in the water due to the activity of bacteria. Consequently, only 1% of organic material is usually collected annually per m^2 of ocean bottom in the world's oceans out of 150 g that is formed.

Now, let us consider what occurs when organic material is collected on the sea bottom. Organisms that are either brought from different continents or are formed directly in the sea are collected rapidly in clay or sandy minerals. Although organic materials contain various substances, the one with the greatest interest for subsequent petroleum formation is "bitumoid". Bitumoid can be extracted from organic material using various solvents such as chloroform, benzene or ether. The main source of bitumoides are lipoid (i.e., fat or a similar compound). The proportion of bitumoides in the sea bottom deposits ranges from 2 to 20% of all organic material. Apart from bitumoides, materials such as hydrocarbons (from 0.1 to 3%) are also available in organic material. Approximately 300 g (and in some cases up to 15 kg) of hydrocarbons are contained in 1 m^3 of minerals formed. The average quantity of dispersed hydrocarbons in minerals is 70-80x10^{12} ton. This exceeds the established volume of hydrocarbons in oil fields (about 2.2x10^{12} ton) by about tenfold. It is therefore evident that the organic material collected as described earlier in this section is sufficient to form the established world petroleum reserves.

Dispersed hydrocarbons in solid minerals and silts in the seas are similar to petroleum hydrocarbons. They are called dispersed petroleum or micro-petroleum.

Mountain minerals are hydrofill, meaning that they are moistened with water instead of petroleum. Thus, in addition to mountain pressure, capillary forces enhance the displacement of petroleum in the solid minerals.

The process of petroleum displacement in the native minerals (i.e. from which it is formed) is referred to as primary migration or emigration. By getting into loose solid minerals (collectors or traps), petroleum begins a new life. Petroleum migration through collectors proceeds as long as it does not encounter a trap (i.e. a layer that is capable of keeping the petroleum as a trapped deposit). Examples of these traps are anticline traps, traps associated salt domes and oil entrapment in a limestone reef. These are shown in Figures 1.2–1.4. Thus, the pre-history of petroleum begins in live organisms from which are synthesized initial biochemical compounds. On the other hand, the history of petroleum begins with the collecting of biological and organic substances in the solid minerals [3].

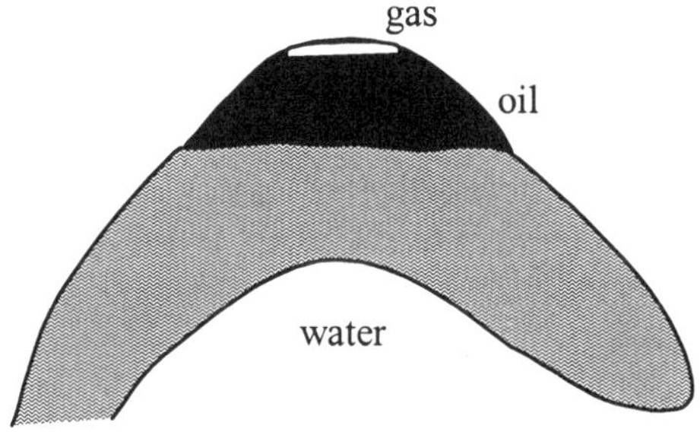

Fig. 1.2: Anticline traps.

Traps associated with salt intrusions are of many types (example: Fig. 1.2 – 1.3); limestone reefs (Fig. 1.4) can also serve as reservoir rocks and give rise to overlying traps of anticlinal form as a result of different compaction. Examples are also known in which the reservoir rock extends to the surface of the earth but oil and gas are sealed in it by clogging of the pores by bitumen or by natural cements [3]. Many reservoirs can display more than one of the factors that contribute to the entrapment of hydrocarbons.

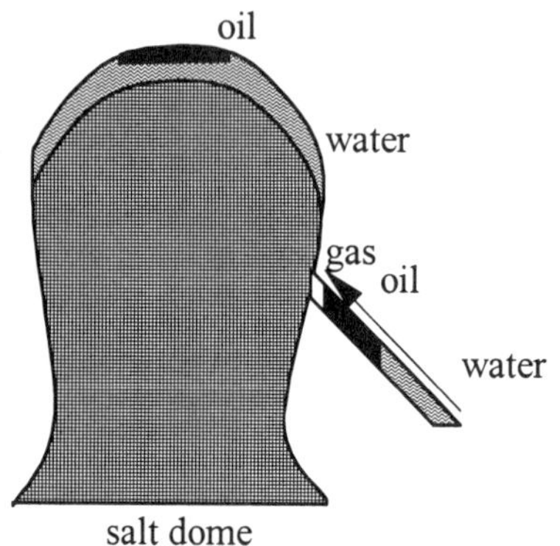

Fig. 1.3: Traps associated with a salt dome.

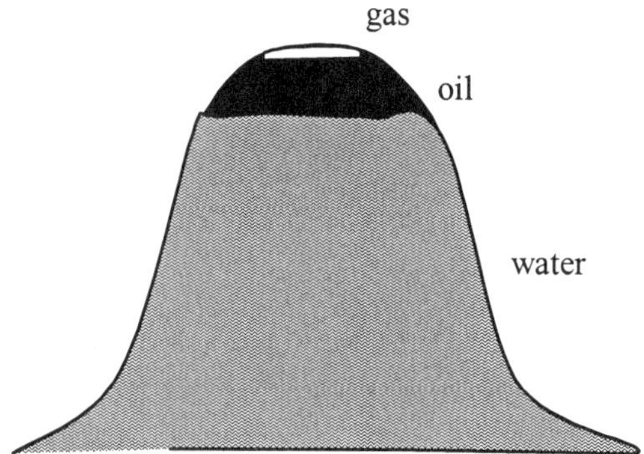

Fig. 1.4: Oil entrapment in a limestone reef.

Detailed discussions on oil and gas formation and modern methods of investigation in this area can be found in references 10-33.

1.1.3 Modern Concept of Formation of Petroleum

The characteristic feature of the modern concept of petroleum formation is based on a new geological idea. Here, there is the representation that there occurs a horizontal movement of separate blocks of the lithosphere, the so-called, "lithosphere-plates". Deep down our planet is a circulation of material according to the so-called "convective movement" [5], which began a long time ago at a depth of about three thousand kilometers into the earth where hot and rather light material started moving upwards. After 15-16 million years, this movement reached the lithosphere – the top and thinnest terrestrial environment. This material spread over and "broke off" on the lithosphere into plates as a result of forces of viscous friction. The plates moved apart from the region of outward flow of material and drifted in a horizontal direction. The original structures were formed as huge failures or "rifts". These were then transformed into the ocean. Today, typical continental rifts exist in East Africa. They are typically filled with water. An example of a modern sea rift that illustrates a subsequent stage of transition of an initial rift structure to the ocean is the Red Sea.

The horizontal movement of the lithosphere plates eventually resulted in the collision of the plates in which one plate was "pushed" under another plate. This created the zone of subduction. Typically, during the immersing of lithosphere-plates, the friction involved generates a considerable amount of heat that results in increasing the temperature of the zone by hundreds of degrees. This process promotes melting of the moved plate and gives rise to the occurrence of volcanic processes. The modern subduction zones are widespread on the coast of the Pacific Ocean and on the eastern part of the Indian Ocean. These processes are accompanied not only by active volcanic processes, but also by strong earthquakes. As a result, the lithosphere is always in continuous movement. What is the relation of the formation of petroleum to these powerful natural phenomena? Formation of petroleum is a very energy-intensive process. It involves the dissociation of various compounds, breaking of chemical bonds between carbon and oxygen, nitrogen and sulfur, etc. These are processes that require significant amounts of energy to be expended. For example, C-C bond scission requires 70-100 kcal/mol whereas C-O bond breaking requires 70-200 kcal/mol. These processes can be initiated and made to proceed actively within the temperature range 100-400°C. Below this temperature range, transformation of dispersed organic material to petroleum will proceed slowly and languidly, and will not completely exploit the potential of the availability of the organic material resource. Chemists have synthesized a product that is practically similar to natural petroleum from natural organic material. This has been made a very rapid process as a result of the high temperature used in the reactor. Hence, if the situation whereby solid minerals with organic material are made to pass through the zone of high temperature can be provided in nature, the formation of petroleum can be facilitated. A required condition for this purpose is that the layer that contains organic material should be located at a minimum depth of 2-3 km. This is where the main stage of petroleum formation proceeds. And what will occur if organic material passes in the zone of rift or subduction? This area is five to six times hotter than the surrounding areas. Hence, the transformation of organic material into liquid petroleum will be facilitated. Practically, it can begin simultaneous with collecting solid minerals to make the trap. That is why zones of rifts and subduction are usually of special interest to geologists when performing oil-prospecting work. Since this knowledge gives them a key to a correct understanding of the genesis of hydrocarbons, they can predict what is likely to occur in such places.

We will now view the processes in the rifts. The process of rift formation precedes a strong increase in temperature of the formation zone ("excitation" of the top layer). This is represented in the structure of modern rifts: thinning of terrestrial layers up to 30-35 km; reduction of asthenosphere depth; strong increase of a thermal flow under the rift; volcano formation; formation of the thermal water sources; and seismicity. All these characterize rifts as extremely active structures in the lithosphere. The mineral pools in the rifts are formed during the initial stage of the destruction of the terrestrial layers. Narrow deflections filled with 4-7 km of organic containing solid minerals exist for short time intervals of 5-20 million years. At the initial stages of collecting of the solid minerals in the rifts, the usual continental river or lake with layers of volcanic formations is formed. Often, the formation of salt complexes is postponed until later. This postponement is connected with postponing salt formation within the deep thermal water. Normal sea minerals are collected during the process of rift formation as well as its transformations from continental layers to sea intercontinental rift (as in the Red Sea). In the central part of the rifts where there is limited water circulation, clay layers enriched by organic material (black clay) usually accumulate. A fast immersion occurs very deep in the earth at the stage of rift formation. This process promotes a substantially abnormally high thermal flow in the rifts. As a result, petroleum formation is facilitated. Therefore, it is possible for formation of hydrocarbons to have already occurred in young superficially located layers. Even the lake minerals containing small quantities of organic material are able to form petroleum. For example, there are numerous petroleum and gas wells in the modern East African rift system. Separate rifts that are filled with water form a system of lakes where gas and light petroleum reserves are found.

There are other kinds of geological events that proceed in zones of subduction, but the result of these processes is the same: the acceleration of the transformation of dispersed organic material to petroleum. The movement zones are two very important areas for petroleum formation phenomena: formation of lenses traps and movement of organic material from the ocean into the trap by means of the displaced plates.

More about modern theories about oil and gas formation and modern investigation in this area can be found in other references [34-37].

1.1.3.1 Crude oil prospecting

The basis for oil prospecting lies on the possibility of obtaining a geological map of the prospecting area. In certain areas (e.g. Iran), one could easily detect possible oil wells by air photography of the earth's surface in the prospecting area. Geological prospecting can be made very exactly. However, it can only enable us to evaluate the structure of the surface complexes of mountain layers. The structure of the mountain layers prospected on the surface does not usually represent the structure of the deeper layers. Geologists use geophysical methods of crude oil prospecting to obtain a deeper insight of what lies below the Earth's surface. There are four popular geophysical methods of crude oil prospecting: seismic prospecting, gravimetric prospecting, magnetic prospecting and electric prospecting.

The seismic method is based on studies of features of the transmission of elastic fluctuations in the terrestrial core. The elastic fluctuations (or seismic waves) can be produced artificially, for example by explosion. The speed of their transmission in each layer varies from 2 to 8 km/s and depends on the density of environment. The higher the density of the layer, the faster seismic waves can be transmitted through it. A fraction of the elastic fluctuations is reflected to a surface (i.e. reflected from the border between two or more layers with different densities), another fraction of seismic waves continues movement but refracted deeper through layers up to a new border between terrestrial layers. Reflected seismic waves can be detected by using special devices called seismic detectors. Researchers then perform an evaluation of the diagrams generated from wave fluctuations of the prospecting surface, including the depth of the maintain layers that reflected the seismic waves, and in some cases, obtain a lithological structure of the layer. Based on these data, the structures of deep layers are clarified, and maps of the underground relief (the so-called structural maps) are made. Based on these maps, the structure of deep terrestrial layers is investigated. The method of reflected waves was first used in Russia in 1923. After then, it became used successfully all over the world. This method is still used by geologists today.

Another method of seismic prospecting is based on detecting the refracted seismic waves obtained at the border between two or more layers under a critical corner. This method is widely applied in the world today. In the practice of seismic crude oil prospecting, other methods, including the method of controlled directed reception and the method of common deep point, are also used. The last method is especially widely applied for prospecting not only anticline traps, but also the zones of their formation. The method of common deep point is carried out by change of a mutual arrangement of the explosion and reception points. In such way, two or more reflected seismic waves from the same underground point can be detected.

The use of explosions as a source of seismic waves is actually somewhat obsolete for geophysicists. Since the 1960s, first in the US and now worldwide, non-explosive methods have been used for generating seismic waves. The most popular of these methods are the method of a falling load, the method based on using vibrators, and methods based on conversion of explosion energy from mechanical power. Today, almost all the seismic prospecting work is carried out without using any explosive sources. Seismic crude oil prospecting in the sea makes use of pneumatic and/or electrical sources of seismic waves.

The gravimetric method is based on investigating the distribution of the gravitational force on the Earth's surface. The acceleration of an object (for example in a mountain area) in a free fall depends on the density of the mountain layers. If the underground is the layer of stone salt having a relatively low density, the acceleration due to free fall decreases, indicating a negative anomaly in the gravitational field. In the case where the layers are composed of a more dense material (granite for example), a positive anomaly in the gravitational field is indicated.

Usually, the gravimetric method is applied in combination with magnetic prospecting. Our planet is a huge magnet. That means the Earth has a magnetic field. The characteristics of the field are influenced by the compositions of the mountain layers constituting the terrestrial core. For example, magma layers are more magnetically active than sands. A magnetic anomaly arises above a place of layer location.

Usually, gravimetric and magnetic methods are carried out before seismic prospecting. Seismic prospecting is carried out based on what information on the gravimetrical and magnetic anomalies is obtained. After detecting anticline traps or any other kind of traps, a detailed seismic investigation of the area is carried out to establish both the exact contours of the trap and the depth of its location. After that, drilling is possible.

There is one more geophysical method. This is the electrical prospecting method developed in France in 1923. This method is based on investigating the Earth's core by measuring the electromagnetic fields either of an artificial or natural origin on a ground. The main idea of the method is that the mountain layers have various electrical properties. For example, petroleum is dielectric, the minerals rich in iron are good electric conductors. Geophysicists investigate the Earth's core by creating an artificial electrical field and studying the electrical resistance of mountain layers. By tracing high-resistance layers, it is possible to identify deep relief anticline traps.

The geological and geophysical methods of crude oil prospecting do not always give the correct answer to the question whether there is an oil or gas deposit in the Earth. As a matter of fact, the presence of traps or collectors is necessary, but it is not a sufficient condition for the accumulation of crude oil deposits. Frequently, it has been observed that after drilling in the prospected area, neither petroleum nor gas is present. This is why it is recommended to carry out geochemical and hydrogeological prospecting after geological and geophysical investigation of the area. Based on results of the geochemical and hydrogeological prospecting, it is possible to confirm the presence of petroleum or gas in traps based on the microconcentration of the hydrocarbons on the Earth's surface in a researched area. Geochemical methods include gas, lumenescic, radioactive, photography and hydrochemical methods.

The gas photography method was first used in Russia in 1929. The main principle of the method is that there is filtration and diffusion of gases through the pores and cracks in the mountain layers of dispersed hydrocarbon gases around any crude oil deposit. Such an anomaly is usually a direct attribute of the crude oil or gas deposits. The disadvantage of the method is that the anomaly can be displaced from a source upwards of the layers.

The lumenescic method is based on an investigation of the bitumen dispersion area. The bitumen content in a layer rises above the crude oil or gas deposits. Samples from the layer are selected from shallow depths, and investigated using ultra-violet light.

The radioactive photography method is based on investigation of distribution of radioactive elements (first of all uranium) above petroleum and gas deposits. The radioactivity above the crude oil deposits is lower than around the deposit. However, radioactive anomalies in surface layers can be due to changed lithologic structure of layers. That is why this method is applied rarely.

With the hydrochemical method, the chemical composition of underground water together with its contents of dissolved gases and organic substances is studied. A large amount of hydrocarbons in the underground water shows a high possibility of the presence of petroleum deposit in this area.

More about oil and gas prospecting and modern investigation in this area can be found in references 38-46.

1.1.3.2 Drilling and crude oil extraction

Practically all the drilling today is carried out according to the rotary principle. A drilling tool screwed in at the lower end of the hollow linkage (either roller chisels or diamond chisels) is shifted in a rotary motion by a turntable installed in the drilling tower. The chisel drills into the Earth's layers. The borehole usually has a diameter of 10 to 70 cm. The borehole begins with the largest diameter at the surface and then decreases with depth.

Heavy bars are installed to increase the load pushing on the chisel and to improve the drilling capacities. The individual parts are lined with steel tubes and sealed against the mountain layers with cement. The layers of material drilled out must be removed from the borehole. The scavenge pump is used to ensure this removal as it maintains a liquid circulation in the drilling borehole. Water is constantly supplied to the chisel as coolant. It ascends the pipe system with constant pressure and thereby carries all detached rock particles forward. During the drilling process, particles that are constantly brought by the flushing water are examined in order to obtain information on the characteristics of the drilled layers. The first pipe system (so-called "preventers") is now installed for protection against uncontrolled oil or gas release. A simplified example of a drilling tower is shown in Figure 1.5.

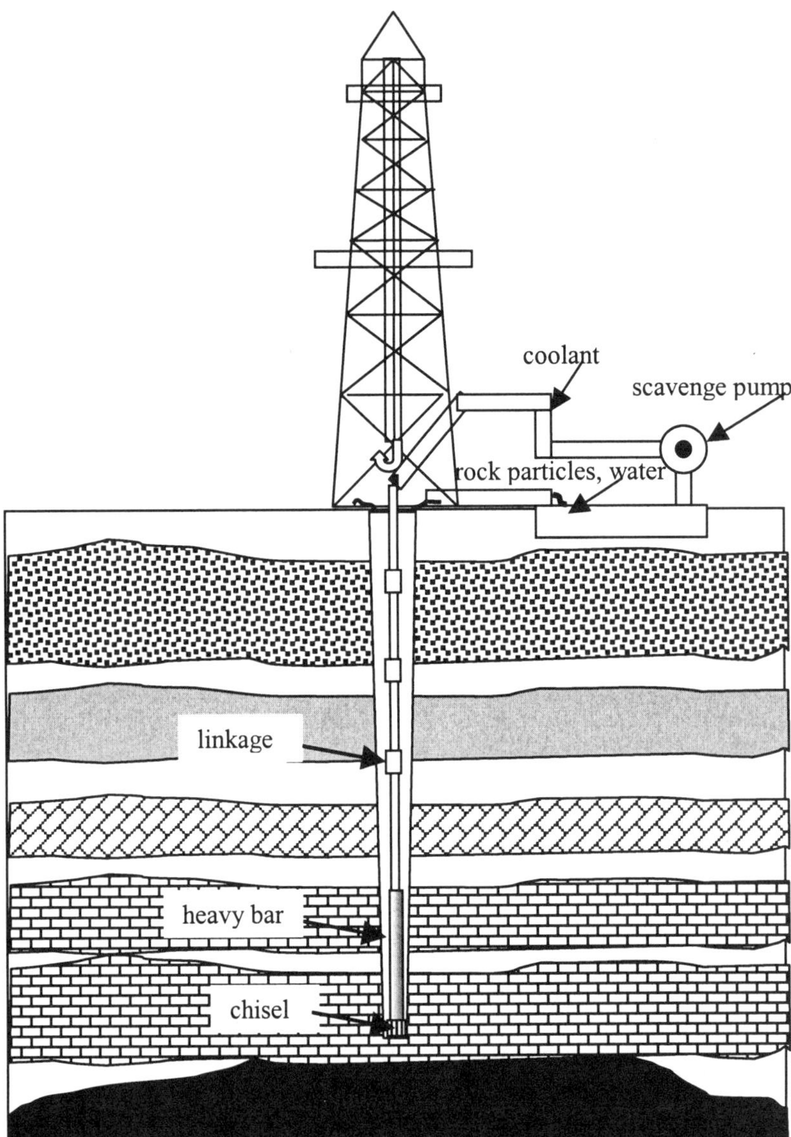

Fig. 1.5: Example of drilling tower.

Crude oil extraction begins after successful drilling. The three most popular extraction methods are:

1. Eruptive extraction. Each crude oil deposit has a natural layer pressure, which increases by up to one bar for every 10 meters of depth. Dissolved gas also flows together with the crude oil from the oil well, the combined flow resulting in pressure depletion in the well. Consequently, the gas begins to exit from the oil accompanied by volume enlargement. The exit of crude oil from the oil well in this case can be compared with the exit of soda water from the bottle when it is opened.
2. Gas elevator extraction. After eruptive extraction has ended naturally, one then sets the oil well under sufficient pressure that will force the oil out, and so extend the period of free flowing out of the oil. Gas elevator extraction has a distinct area of application. Frequently, one prefers to pump to gas elevators during extraction of oil from larger depths (approximately between 2500 and 3500 meters).
3. Pumping extraction. Pumping is the most frequently used artificial extraction method. The most important feature of this extraction method is the use of a pump. The pump consists of three sections: the deep pump, the pump linkage and finally the drive unit, which is represented by the pump support (so-called horse head) with the driving motor. The usual stroke rate for this pump varies from a few strokes up to 20 strokes per minute. An example of such a pump is shown in Figure 1.6.

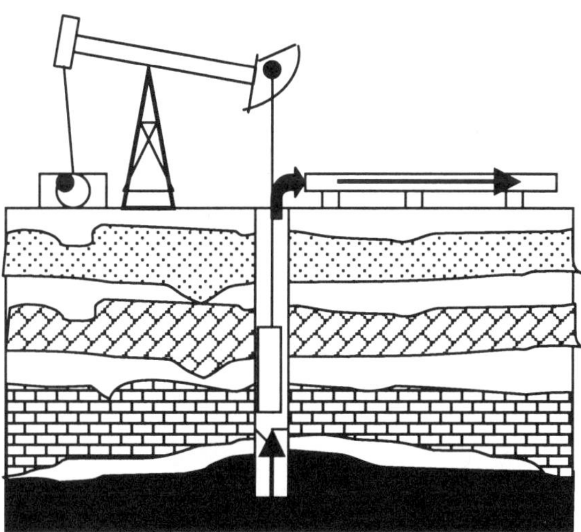

Fig. 1.6 Example of pumping extraction.

A special problem in crude oil extraction arises if high viscous petroleum or bituminous petroleum is being extracted. There is a significant number of oil wells in some places in the world, especially in Canada, with viscous and paraffinic crude oils, which are remote from practical power sources. Most of such oil wells are concentrated in deposits with porous traps. The most popular method for extraction of such oils is the thermal method.

In this method, the oil deposit is opened for extraction by the opening of boreholes, which are located in a uniform triangular grid formed by a thirteen-borehole system with six boreholes in each of the two concentric circles. These thirteen boreholes are located as follows: surrounding one central borehole is a circle (i.e. first concentric circle) of boreholes consisting of six boreholes, and a further six-borehole circle forming the second concentric circle. Thirteen boreholes are thus located so that each of the boreholes is located from the nearest ones by an identical distance. The heat-medium (for viscosity decrease of the petroleum) is carried out cyclically, with each cycle consisting of three stages.

In the first stage, the introduction of the heat-medium is conducted simultaneously through the central input borehole and every second extraction borehole of an external ring. Crude oil is extracted from all the other boreholes.

In the second stage, the introduction of the heat-medium is carried out through the central borehole as well, but the role of boreholes in the external ring changes: the heat-medium input boreholes now become extraction boreholes and the extraction boreholes now become heat-medium input boreholes. The amount of heat-medium introduced in the second stage is the same as that in the first stage.

In the last stage, only the central borehole is used as the heat-medium borehole, and all the other boreholes play the role of extraction boreholes.

More drilling and extraction of crude oil can be found in reference 47.

1.1.4 Consequences of Intensive Extraction and Processing of Oil [6-8]

Initially, the adverse effects of intensive petroleum extraction were not of any prime consideration. The key was to extract as much petroleum as possible. However, about fourteen years into the twentieth century, some indications of these adverse effects had already appeared.

It happened in the oil well in Wilmington (California, USA). This oil well is located between the southwest areas of Los Angeles and a gulf where Long Beach reaches the coastal quarters of the same resort city. The area of this petroleum pool is 54 km^2. The oil well was drilled in 1936. In 1938, it became the center of oil extraction in California. By 1968, almost 160 million tons of petroleum and 24 billion m^3 of gas were extracted from this oil well.

The location of the oil well at the center of industrial and densely populated areas of southern California, and also its proximity to the large oil refineries in Los Angeles, was considered very crucial to the economic development of the whole of California. As a result, a very high level of extraction from this well (as compared with other petroleum wells in North America) was constantly supported from the beginning of the operation of the oil well till 1966.

In 1939, the inhabitants of the cities of Los Angeles and Long Beach observed an appreciable concussion of the ground surface. The lowering of the ground above the oil pool thereby began. The intensity of this process amplified in the next fourteen years following when it started. The overall process resulted in what could be considered as an elliptic rift. After sixteen years, the amplitude of the lowering had already reached 8.7 m. The horizontal displacement with this amplitude was up to 23 cm, directed towards the center of the area. Movement of the ground was accompanied by earthquakes. Five strong earthquakes were registered in the period from 1949 till 1961. In a literal sense, the earth fell from under the feet. Ports, pipelines, urban structures, highways, bridges and petroleum wells collapsed. About 150 million dollars was spent for reparation. In 1951, the rate of sinking of the ground surface achieved a maximum value of 81 cm per year. There was a threat of flooding. Frightened by these events, the urban authorities of Long Beach stopped the operation of the oil wells in order for them to develop a good method to solve the problem.

In 1954, it was shown that the most effective method to mitigate this problem was water flooding (i.e., the input of water in the layer). Water input had other advantages as well. For example, there was an increase in oil extraction. The first stage of work for the flooding of the oil layer started in 1958, from which time a southern part of the productive layer was input 60 thousand m^3 of water per day. In the ten years that followed, the rate of flooding of this layer increased to 122 thousand m^3 per day. The lowering of the ground surface practically stopped. At the end of the twentieth century, the rate of lowering of the center of the zone did not exceed 5 cm per year. In some areas, there was even an elevation of the surface of about 15 cm. The oil wells are again in operation. Now, for each ton of petroleum extracted, about 1600 liters of water is needed. The maintenance of the layer pressure gives up to 70% of daily petroleum extraction in the old areas of Wilmington. Generally, 13,700 ton per day of petroleum is extracted from the oil wells.

Reports appeared in 1999 about the lowering of the Northern Sea, close to oil wells in Ecofisc, after the extraction of about 172 million tons of petroleum and 112 billion m^3 of gas. It was accompanied by deformations of well trunks and sea platforms. The consequences are difficult to predict, but their catastrophic character is obvious.

Lowering of the ground accompanied by earthquakes also occurred in old oil-extracting areas of Russia. It was especially strongly felt in Starogroznenskij oil wells. Weak earthquakes that resulted from intensive petroleum extraction were felt in this city in 1971. The earthquake was of intensity of 7 M at the epicenter and was located 16 km from the city of Groznyj. The aftermath was that owners of homes and office buildings had to be compensated. Workers who were displaced from working in the oil wells also had to be settled. Lowering of the ground occurred in the old oil wells in Azerbaijan. This was considered to be due to horizontal motions. These horizontal motions were responsible for breaking of pipes in the operational petroleum wells.

An earthquake was registered in April, 1989 in Tataria with an intensity up to 6 M. In the opinion of the local experts, there was a direct connection between amplification of petroleum extraction from oil wells and activation of weak earthquakes. Cases of breakage of oil well trunks and columns are on record in the Tataria example. Earthquakes in this area are especially dangerous since Tataria nuclear power station is located in this area. In all these cases, one effective measure to mitigate the problem is water flooding. Forcing water into the productive layer compensates for the extracted petroleum.

1.1.4.1 Dangerous fogs

Another danger in petroleum lies in the use of petroleum and gas as a fuel. During the combustion of these materials, enormous amounts of carbon oxides (such as carbon dioxide (CO_2)), various sulfur compounds (such as sulfur dioxide (SO_2)), nitrogen oxides (such as nitric oxide (NO) and nitrogen dioxide (NO_2)), etc. are released into the atmosphere. In the last half of the twentieth century, the contents of CO_2 in the atmosphere has increased by almost 288 billion tons, and more than 300 billion tons of oxygen has been used up for combustion processes involving various kinds of fuel, including stone coal. Thus, starting from the first fires of primitive man to the present, the atmosphere has lost about 0.02% of oxygen whereas the content of carbon oxides has increased by 12%. Annually, mankind burns 7 billion tons of fuel, for which more than 10 billion tons of oxygen is used up, and up to 14 billion tons of CO_2 is released into the atmosphere. In the future, these values will grow because of the general increase in the production and combustion of combustible minerals. It is predicted that in 2020, about 12,000 billion tons of oxygen (0.77%) will disappear from the atmosphere as a result of being used up for combustion processes. This means that in the next 100 years, the composition of the atmosphere will be essentially changed, probably, in an adverse direction.

It is feared that reduction in the quantity of oxygen and the growth in the content of CO_2 will cause adverse changes in the climate. The molecules of CO_2 allow short wave solar radiation to penetrate the atmosphere of the Earth and retain infrared radiation which penetrates into the terrestrial surface. This gives rise to the so-called greenhouse effect, resulting in an increase in the average temperature of the planet. It is indicated that the change in climate from 1880 till 1940 is substantially related to this effect. It seems that the climate will progressively change due to the greenhouse effect. However, other human influences on the atmosphere may help to neutralize the greenhouse effect.

Mankind contributes huge quantities of dust and other microparticles into the atmosphere. These particles shield solar beams and reduce the heating action of CO_2. According to the American expert K. Frazer, the turbidity of the atmosphere above Washington in the period from 1905 to 1964 has increased by 57%.

The transparency of the atmosphere above the Pacific Ocean was decreased by 30% from 1957 till 1967.

Atmospheric pollution by itself introduces another problem: it reduces the quantity of solar radiation that reaches the Earth's surface. According to data released from studies of the oceans and the atmosphere above the US by a US health agency, solar radiation in the period from 1950 to 1972 decreased by 8% during the fall season, and increased by only 3% in the spring. On the average, solar radiation has fallen by 1.3% since 1964. This is equivalent to the loss of approximately 10 minutes of daylight per day. This apparent triviality can have serious consequences on the Earth's climate.

In 1975, the atmospheric pollution above the United States resulted in an absolutely unexpected phenomenon. In the area of Boston (Massachusetts), it was established that there was a large increase in the quantity of ozone in the atmosphere – 0.127 ppm, whereas the established USA EPA safety limit is 0.08 ppm. It is known that ozone is formed in the atmosphere during the interaction of hydrocarbons with oxygen. A high concentration of ozone is more poisonous than charcoal gas. On August 10, 1975, the Department of Public Health Services of the state issued an "ozone-alarm", which lasted till August 14, 1975. This was already the second alarm for one year.

Other notable contributors to atmospheric pollution include jet planes, machines and factories. For example, modern jets need to use 35 tons of oxygen to enable them to cross the Atlantic Ocean. Also, the process of flying leaves "traces" behind thereby increasing the cloudiness of the atmosphere. Cars, whose worldwide total is already more than 500 million, pollute the atmosphere very significantly. They use fossil fuel and emit CO_2, SO_x, NO_x, etc. into the atmosphere. In the US, automobiles contribute up to half of the air pollution. This type of statistic led a US senator, E. Muskie, to declare in 1976 that 15 thousand men and women die each year in the US because of diseases caused by air pollution. There are strong efforts to seek to design automobile engines that can work with other types of fuel. For example, electric cars are no longer dreams of the past. There are demonstrations of various types of electric cars in many countries of the world. However, their commercial application worldwide has been hampered by the low capacity of the accumulators.

Petroleum fueled electrical power plants are also a major contributor to air pollution. Such power plants emit about 500 tons per day of sulfur into the environment in the form of sulfuric anhydride. This reacts with water resulting in the immediate formation of sulfuric acid. A French journalist, M. Ruze, has presented data to show that a French thermal power plant belonging to Electricite de France emits about 33 tons of sulfuric anhydride into the atmosphere on a daily basis. This can result in a daily production of about 50 tons of sulfuric acid. The aftermath of acid production is acid rain, the adverse effect of which covers the power plant and surrounding territory up to a radius of 5 km. Such rains have high chemical activity. They corrode even cement and marble.

Old monuments also suffer due specifically to atmospheric pollution. For example, the Athenian Acropolis, which had already seen 2,500 destructive earthquakes and fires, is today being threatened by another danger – atmospheric pollution. Atmospheric pollution has gradually destroyed the surface of the marble. This destruction is due to a combination of various processes. Smoke released into the air from industrial enterprises in Athens and wetted by droplets of water find their way on the marble. By the morning, evaporation of the water takes place, leaving behind on the marble an uncountable set of rifts that make the marble hardly appreciable. According to a Greek archeologist, Professor Narinatos, the monuments of ancient Ellada have suffered more from atmospheric pollution in the last 20 years than in the last 25 centuries of wars and invasions. To keep these invaluable creations of the ancient architects for future generations, the experts decided to cover a part of the monuments with a special blanket made from plastic.

Atmospheric pollution resulting from the release of various harmful gases and solid particulates has the result that the air in large cities has become unsuitable and even dangerous for human life. For example, in some cities of Japan and Germany, policemen on the streets breathe oxygen from special cylinders. This opportunity also exists for pedestrians for a fee. In the streets of Tokyo and some other cities in Japan, oxygen cylinders are provided for children so that they can get fresh air on their way to school. Japanese businessmen have opened special bars where humans can get non-alcoholic drinks and fresh air. It should be noted that in the last few years, conditions have improved considerably.

Another danger to human life is caused by smoke that is frequently emitted in large cities. The largest tragedy took place in London in 1952. On the morning of December 5, people in London could not see the sun. Extraordinarily dense clouds formed from the mixture of smoke and fog that hung above the city for 4 days. According to the official data, this took the life of four thousand people, and worsened the health conditions of thousands of others. Such smog has worsened the health conditions of people in other cities of Western Europe, America and Japan. In the Brazilian city of Sao Paulo, the level of air pollution exceeds three times the maximum allowable limits, and in Rio de Janeiro, this exceeds two times the limit. The usual diseases in these cities include irritation of the eyes, allergic diseases, and chronic bronchitis. It is also because of smog formation that the Japanese city of Nagoya has received the name "The Japanese smog capital".

Tokyo got third place among Japanese cities with the number of diseases caused by environmental pollution. More than four thousand patients were registered in this city in 1975. Also, in October of the same year, there was serious threat of poisoning in this huge city with a population of almost 12 million people. The concentrations of various harmful gases in the city had exceeded the allowable levels. Tokyo authorities had to order all factories to reduce the consumption of fuel by 40%. The inhabitants (especially children) were advised to stay indoors.

Even the plants are also affected by smog. For example, the green zone of Tokyo has been reduced by 12% in the last 10 years.

As a protective measure, the University of Kentucky (USA) has designed a special mini-gas mask against concentrations of various gases exceeding the allowable limits. If air contamination or pollution reaches a dangerous level, a tiny bulb flashes on the device.

1.1.4.2 Black oceans

From 2 to 10 million tons of petroleum is released annually into the world's ocean. One liter of petroleum deprives about 40 thousand liters of sea water of the oxygen that is used to sustain living inhabitants such as fish. Also, one ton of petroleum can pollute about 12 km^2 surface of the ocean.

There are many sources of petroleum pollution in the seas and oceans. These include failures of tankers and drilling platforms as well as dumping of ballast and clearing waters.

Perhaps, the first catastrophe that stirred worldwide interest in this issue was the one that took place in 1967. The supertanker "Tory Canyon" sank at the coast of Western Europe, and 120 thousand tons of petroleum poured into the sea. A huge petroleum slick painted the coastal waters of France and England. Approximately fifty thousand birds died (i.e. almost 90% of the see bird population in these areas).

In 1974 there was the failure of the American tanker "Transheron", which had on board 25,000 tons of petroleum. About 3,500 tons of petroleum flowed out from the holes in the tanker in only the first week. A huge petroleum slick covered the area of ten square kilometers and moved slowly towards the coastal city of the South Indian State of Kerala.

450 tons of petroleum poured from the Gulf Oil tanker "Afran Zodiac" into the Gulf of Bantry (Ireland) in January of 1976. The whole northern part of the gulf was under its cover.

In February of 1976, there was a fire onboard the tanker "San-Peter" carrying 33 thousand tons of petroleum as it was navigating its way from Peru to Colombia. The vessel sank and the petroleum content poured into the sea. The seamen from Colombia tried unsuccessfully for ten days to clean the waters in the area of the disaster.

The supertanker "Olympic Bravery", property of the company owned by the Greek magnate A. Onassis, sank at the coast of Great Britain in 1976. A mixture of petroleum and sand flooded the coast. The British government was compelled to involve naval forces in clearing the coast. However, it was not before irreparable damage was done to vegetation and animals.

About 20 million liters of petroleum was released into the waters in the area of the Hawaiian Islands in 1977 as a result of the disaster with the tanker "Irins Challenger". In the same year, 90 thousand tons of petroleum was released into the waters of the northern part of the Pacific Ocean as a result of the fire on board the tanker “Hawaiian Patriot”.

The year 1978 was marked by the largest tanker disaster on the coast of Great Britain. The American supertanker “Amoko Cadiz” sank on the reefs, and about 230 thousand tons of petroleum poured into the sea.

The collision of tankers “Atlantic Empress” and “Idgen Captain” in the Caribbean Gulf was the largest disaster in 1979. About 300 thousand tons of petroleum poured into the sea.

In the November storm of 1981, the Greek tanker “Globe Asini” had a disaster on the wave protector off the Port of Klaipeda. About 10 thousand tons of petroleum poured into the sea.

In August of 1983, the tanker “Castillo de Believer” had a disaster near the European Atlantic coast. The vessel sank, and about 250 thousand tons of petroleum was released into the ocean.

The tanker “Baia Paraiso” with one thousand tons of diesel oil on board sank at the coast of the Antarctic continent in January of 1989. Another tragedy happened in the Arctic waters of Alaska two months later. The tanker “Exxon Valdez” sank at the reef because of the fault of the captain. More than 40 thousand tons of petroleum flowed out from the hole into the waters. A petroleum slick covering an area up to 800 km^2 was formed. The area in the Strait of Prince William was declared "a zone of disaster". The US Navy was involved with the cleanup. Nevertheless, there was the "potential for ecological disaster" with consequences that are difficult to foresee, according to the Washington Post.

At the end of March of 1989, the Dutch River tanker ran aground in the area of Bad-Honnefa. Approximately one thousand tons of petroleum poured into the river. Petroleum film covered the river up to an area of 7 km^2.

In April of 1989, the Indian tanker "Kanchendgunga" ran aground at the reef in the Red Sea in the territorial waters of Saudi Arabia 5 km from the port of Jeda. More than 10 thousand tons of petroleum flowed out into the sea.

The sad list of tanker failures could go on, but their share in petroleum pollution of seas and oceans is not big. Three times more petroleum is released due to dumping of the water used in washing tanker tanks into the sea or oceans. Also, four times more pollution comes from the waste water of petrochemical factories; and almost the same amount of petroleum is released by the disasters that occur at sea platforms.

Now, the question arises: How can the ocean be rescued?

Fortunately, there are good methods. Some of these methods involve the application of dispersing additives - special substances - that adsorb petroleum; another is the treatment of petroleum slicks with iron powder and the subsequent collection of iron particles with a magnet. A promising method, however, lies in biological protection: the use microbes. These micro-organisms (already designed in the US and elsewhere for this purpose) are capable of splitting molecules of hydrocarbons.

Russian scientists have established that any bio-organisms that inhabit the sea are not affected by petroleum pollution. An example is cardium. This plays an important role in clearing sea water of petroleum by extracting for itself both food and oxygen from the pollutant. Nature has already designed for the clearing of seas and oceans following natural penetration of petroleum in the ocean. The penetration of oil from underground is seen, for example, on the coasts of California, Australia, Canada, Mexico and Venezuela as well as in the Persian Gulf. In one part of the bottom of the Californian Gulf, in the Strait of Santa Barbara, there is a natural outflow of petroleum from underground. It is supposed that this oil penetration had been taking place for the past ten thousand years. However, it was noticed for the first time in 1793 by the English seafarer D. Vancouver. According to US scientists, the annual penetration of petroleum into the world oceans from natural infiltration is approximately 200 thousand tons. This is about 6% from the total volume of petroleum that penetrates into the seas and oceans on the planet from anthropogeneous sources. Suffice it to say that during the disaster of the tanker "Tory Canyon", as much petroleum that will penetrate into water from the Californian oil wells for 28 years poured into the ocean. This was too large an amount of petroleum pollution to be mitigated by natural cleaning alone.

1.2 GENERAL PROPERTIES AND CLASSIFICATION OF PETROLEUM: COMPARISON OF PETROLEUM FROM DIFFERENT COUNTRIES

1.2.1 Fractions and Chemical Composition of Petroleum

Petroleum is a complex mixture of various organic compounds. It consists of different hydrocarbons and heteroatomic compounds. It is technically impossibly to separate petroleum into individual compounds. In any case, it is unnecessary to separate the petroleum to the component level in order to obtain a technological or industrial classification.

A very important petroleum property is its fractional composition. This property is determined in the laboratory by slowly heating the oil and separating it into fractions having specified boiling ranges. Every fraction is characterized by the temperature at which boiling begins as well as the temperature boiling ends.

In the industrial method, fractionation is achieved by the method of rectification. Using this method, the fractions with boiling point up to 350°C are separated at atmospheric pressure. These are called the light fractions. Usually, during atmospheric rectification, the following individual fractions are obtained:

- Boiling begins –140°C – gasoline fraction
- 140–180°C – heavy naphtha

- 180–240°C – kerosene fraction
- 240–350°C – diesel fraction

The residue after atmospheric distillation is called "atmospheric residue". This fraction, with a boiling point over 350°C, is usually distillated further at a low pressure or in a vacuum. This residue can be classified in two different ways depending on the intended application:

For further processing to fuel fractions:

- 350–500°C – vacuum gas oil
- over 500°C – vacuum residue

For further processing to lubricating oils:

- 300–400°C – light oil fraction
- 400–450°C – medium oil fraction
- 450–490°C – heavy oil fraction
- over 490°C – vacuum residue

All the fractions obtained from atmospheric residue are called "heavy fractions". On the other hand, the products obtained after secondary processing of the heavy fractions are considered to be light fractions if boiling of the fraction ends at < 350°C, and to be heavy fractions if boiling begins at > 350°C.

The amount of light fractions can be very different for oils from different oil wells. For example, it can be under 20% for some heavy oils from Alberta (Canada) and over 60% for some light oils form western Siberia (Russia). Typical analysis of a Canadian petroleum from oil wells in McMurray resulted in the following fractions:

- Gasoline and naphtha – 2.8%
- Kerosene – 0%
- Diesel – 19.0%
- Light oil – 4.3%
- Medium oil – 8.5%
- Heavy oil fraction – 13.2%
- Vacuum residue – 49.5%

The chemical composition of petroleum can be very different too. The main constituents of petroleum are:

- carbon (83 – 87%)
- hydrogen (11.5 – 14%)
- hetero-atoms (1 – 5.5%)

Table 1.1 shows the chemical composition of oils from the US, Canada, and Russia.

Even though the chemical composition of petroleum varies, almost all the hydrocarbons found in petroleum do not include alkenes. On the other hand, many oils with high amounts of paraffinic or naphthenic compounds or arenes are known.

It is necessary to make a chemical classification of petroleum since chemical properties are very crucial in selecting the right method for processing the oil. There are many classifications with regard to fractions and chemical compositions of petroleum. In this chapter we will present the basis for these classification methods.

Table 1.1: Ultimate analysis for crude oils.

Origin	Composition (wt.%)				
	Carbon	Hydrogen	Nitrogen	Oxygen	Sulfur
USA	86.6	11.8			
	83.5	13.3			
	85.5	14.2			
	83.6	12.9		3.6	
Canada	86.9	12.9			
	83.4	10.4	0.5	1.2	4.5
	82.8	11.8	0.3	1.7	3.4
Russia	85.3	11.6			
	86.1	12.8		0.9	0.2
	86.3	12.9		0.6	0.2

1.2.2 Chemical Classification of Petroleum

The chemical classification of petroleum that distinguishes between oils of a paraffin base from those of an asphaltene base was introduced into petroleum chemistry to distinguish the oils that separate paraffin on cooling from those that separate asphaltenes. The presence of paraffins is usually reflected in the paraffinic nature of the constituent fractions whereas a high asphaltic content corresponds with the naphthenic properties of the fractions. This could lead to the misconception that paraffin-base petroleum consists mainly of paraffins and that asphalt-base petroleum consists mainly of cyclic (or naphthenic) hydrocarbons. In order to avoid confusion, a mixed base has been introduced for those oils that leave a mixture of asphaltic petroleum and paraffins as residue from nondestructive distillation. A fourth class has also been suggested, the hybrid base; it includes asphaltic oils that contain a small amount of paraffins. A simplified scheme has been proposed by Speight [3] with paraffinic, naphthenic, aromatic, and asphaltic petroleums as extremes (Fig. 1.7). It is indeed possible to characterize petroleum semi-quantitatively in this manner.

An attempt to give the classification system a quantitative basis suggested that petroleum should be called asphaltic if the distillation residue contained less than 2% of parafins and paraffinic if it contained more than 5%. A division according to the chemical composition of the 250 to 300°C fraction has also been suggested (Table 1.2), but the difficulty in using such a classification is that in the fractions boiling above 200°C, the molecules can no longer be placed in one group because most of them are typically of a mixed nature.

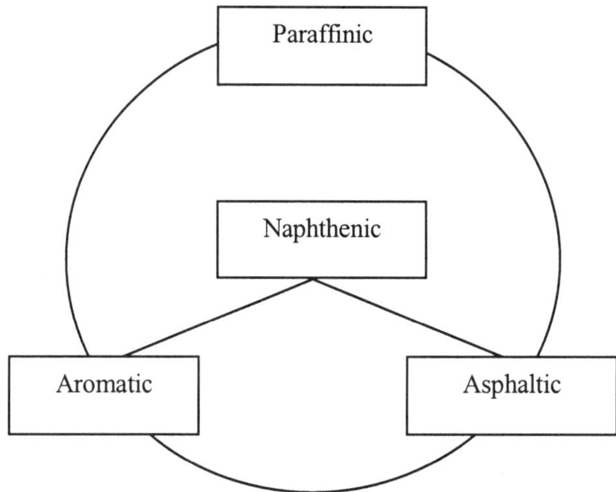

Fig.1.7: Composition diagram for petroleum.

Purely naphthenic or aromatic molecules occur very seldom; cyclic compounds generally contain paraffinic side chains, and often even aromatic and naphthenic rings occur side by side. More direct chemical information is often desirable and can be supplied by means of the correlation index (CI).

This index was developed by the U.S. Bureau of Mines. It is based on the plot of specific gravity at 48.64°C versus the reciprocal of the boiling point in degrees Kelvin (K = T°C + 273) for pure hydrocarbons for which the line described by the constants of the individual members of the normal paraffin series is given a value of CI = 0, and a parallel line passing through the point for the values of benzene is given as CI = 100. The following empirical equation (1.2) has been derived for estimating CI:

$$CI=473.7d - 456.8 + 48.640/T \tag{1.2}$$

where T, in the case of a petroleum fraction, is the average boiling point, determined by the standard Bureau of Mines distillation method
d is the specific gravity.

Table 1.2: Petroleum classification according to chemical composition.

Class of petroleum	Composition of 250-300°C fraction, wt.%				
	Par.	Naphth.	Arom.	Wax	Asph.
Paraffinic	46-61	22-32	12-25	1.5-10	0-6 - 6
Paraffinic-naphthenic	42-45	38-39	16-20	1-6	-6
Naphthenic	15-26	61-76	8-13	trace	0-6 - 6'
Paraffinic-naphthenic-aromatic	27-35	36-47	26-33	0.5- 1	0 - 10
Aromatic	0- 8	57-78	20-25	0 - 0.5	0-20

Thus, values for the index between 0 and 15 indicate a predominance of paraffinic hydrocarbons in the fraction; values from 15 to 50 indicate a predominance of either naphthenes or mixtures of paraffins, naphthenes, and aromatics; values above 50 indicate a predominant aromatic character. Although the correlation index yields useful information, it is in fact limited to distillable materials and, when many petroleum samples are to be compared, the analysis of results may be cumbersome.

It is also possible to describe a crude oil by an expression of its chemical composition on the basis of the correlation index figures for its middle portions.

1.2.3 Classification by Density

Density has been the principal and often the only specification of petroleum products and was taken as an index of the proportion of gasoline and kerosene present. As long as only one kind of petroleum is in use the relations are approximately true. However, since a wide variety of crude oils having various other properties occur in nature and have come into use, the significance of density measurements has disappeared. Nevertheless, petroleum samples having other properties that are similar can still be rated by gravity as can gasoline and naphtha within certain limits of other properties. The use of the density values has been advocated for quantitative application using a scheme of the American Petroleum Institute (API) based on the gravity of the 250 to 275°C (at the pressure 1 bar) and the 275 to 300°C (50 mbar) distillation fractions. Indeed, analysis of petroleum from different sources worldwide showed that 85% fell into one of the three classes: paraffin, intermediate, or naphthene base. It has also been proposed to classify heavy oils according to characterization gravity, defined as the arithmetic average of the instantaneous gravities of the distillates boiling at 177°C, 232°C and 288°C vapor line temperature at 33 mbar pressure in a true boiling point distillation.

In addition, a method of petroleum classification has been developed that is based on other properties as well as the density of selected fractions. The method consists of a preliminary examination of the aromatic content of the fraction boiling up to 145°C as well as that of the asphaltene content, followed by more detailed examination of the chemical composition of the naphtha (b.p. <200°C). For this examination, a graph (a composite of curves expressing the relation between percentage distillate from the naphtha, the aniline point, refractive index, specific gravity, and the boiling point) is used. The aniline point after acid extraction is included in order to estimate the paraffin-naphthene ratio.

1.2.4 Characterization by Viscosity-Gravity Constant (vgc)

This parameter, along with the Universal Oil Products (UOP) characterization factor, has been used, to some extent, as a means of classifying crude oils. Both parameters are usually employed to give an indication of the paraffinicity of the petroleum. Both have been used, if a subtle differentiation can be made, as a means of petroleum characterization rather than for petroleum classification.

Nevertheless, the viscosity-gravity constant is one of the indexes proposed to characterize oil types. For heavy oils, the low-temperature viscosity is difficult to measure. The viscosity-gravity constant for such type of petroleum is calculated by the formula

$$VGS = d - 0.24 - 0.022 \log (v - 35.5)/0.755 \qquad (1.2)$$

where d is the specific gravity at 48.64°C

v is the Saybolt viscosity at 99°C

The viscosity-gravity constant is of particular value in indicating petroleum of a predominantly paraffinic or cyclic composition. The lower the index number, the more paraffinic the stock; for example, naphthenic lubricating oil distillates have a vgc of 0.876 while the raffinate obtained by solvent extraction of lubricating oil distillate has a vgc of 0.840.

The UOP characterization factor is perhaps one of the more widely used of the derived characterization or classification factors and is defined by the formula

$$K = T_B^{1/3}/d \tag{1.3}$$

where T_B is the average boiling point in degrees Rankine (degrees Fahrenheit + 460) and d is the specific gravity at 48.64°C.

This factor has been shown to be additive on a weight basis. It was originally devised to show the thermal cracking characteristics of heavy oils; thus, highly paraffinic oils have K of from 12.5 to 13.0 while naphthenic oils have K of from 10.5 to 12.5.

1.2.5 Technological Classification of Petroleum

According to technological petroleum classification, the oil can be classified as:

- low sulfur oil containing not more than 0.5% of the sulfur, whereby the gasoline fraction contains less than 0.1% sulfur and diesel fraction less than 0.2%.
- sulfur petroleum containing over 0.5% but under 2% of the sulfur, whereby the gasoline fraction contains less than 0.1 % sulfur and the diesel fraction less than 1.0%.
- high sulfur petroleum containing over 2% of sulfur.
- low paraffinic petroleum containing under 1.5% of paraffins. This type of oil can be used for production of jet and winter diesel fuels without deparaffinization.
- medium paraffinic petroleum containing over 1.5% and under 6% of paraffins. This type of oil can be used for production of jet and summer diesel fuels without deparaffinization.
- high paraffinic petroleum containing over 6% of paraffins. This type of oil can be used for production of diesel and jet fuels only after deparaffinization.

1.3 PRODUCTS FROM CRUDE OIL [2-4]

The list of products from petroleum is endless. Oil products fuel planes, trains, cars, trucks, buses, and so on. Oil is also used to heat homes. Chemicals made from oil are used to make products that range from makeup, toys, fabrics, sneakers and football helmets to aspirin, toothpaste, deodorant, clothes, hair dryers and lipstick to name just a few. Plastics made from oil are widely used in everything from compact discs and video cassette recorders, to computers, television sets, and telephones. In this chapter we will focus on petroleum fuels since it is presently the most important power source.

Crude oil contains a wide range of hydrocarbons and other compounds containing sulfur, nitrogen, etc. In the refinery, petroleum is distilled into various fractions. Depending on the desired final products, these fractions are further processed and then blended to yield a wide variety of products.

Typical final products are:

- gases for chemical synthesis and fuel, liquified gases
- aviation and automotive gasoline
- aviation (jet) and lighting kerosene
- diesel fuel
- distillate and residual fuel oils
- lubricating oil base grades
- paraffin oils and waxes

Many of the common processes in the refinery are intended to increase the yield of blending feedstocks for gasoline. As such, most modern fuels are represented by fuel fractions compounded from the products of many different processes.

Typical modern refinery processes for gasoline components include:

- Catalytic cracking
- Hydrocracking
- Isomerization
- Reforming
- Alkylation

1.3.1 Gasoline

In the late 19th century, the most suitable fuels for automobile use were coal tar distillates and the lighter fractions from the distillation of crude oil. During the early 20th Century, the oil companies were producing gasoline as a simple distillate from petroleum. On the other hand, automotive engines became vastly improved and these required a more suitable fuel. Typical gasoline products in the 1920s had octane numbers (ratings) in the range 40-60. Tetraethyl lead was often used to enhance the octane number.

Because sulfur in gasoline inhibited the octane-enhancing effect of the alkyl lead, there was a restriction on the sulfur content of thermally cracked refinery streams for gasoline. By the 1930s, it was determined that larger hydrocarbon molecules had major adverse effects on the octane number of the gasoline, and specifications consistent with the desired properties were developed.

The increase in the compression ratios of cars started in the 1950s. The consequence was that such car engines required fuels with higher octane numbers. Thus, octane number, lead levels, and vapour pressure increased whereas sulfur and olefin contents decreased. Some new refining processes (such as hydrocracking), designed to provide hydrocarbon components with good lead response and octane number, were introduced. Minor improvements were made to gasoline formulations to improve yields and octane number until the 1970s, when unleaded fuels were introduced to protect the catalysts that were also being introduced for environmental reasons. From 1970 until 1990, there was a slow but progressive change in the gasoline produced. Lead was being phased out. Lead levels decreased, octane number initially decreased as well, but vapor pressures continued to increase while sulfur and olefins remained constant. The aromatic content increased. In 1990, the US Clean Air Act started to force the implementation of major compositional changes in gasoline. This resulted in the plummeting of the vapor pressure and the increase in the oxygenate levels. These changes will continue into the 21st century as a way to minimize the polluting effect of the use of gasoline as fuel in automobile engines.

The move to unleaded fuels continues worldwide. However, several countries have increased the aromatics content (by up to 50%) to replace alkyl lead. These highly aromatic gasolines can lead to damage of elastomers and increased levels of toxic aromatic emissions if is used without catalysts.

Gasoline as a fuel is composed of a mixture of various hydrocarbons, which can be burnt to form water (H_2O) and CO_2. If combustion is not complete, carbon monoxide (CO) is also formed.

The following main groups of hydrocarbons are contained in gasoline:

- saturated hydrocarbons or alkanes
- unsaturated hydrocarbons or olefins
- naphthene or cyclic hydrocarbons
- aromatics
- oxygenates
- other hetero-atom compounds

Gasoline contains over five hundred types of hydrocarbons that have between 3 to 12 carbon atoms in their structure. Gasoline has a boiling range from 30 to 180°C at atmospheric pressure. The boiling range is narrowing as the initial boiling point is increasing, and the final boiling point is decreasing. Both these changes are for environmental reasons.

Saturated hydrocarbons have the following major properties:

- Are thermally and chemically stable, it is the major component of leaded gasolines.
- Tend to burn in air with a clean flame.

- The octane number depends on branching and number of carbon atoms. Generally octane number is low for n-alkanes and high for the iso-alkanes (iso-octane is assigned the octane number of 100).

The cyclic hydrocarbons all have the positive properties of the alkanes with regard to thermal and chemical stability as well as good environmental properties. In addition, they have a higher octane number in comparison to alkanes.

Alkenes or unsaturated hydrocarbons have the following major properties:

- Are chemically unstable.
- Tend to be reactive and toxic, but have high octane numbers.

Arenes or aromatics are characterized as follows:

- Gradually being reduced to less than 20% in the US.
- Tend to be more toxic than other hydrocarbons, but have the highest octane ratings.
- Some countries are increasing the aromatic content (up to 50% in some super unleaded fuels) to replace the alkyl lead octane enhancing additive.

Polynuclear aromatics are high boiling compounds, and are only present in small amounts in gasoline. The simplest and least toxic polynuclear arene, naphthalene, is present in only trace amounts in traditional gasoline, and in even lower levels in reformulated gasoline. The larger multi-ringed polynuclear arenes are highly toxic, and are not present in gasoline.

Oxygenates contain oxygen that does not contribute to the energy content, but because of their structure, provides a reasonable antiknock value. Thus, they are good substitutes for aromatics. They also reduce the tendencies of the formation of toxic gases. Most oxygenates used in gasoline are either alcohols or ethers that contain 2 to 6 carbon atoms per molecule. Alcohols have been used in gasoline since the 1930s whereas ether (such as methyl tertiary butyl ether (MTBE)) was first used in commercial gasoline in Italy in 1973 and in the US in 1979.

Oxygenates can be produced from fossil fuels or from biomass. MTBE is produced by reaction of methanol with iso-butylene in the liquid phase over an acidic ion-exchanger resin catalyst at 100°C. MTBE production has increased at the rate of 10 to 20% per year.

Oxygenates are added to gasoline to reduce emissions. However, they can only be effective if the hydrocarbon fractions are modified to utilize the oxygenate's octane number and volatility properties. If the hydrocarbon fraction is not correctly modified, oxygenates can even increase smog-forming and toxic emissions both of which are undesirable. It should also be noted that oxygenates do not necessarily reduce all toxins.

Initially, oxygenates were added to hydrocarbon fractions that were just slightly modified unleaded gasoline fractions. These were known as "oxygenated" gasolines. Since 1995, the hydrocarbon fraction has been significantly modified resulting in "reformulated gasoline".

Oxygenates added to gasoline function in two ways. First, they have a high blending octane number, and so can replace high octane aromatics in the fuel. These aromatics are responsible for disproportionate amounts of CO and hydrocarbons emissions during combustion. Oxygenates cause engines without sophisticated engine management systems to move to the lean side of stoichiometry, thus reducing emissions of CO (2% oxygen can reduce CO by 16%) and hydrocarbons (2% oxygen can reduce hydrocarbons emissions by 10%). Other researchers have observed that similar reductions occur when oxygenates are added to reformulated gasoline and used on older and newer vehicles. They have also shown that, unfortunately, NO_X levels may increase, as may some regulated toxins.

On the other hand, on vehicles with engine management systems, the fuel volume will be increased to bring the stoichiometric combustion to the preferred optimum setting. It is to be noted that oxygen in the fuel cannot contribute energy. Consequently, the fuel has less energy content. Thus, for the same efficiency and power output, more fuel has to be burned. Therefore, the slight improvements in combustion efficiency that oxygenates provide on some engines usually do not completely compensate for the energy loss due to the presence of oxygen.

There are some other properties of oxygenates that have to be considered if they are to be used as fuels. These relate particularly to (i) their ability to form very volatile azeotropes that cause the fuel's vapor pressure to increase, (ii) the chemical nature of the emissions, and (iii) their tendency to separate into a separate water-oxygenate phase when water is present. These problems are solved more successfully in the reformulated gasoline than in the original oxygenated gasoline.

Table 1.3 shows some hydrocarbons in gasoline and their corresponding octane numbers.

The main characteristics of gasoline are:

- Vapor pressure and distillation classes. There are six different classes according to location and/or season. As gasoline is distilled, the temperatures at which various fractions are evaporated are recorded. Specifications define the temperatures at which various percentages of the fuel are evaporated. Distillation limits include maximum temperatures at which 10% is evaporated (50-70°C), 50% is evaporated (110-121°C), 90% is evaporated (185-190°C), and the final boiling point (225°C). A minimum temperature for 50% evaporated (77°C), and a maximum amount of residue (2%) after distillation. Vapor pressure limits for each class (54, 62, 69, 79, 93, 103 kPa, respectively) are also specified.
- Vapor lock protection classes. There are five classes for vapor lock protection. This classification depends on location and/or season. The limit for each class is a maximum vapor-liquid ratio of 20 at one of the specified testing temperatures of 41, 47, 51, 56, or 60°C.

- Antiknock index. Limits are not specified, but changes in engine requirements according season and location are discussed.
- Lead content
- Copper strip corrosion. This determines the ability to tarnish clean copper, indicating the presence of any corrosive sulfur compounds.
- Maximum sulfur content. Sulfur adversely affects catalysts and fuel hydrocarbon lead response. Sulfur may also be emitted as polluting sulfur oxides.
- Maximum solvent washed gum limits the amount of gums present in fuel at the time of testing to 5 mg/100 mLs. The results do not correlate well with actual engine deposits caused by fuel vaporization.
- Minimum oxidation stability. This ensures that the fuel remains chemically stable, and does not form additional gums during periods in the distribution systems (can be up to 3-6 months). The test is conducted by heating the sample with oxygen inside a pressure vessel. The time required for a significant oxygen uptake is measured to indicate minimum oxygen stability.
- Water tolerance. This is the highest temperature that causes phase separation of oxygenated fuels. The limits vary according to location and month.

An understanding of the reasons for detonation makes it obvious that there are two ways to improve the octane number of gasoline. It is possible to change the composition of gasoline. However, this is a difficult and expensive method. It is also possible to add additives that destroy peroxides. However, any "knock" caused by the fuel will rapidly mechanically destroy the engine. The problem is to identify economic additives which could be added to gasoline to prevent knock effects. Originally, iodine was the best antiknock additive available, but was not a practical gasoline additive, and was used as the benchmark. Later, tetra ethyl lead was added to the gasoline to improve the octane number.

Tetra ethyl lead $(C_2H_5)_4Pb$ is an organometallic compound, which is well soluble in hydrocarbons in the gasoline. At the temperatures of 200-250°C, this substance is cracked to lead and four ethyl radicals $C_2H_5\cdot$. All the compounds formed prevent the formation of explosive compounds in gasoline, or promote their fast destruction. However, the application of pure tetra ethyl lead was impossible. The formed lead deposits as a residue on the walls of the cylinder, and further makes the working of the engine impossible. This is why, in practice, tetra ethyl lead is mixed with various alkyl-halogens. At high temperature, alkyl-halogens are cracked and react with lead to form volatile salts, which leave the engine together with the exhaust gases.

Table 1.3: The average gasoline composition.

15% n-paraffins	RON	MON
n-pentane	62	66
n-hexane	19	22
n-heptane	0	0
n-octane	-18	-16
n-decane	-41	-38
n-dodecane	-88	-90
n-tetradecane	-90	-99
30% iso-paraffins		
2-methylpropane	122	120
2-methylbutane	100	104
2-methylpentane	82	78
3-methylpentane	86	80
2-methylhexane	40	42
3-methylhexane	56	57
2,2-dimethylpentane	89	93
2,2,3-trimethylbutane	112	112
2,2,4-trimethylpentane	100	100
12% cycloparaffins		
Cyclopentane	141	141
Methylcyclopentane	107	99
Cyclohexane	110	97
Methylcyclohexane	104	84
35% aromatics		
Benzene	98	91
Toluene	124	112
ethyl benzene	124	107
meta-xylene	162	124
para-xylene	155	126
ortho-xylene	126	102
3-ethyltoluene	162	138
1,3,5-trimethylbenzene	170	136
1,2,4-trimethylbenzene	148	124
8% olefins		
2-pentene	154	138
2-methylbutene-2	176	140
2-methylpentene-2	159	148
cyclopentene	171	126
1-methylcyclopentene	184	146
1,3 cyclopentadiene	218	149
dicyclopentadiene	229	167
Oxygenates		
methanol	133	105
ethanol	129	102
iso-propyl alcohol	118	98

Adding tetra ethyl lead appeared to be quite an effective method to improve antidetonating properties of gasoline. The additive added in a very small amount in gasoline allows the increase of the octane number by about 5 - 10 units. Unfortunately however, lead and tetra ethyl lead are very poisonous substances. If contacted with the skin, they can filter into the blood. The lead compounds formed in engines falls on the ground and near vegetation. This can produce tragic consequences for the ecology.

The alkyl leads rapidly became the most cost-effective method of enhancing octane. Up until the late 1960s, alkyl leads were added to gasoline in increasing concentrations to improve octane number. However, in later years, the use of tetra ethyl lead was not encouraged. Presently in most countries of the world, the use of tetra ethyl lead as an additive to gasoline is prohibited by law.

However, the need for gasoline with high octane number is ever increasing. The next possible way to improve the octane number of gasoline is to mix it with other compounds having high octane numbers. In the table above is shown that oxygenates have very high octane number (over 100). It is known that in the US, the so-called gasohol, a mixture of gasoline and alcohol, was used. In Europe syntin was produced from CO_2 and hydrogen. The product of this process includes a mixture of different types of alcohol.

However, the most effective and ecologically sound additive to gasoline nowadays is methyl tetra butyl ether ($(CH_3)_3COCH_3$). This additive is well known as MTBE. The octane number of MTBE is 135. The use of methyl tetra butyl ether as the additive fortunately solved a problem that resulted from using alcohols as the additive. In contrast to alcohols, MTBE is not soluble in water. It is known that water in different amounts can be found in industrial reservoirs. Even gasoline contains small traces of water. All alcohols are well soluble in water such that alcohol can stay in water at the bottom of the reservoir during storage. MTBE is insoluble in water and so this problem does not arise. Another problem with alcohol is that light alcohols (for example methanol) have a much lower heat of combustion than gasoline. It means that the need of fuel in tanks of the cars should be increased. MTBE has approximately the same fuel characteristics as gasoline. Moreover the presence of oxygen in its structure essentially improves the combustion process of gasoline in cylinders and reduces the contents of products of incomplete combustion in the exhaust gas.

There is only one disadvantage of using MTBE. There is a shortage in the production of MTBE in comparison to the need for high quality fuel for modern cars. One more method to improve the antidetonating properties of gasoline was introduced in the 1970s. In this method, the properties of gasoline were not changed. Instead the working mechanism of the engine was changed. It was introduced to make the engine capable of working with two different gasolines. Because the highest danger of detonation in engine occurs only during the forced working regime, it is rational to use two different gasolines in one engine: the high quality gasoline for the forced regime of the engine and the gasoline with low octane number for the stable working regime. It was shown that normally the engine works about 20% of the general working time in forced regime. This means that by using two tanks in a car, one can save up to 80% of the high quality gasoline by partly using low quality fuel. However, this method did not become popular and therefore did not result in industrial production of cars with dual tanks.

It is useful at this stage, close to the end of the chapter on gasoline, to mention additional properties of gasoline of importance:

- Volatility. This affects evaporative emissions and derivability. It is a property that is changed with location and season. For example, fuel for mid-summer Arizona would be difficult to use in mid-winter Alaska. For this reason, the US is divided into zones, according to altitude and seasonal temperatures. The fuel volatility is adjusted accordingly. Incorrect fuel may result in difficult starting in cold weather, carburetor icing, vapor lock in hot weather, and crankcase oil dilution. Volatility is controlled by distillation and vapor pressure specifications. The higher boiling fractions of the gasoline have significant effects on the emission levels of undesirable compounds (hydrocarbons and oxygen containing compounds), and a reduction of 40°C in the end boiling point will reduce the levels of benzene, butadiene, formaldehyde and acetaldehyde by 25%, and will also reduce hydrocarbons emissions.
- Combustion Characteristics. Gasoline contains a wide variety of hydrocarbons. Thus, the only significant variable that characterizes different gasoline grades is the octane number. Most other properties are similar. There are only slight differences in combustion temperatures (most are around 2000°C in isobaric adiabatic combustion). It should also be noted that the actual temperature in the combustion chamber is also determined by other factors, such as load and engine design. The addition of oxygenates changes the pre-flame reaction pathways and also reduces the energy content of the fuel. The level of oxygen in the fuel is regulated according to regional air quality standards.
- Stability. Motor gasoline may be stored up to six months. As such, they must not form gums during storage. Gums formed may precipitate. Gums are formed due to oxidation and polymerization reactions involving unsaturated hydrocarbons. Antioxidants and metal deactivators are added to prevent gum formation due to these reactions.
- Corrosiveness. Sulfur in the fuel facilitates corrosion. The combustion of sulfur containing fuels results in the formation of corrosive gases that attack the engine, the exhaust as well as the environment. Sulfur also adversely affects the alkyl lead octane response. Furthermore, it adversely affects the exhaust catalysts. However, the monolithic catalysts usually recover when the sulfur content of the fuel is reduced. In this case, sulfur is considered to be an inhibitor rather than a catalyst poison. The copper strip corrosion test and the sulfur content specification are used to ensure fuel quality. The copper strip test measures the active sulfur whereas the sulfur content reports the total sulfur present.

It has to be noted, that aviation gasoline is required to be an all hydrocarbon product. Its components must be chemicals that contain only carbon and hydrogen atoms. The use of oxygenated or other oxygen containing compounds, such as alcohols or ethers, is not allowed. Only a few select additives are permitted and their use is strictly controlled and limited. The primary ingredient in aviation gasoline is isooctane.

More about gasoline can be found in references 51-53.

1.3.2 Jet Fuel (Kerosene)

Kerosene is a hydrocarbon fraction that typically boils between 170-270°C (for narrow kerosene fraction or Jet A1), 100-250°C (for wide kerosene fraction or JP-4), or 170-315°C (for heavy kerosene, Russian standards). It contains about 20% aromatics. However, the aromatic content is usually reduced for high quality lighting kerosene, as the aromatics reduce the smoke point. The other constituents include 20 to 60% aliphatic compounds, 20 to 60% naphthenes and up to 1% unsaturated compounds. The major use for kerosene today is as aviation turbine jet fuels. Special properties are required for this application, including high flash point for safe refueling (38°C for Jet A1), low freezing point for high altitude flying (-47°C for Jet A1), and good water separation characteristics.

The quantity and quality of kerosene vary with the type of crude oil; some crude oils yield excellent kerosene, but others produce kerosene that requires substantial refining and hydrotreatment. Kerosene is a very stable product, and additives are not required to improve the quality. Apart from the removal of excessive quantities of aromatics, kerosene fractions may need only hydrotreatment if hydrogen sulfide is present. The kerosene fraction from shale oil is like the gasoline fraction, which generally contains high amounts of nitrogen and oxygen. However, hydrotreatment will remove most of the nitrogen, but catalyst degeneration can be quite severe.

Jet fuels must burn cleanly and remain fluid and free from wax particles at the low temperatures experienced in high-altitude flight. The conventional freeze-point specification for commercial jet fuel is –50ºC. The fuel must also be free of any suspended water particles that might cause blockage of the fuel system with ice particles. Special-purpose military jet fuels have even more stringent specifications.

The quality of this fuel varies widely. This is why there is no universal jet fuel. Instead, there very many different types of jet fuels. The main types for North America and Russia are:

- Jet-A: This is a narrow cut kerosene product. This is the standard commercial and general jet fuel available in the United States. It usually contains no additives but anti-icing chemicals may be added.
- Jet-A1: This is identical to Jet-A with the exception of its freezing point. It is used outside of the US and is the fuel of choice for long haul flights where the

fuel temperature may fall to near the freezing point. It often contains a static dissipator additive.

- Jet-B: This is a wide cut kerosene with lighter gasoline type naphtha components. It is used widely in Canada. It contains a static dissipator and has a very low flash point;
- JP-4: This is a military designation for a fuel like Jet-B but contains a full additive package including corrosion inhibitor, anti-icing and static dissipator.
- JP-5: This is another military fuel. It has a higher flash point than JP-4 and was designed for use by the US Navy on board aircraft carriers. It contains anti-ice and corrosion inhibitors.
- JP-8: This is like Jet-A1 with a full additive package.
- T-5: This is a Russian heavy jet fuel without any additives and had been used for civil aviation. It has not been used any more since 1971.
- RT: This is a Russian hydroteated jet fuel with antioxidation additive ionol;
- T6: This is a Russian thermo-stabilized jet fuel made from the east sulfuric petroleums with antioxidation additive ionol.
- T7: This is the same as the RT jet fuel without any additives.
- T8W: This is a Russian hydrotreated jet fuel without any additives.

1.3.3 Diesel

Diesel fuel is derived from petroleum. Diesel, gasoline and jet fuel are different cuts from the refining of petroleum. The difference is that diesel contains heavier hydrocarbons with a higher boiling point than gasoline and jet fuel. The term diesel fuel is therefore generic; it refers to any fuel mixture developed to run a diesel-powered vehicle, i.e. engines with compression ignition engines.

Diesel is a hydrocarbon fraction that typically boils between 250-380°C. Diesel engines use the cetane (n-hexadecane) rating to assess ignition delay. Normal alkanes have a high cetane rating, (n-C_{16}=100) representing short ignition delays. On the other hand, aromatics (alpha-methylnaphthalene = 0) and iso-alkanes have low ratings representing long ignition delays. Because of the size of the hydrocarbons, the low temperature flow properties control the composition of diesel. Consequently, additives are usually added to prevent filter blocking in cooler temperatures. There are usually summer and winter diesel grades. Environmental legislation is reducing the amount of aromatics and sulfur permitted in diesel. Their emission as well as those of small particulates are considered as possible carcinogens and are also known to cause other adverse health effects.

Cetane number is one of the most widely known quality characteristics of diesel fuel. It is important to not confuse cetane number with cetane index.

The fuel in a diesel engine is ignited by the high-temperature and high-pressure air created in the cylinder as the piston nears the end of the compression stroke. In contrast, fuel in gasoline engines is ignited by a spark plug.

Cetane number is therefore a measure of the power with which diesel fuel is ignited during the compression stroke. The number is determined using a specified laboratory test engine. Cetane index is calculated using an equation involving the density and the distillation curve of the fuel. Consequently cetane index cannot be increased and improved by cetane-improving additives because the equation does not account for the type and amount of cetane-improving additive in the fuel. However, cetane index can be modified through changes in the density or distillation curve of the fuel.

The time between the beginning of fuel injection and the start of combustion is called "ignition delay". As stated earlier, higher cetane number fuels result in shorter ignition delays, providing improved combustion, lower combustion noise, easier cold starting, faster warm-up, less smoke, and in many engines, reduction of emissions.

In most diesel engines, the ignition delay is shorter than the duration of injection. Under these circumstances, the total combustion period can be divided into the following four stages:

- Ignition delay
- Rapid pressure rise
- Constant pressure or controlled pressure rise
- Burning on the expansion stroke

Rapid pressure rise results from the large number of ignition points and the accumulation of fuel during the ignition delay period. Following this stage, the rate of combustion can be controlled to a much greater degree by controlling the injection rate since the fuel is being injected into the flame. Because the rapid pressure rise represents uncontrolled and inefficient combustion, it is necessary to limit the ignition delay to a minimum. This limitation can be accomplished mechanically by the selection of a spray pattern configuration properly tailored to the combustion chamber.

The nature of the fuel is also an important factor in reducing ignition delay. Physical characteristics such as viscosity, density, and medium boiling point are influencing parameters. Hydrocarbon composition is also important as it affects both the physical and combustion characteristics of the fuel. Ignition delay characteristics of diesel fuel are of primary importance since they directly influence the interval of uncontrolled combustion during the injection and, as such, the overall engine performance.

The next important parameter of diesel fuel is stability or storage stability. As fuel ages, it can become unstable and form insoluble particulates that accumulate and eventually end up on the fuel filter. For the most part, instability involves the chemical conversion of precursors to compounds of higher molecular weight with limited fuel solubility. The precursors are certain nitrogen and/or sulfur containing compounds, organic acids, and reactive olefins. The conversion process often involves oxidation of the precursors. Certain dissolved metals, especially copper, contribute by functioning as oxidation catalysts. Fuel solvency also plays a role because the development of insoluble compounds is always a function of both the presence of higher molecular weight compounds and the capacity of the fuel to dissolve them. Diesel fuel is increasingly used as a coolant for high-pressure fuel injection systems in which the fuel can be thermally stressed. Sometimes, this can cause the fuel to degrade and form insoluble materials that can restrict fuel flow through filters and injection systems. Three tests are routinely used to evaluate fuel stability in the USA. These are ASTM D 2274, the Octel F21-90, and the Octel F21-180. ASTM D 2274 is an accelerated oxidation stability test. Oxygen is bubbled through a fuel sample for 16 hours, after which the fuel is filtered to collect any insoluble materials. Fuels that have insoluble materials of less than 15 mg/liter are considered to be stable.

Both Octel tests are thermal stability tests. The difference between the two is the length of time the fuel is thermally stressed. A sample is subjected to a 150°C bath for either 90 or 180 minutes, filtered to collect insoluble materials, and measured by light reflectance. ASTM has developed the 180-minute test as a standard because it works better than the 90-minute test.

The storage stability is very important for bio-stability as well, especially for states with a warm climate. Microorganisms in diesel fuel have increasingly become a concern for fuel users. Prevention is important in fuel storage facilities. These bugs grow wherever water comes into contact with fuel, feeding at the interface, but living in the water.

The best way to prevent the problem is to keep water out of the fuel system. This is, however, practically impossible. The storage tanks cannot always keep the fuel from the humidity from the air.

In order to make diesel work most efficiently in engines year round, the low temperature properties of seasonal diesels are adjusted throughout the year according to historical temperature data of the region where the fuel is used. Generally all the diesel fuels can be divided into two groups:

- diesel fuel for summer season
- diesel fuel for winter season

For this classification of diesel fuel, the low temperature properties of the fuel are especially important. At low temperatures, wax crystals can be formed in the diesel fuel. These wax crystals can collect on and plug fuel filters in a vehicle's fuel system causing the engine to stumble or stall. The temperature at which this occurs is called the low temperature operability limit of the fuel and vehicle. Both the fuel system design and the fuel properties are important factors in determining this minimum temperature for acceptable operation.

The next quality parameter of diesel fuel is the ability to keep the fuel injector clean. The fuel injector is the heart of a diesel engine. These precision components meter fuel to a high degree of accuracy. Correct engine behavior depends on the injector functioning properly.

Proper injector functioning is essential for optimum performance. As such, it is beneficial to keep the nozzles operating at their optimum. Build up of carbon on the injector can disrupt the spray pattern of the fuel being injected into the cylinder. This can lead to incomplete combustion, which, in turn, can cause increased emission and high fuel utilization.

Many various additives are used to improve the quality of diesel fuel. These additives are used for a variety of purposes that can be classified into four major categories:

- Engine performance
- Fuel handling
- Fuel stability
- Contaminant control

The engine performance additives improve engine performance. Different members of this class are used to improve engine performance in different time frames. Any benefit that is provided by a cetane number improver is immediate, whereas that provided by detergent additives or lubricity additives is typically seen over the long term.

Cetane number improvers or diesel ignition improvers reduce combustion noise and smoke. The magnitude of the benefit varies among engine designs and operating modes, ranging from no effect to readily perceptible improvement.

2-Ethylhexyl nitrate (EHN) is the most widely used cetane number improver. It is also sometimes called, "octyl nitrate". EHN is thermally unstable and decomposes rapidly at the high temperatures in the combustion chamber. The products of decomposition help initiate fuel combustion and, thus, shorten the ignition delay period as compared with that of the fuel without the additive.

The increase in cetane number from a given concentration of EHN varies from one fuel to another. It is greater for a fuel whose natural cetane number is already relatively high. The incremental increase gets smaller as more EHN is added. Thus, there is little benefit to exceed a certain concentration. EHN typically is used in the concentration range of 0.05 to 0.4 mass % and this may yield a 3 to 8 cetane number benefit. A disadvantage of EHN is that it decreases the thermal stability of the fuels.

Other alkyl nitrates as well as ether nitrates and some nitroso compounds, have also been found to be effective cetane number improvers. However, they are not currently used commercially. Di-tertiary butyl peroxide has recently been introduced as a commercial cetane number improver.

Injector cleanliness additives and crankcase lubricant can form deposits in the nozzle area of injectors. The extent of deposit formation varies with engine design, fuel composition, lubricant composition, and operating conditions. Excessive deposits upset the injector spray patterns and these hinder the fuel-air mixing process. In some engines, this leads to increased emissions and fuel utilization. Ashless polymeric detergent additives can clean the fuel injector from deposits.

These additives are composed of a polar group that bonds to deposits and deposit precursors and a non-polar group that dissolves in the fuel. Thus, the additive can redissolve deposits that already have formed and reduce the opportunity for deposit precursors to form deposits.

Lubricity additives are used to compensate for the poor lubricity of severely hydrotreated diesel fuels. They contain a polar group that is attracted to metal surfaces, causing the additive to form a thin surface film. The film acts as a boundary lubricant when two metal surfaces come in contact. Two additive chemicals, fatty acids and esters, are typically used for this purpose.

Smoke suppressants are combustion catalysts. This class of additives is usually composed of organometallic compounds. Adding these compounds to fuel can reduce the emissions of pollutants that result from incomplete combustion. Smoke suppressants that are based on organometallic compounds of iron, cerium, platinum are in common use.

The following additives belong to the fuel handling additives:

Antifoam additives prevent formation of foam. Diesel fuels tend to be prone to foam formation when they are pumped into the tanks. Most antifoam additives are organosilicone compounds and are typically used at very low concentrations (about 10 ppm).

De-icing additives prevent ice formation from the free water in diesel fuel. If free water in diesel freezes, the resulting ice crystals can plug fuel lines and filters. Low molecular weight alcohols or glycols can be added to diesel fuel to prevent ice formation. The alcohols/glycols preferentially dissolve in the free water, giving the resulting mixture a lower freezing point than that of pure water.

Low temperature operability additives lower a diesel fuel's pour point and improve its cold flow properties. Most of these additives are polymers that interact with the wax crystals that form in diesel fuel when it is cooled below the pour point. The added polymers mitigate the adverse effect of wax crystal formation on fuel flow by modifying their size, shape, and degree of agglomeration. The polymer-wax interactions are fairly specific. As such, a particular additive generally will not perform equally well in all fuels. Unfortunately, the best additive for a particular fuel cannot be predicted; it must be determined experimentally.

Drag reducing additives are used by pipeline companies to increase the volume of product they deliver. These high molecular weight polymers reduce turbulence in fluids flowing in a pipeline. This can increase the maximum flow rate by up to 40%. When the additive modified product passes through a pump, the additive is broken into smaller molecules that have no effect on product performance in engines.

Fuel stability additives prevent the formation of gums that can lead to injector deposits or particulates that can plug fuel filters and the fuel injection system. The need for a stability additive varies widely from one fuel to another. Stability additives typically work by blocking one step in a multi-step reaction pathway of gum formation. Because of the complex chemistry involved, an additive that is effective in one fuel may not work as well in another. If a fuel needs to be stabilized, it should be tested to select an effective additive and treatment rate. The best results are obtained when the additive is added immediately after the fuel is manufactured.

Antioxidants prevent the oxidation of the fuel. Antioxidants work by interrupting the chain in free radical chain reaction by blocking the free radicals. Phenylenediamines are the most commonly used antioxidants.

Stabilizers prevent acid-base reactions in the diesel fuel. Basic amines are usually used as stabilizers. They react with weakly acidic compounds with the formation of products that remain dissolved in the fuel and do not react further.

Metal deactivators prevent the catalytic influence of the metals on reactions that can lead to its instability in diesel fuel. They are typically used in concentrations of up to 15 ppm.

Dispersants disperse the formed insoluble products in diesel fuel, preventing them from clustering into aggregates large enough to plug fuel filters or injectors. Dispersants are typically used in concentrations of up to 100 ppm.

Biocides prevent the growth of bio-organisms. The best choice is an additive that dissolves in both the fuel and the water such that it can attack the microbes in both phases. Biocides are typically used in concentrations of up to 600 ppm. A biocide may not work if a heavy biofilm has accumulated on the surface of the tank or other equipment, because then it does not reach the organisms living deep within the film.

Demulsifiers aid in the separation of fuel from water. Normally, hydrocarbons and water separate rapidly and cleanly. But if the fuel contains polar compounds that behave like surfactants and if free water is present, the fuel and water can form a stable emulsion. Any operation that subjects the mixture to high shear forces, like pumping the fuel, can stabilize the emulsion. Demulsifiers are surfactants which can destabilize the emulsions and allow the fuel and water phases to separate. Demulsifiers are used in concentrations of up to 30 ppm.

Corrosion inhibitors prevent the oxidation of the metal pieces in diesel engines. Since most petroleum pipes and tanks are made of steel, the most common corrosion is the formation of rust in the presence of water. The fuel is thus contaminated with rust particles, and these can plug fuel filters as well as increase fuel pump and injector wear. Corrosion inhibitors attach to metal surfaces and form a film that prevents attack by corrosive agents. They typically are used in concentrations of up to 15 ppm.

Modern diesel fuel can be produced from sources other than petroleum. Recently, attention has been focused on biodiesel. This is a cleaner-burning fuel made from natural and renewable sources such as vegetable oils. Just like petroleum diesel, biodiesel can operate in combustion-ignition engines. Essentially no engine modifications are required, and also biodiesel maintains the payload capacity and range of diesel.

The use of biodiesel in a petroleum diesel engine results in substantial reduction of unburned hydrocarbons, carbon monoxide, and formation of particulate matter, i.e. in reduction of emission of pollutants. Emissions of nitrogen oxides are either slightly reduced or slightly increased depending on the duty cycle and testing methods. The use of biodiesel decreases the solid carbon fraction of particulate matter (since the oxygen in biodiesel enables more complete combustion to CO_2) and eliminates sulfate fraction (as there is no sulfur in the fuel), while the soluble and hydrocarbon fraction stay the same or are increased. Therefore, biodiesel works well with new technologies such as catalysts (which reduce the soluble fraction of diesel particulate but not the solid carbon fraction), particulate traps, and exhaust gas recirculation (leads potentially to longer engine life due to less carbon).

Biodiesel has physical properties that are very similar to those of petroleum diesel. As mentioned earlier, emission properties, however, are better for biodiesel than for petroleum diesel.

Biodiesel is produced by a process called “trans-esterification”. Vegetable oil or animal fat is first filtered, then pre-processed with alkali to remove free fatty acids. It is then mixed with an alcohol and a catalyst. The oil's triglycerides react to form esters and glycerol, which are then separated from each other and purified.

Much of the current interest in biodiesel production comes from vegetable oil sources such as soybean (in the USA), canola oil (in Canada) and rapeseed oil (in Europe) because the vegetable oil producers are faced with excess production capacity, product surpluses, and declining prices.

Waste animal fats and used frying oils (known as 'Fellow grease') are also potential feedstocks. These are cheaper than soybean oil and are being considered as a way to reduce feedstock costs.

1.3.4 Residual Fuel

Many marine vessels, power plants, commercial buildings and industrial facilities use residual fuels or combinations of residual and distillate fuels for heating and processing. The two most critical specifications of residual fuels are viscosity and low sulfur content for environmental control.

Furnace oil is one kind of residual fuel. These are blended with other suitable gas oil fractions in order to achieve the viscosity required for convenient handling. As a residue product, fuel oil is the only refined product of significant quantity that commands a market price lower than the cost of crude oil.

Because the sulfur contained in crude oil is concentrated in the residue material, residual fuel sulfur levels naturally vary from less than 1 to over 6%. The sulfur level is not a critical factor for the combustion process as long as the flue gases do not impinge on cool surfaces. However, residual fuels may contain large quantities of heavy metals such as nickel and vanadium; these produce ash upon burning and can foul the burner systems. Such contaminants are not easily removed and usually lead to lower market prices for fuel oils with high metal contents.

In order to reduce air pollution, most industrialized countries have restricted the sulfur content of residual fuel. Such regulations have led to the design and construction of residual desulfurization units or cokers in modern refineries to produce fuels that meet these restrictions.

The standardized properties of residual fuels are:

- Density at 15°C ranging from 0.975 to 1.01 g/cm^3.
- Kinematic viscosity at 100°C ranging from 10 to 55 mm^2/s. The viscosity of an oil is a measure of its resistance to flow. Over the years different units have been used for viscosity (Engler Degrees, Saybolt Universal Seconds, Saybolt Furol Seconds and Redwood No. 1 Seconds). Nowadays, a majority of residual fuels are traded internationally on the basis of viscosity measured in centistokes (one centistoke = 1 mm^2/sec). The value for viscosity should be given together with the temperature at which it is determined. Accepted temperatures for viscosity determination of residual fuels are 40°C for distillate fuels and 100°C for vacuum residues. If a fuel contains an appreciable amount of water, testing for viscosity determination at 100°C is impossible. In this case, viscosity is determined at a lower temperature (usually 80-90°C). It is then recalculated to obtain viscosity at 100°C. Due to the variability of the composition of residual fuels, calculations for viscosity at another temperature from that measured at one temperature may not be accurate. Therefore, such calculated results should be treated with caution. The viscosity result is used first to ensure that the fuel is correctly heated in storage and made ready for pumping. The maximum viscosity for efficient pumping is considered to be around 600 cSt. However, lower viscosity should be maintained if long pipelines are installed.
- Flash point has to be over 60°C for residual fuels. The flash point is considered to be a useful indicator of the fire hazard associated with the storage of residual fuels. The flash point is determined by using the Pensky Martens closed cup tester. It should be remembered that results can be affected if the fuel sample contains a significant amount of water. There is one difficulty encountered by laboratories, which are requested to carry out a wide range of tests on one sample of residual fuel, including flash point. To ensure that a fuel sample is fully homogeneous, it should be heated for a period of time and shaken or preferably homogenized before it is split to perform individual tests. This preheating and homogenization may release some of the vapors from the sample thereby affecting the flash point result.
- Pour point varies from 0 to 45°C depending on the kind of fuel. Pour point is defined as the lowest temperature at which an oil will continue to flow when it is cooled under prescribed standardized conditions. If fuels are held at temperatures below their pour points, wax will begin to crystallize. This wax causes blocking of the filters and can deposit on heat exchangers. Generally, high viscosity fuels need to be heated well above the pour point to achieve the desired pumping viscosity. However, these fuels should be stored at a tem-

perature of around 10°C above the pour point if wax deposition is to be avoided. Some waxy residual fuels have pour points between 40 and 45°C.

- Carbon residue varies from 10 to 22% depending on the residual fuel. The carbon residue provides information on the coke or carbonaceous deposits which result from combustion of the fuel. Fuels which are rich in carbon prove more difficult to burn fully, resulting in increased deposits in the combustion and exhaust spaces. Fuels produced from thermal cracking residues show a higher carbon/hydrogen ratio and, hence, high percentage carbon residue. Carbon residue can only be an indicator of potential deposit-forming tendency of the fuel. Thus, operating experience with different fuels at different engine loadings and conditions should be recorded to determine individual engine tolerance levels to high carbon fuels.
- Ash ranges from 0.1 to 0.2%. The ash content of a fuel is a measure of the amount of inorganic noncombustible material it contains. Some of the ash-forming constituents occur naturally in crude oil; others are present as a result of refining or contamination during storage or distribution. For instance, it could be due to the presence of compounds of the following elements: vanadium, sodium, calcium, magnesium, zinc, lead, iron, nickel.
- Water content varies from 0.5 to 1%. Water is introduced into residual fuels by poor storage. The standards allow water up to a maximum of 1% in residual fuels. However, the majority of fuel deliveries have water contents below 0.5%. The problems with high water levels in fuel can be complex and include sludging of fuel tanks, filter blockage, corrosion of fuel injection equipment, exhaust valve corrosion, etc.
- Sulfur ranges from 3.5 to 5%. The main problem resulting from the use of high sulfur fuels is corrosion. During the combustion process, sulfur dioxide (SO_2) and sulfur trioxide (SO_3) are produced. Depending upon engine conditions such as excess air, temperature and pressures, these gases will convert to sulfurous acid (H_2SO_3) and sulfuric acid H_2SO_4), to some extent. These acids are formed when the gases exist below their dew points. For the formation of sulfurous acid, the favorable temperature should range between 50 and 60°C whereas for sulfuric acid, it should range between 110 and 150°C. These acids cause corrosion in the low temperature zones of engines and boilers. Hence, their effect is often referred to as "cold end corrosion".
- Vanadium content ranges from 150 to 600 mg/kg. The vanadium content of fuels from different countries varies considerably and is directly related to the crude oil source. Problems associated with high vanadium have been addressed under ash, sodium and sulfur. These problems are largely overcome by good engine design, and correct fuel treatment. To prevent these problems additives are widely used.

1.3.5 International Standards for Fuels

MS ISO 8216/0 — 8216/4 have been developed as standards for fuel classification in the framework of international standardization.

MS ISO 8216/0 establishes the general classification of petroleum fuels (fuels of class F). There are five categories of products which are included in class F. These categories depend on the type of fuel.

A detailed classification of fuel groups with regard to conditions of application, type, properties and characteristics establishes the groups of products for each category and these are provided by separate parts of ISO 8216. Thus, parts designated as ISO 8216/1, ISO 8216/2, ISO 8216/4 which have been developed are based on these standards. The Russian standards are included in the international standards system:

- ISO 8216/0-86 - Classification. Part 0. This is a general classification of petroleum fuels (Class F). The Russian equivalent for this standard is GOST 28577.0-90.
- ISO 8216/1-86 - Classification. Part 1. This covers the categories of fuels for sea engines. The Russian equivalent for this standard is GOST 28577.1-90.
- ISO 8216/2-86 - Classification. Part 2. This covers the categories of fuels for gas turbines for applications in industry and for sea engines. The Russian equivalent for this standard is GOST 28577.2-90.
- ISO 8216/4-86 - Classification. Part 3. Group 1. This covers liquefied petroleum gases. The Russian equivalent for this standard is GOST 28577.3-90.

Table 1.4 shows the international classification of petroleum fuels.

The fuel classification used for jet planes deserves special attention. In a majority of countries, the quality for these fuels is defined in the basic specifications developed in ASTM and DERD (the British Defense Ministry). The national specifications of the countries of manufacture of the jet fuel have insignificant deviations from the ASTM and DERD specifications.

To avoid confusion in interpreting fuel marks for delivery and storage of jet fuels, their basic manufacturers and the suppliers have developed a general specification covering the restrictive requirements of the basic specifications. These are used in the countries in which consumption of the jet fuels is greatest. This document is known as "The Requirements for the quality of jet fuel for common used systems" or AFQRJOS.

In the USA, the requirement for quality of jet fuels (list of the physical-chemical parameters and operational properties) for civil aircraft are made out as the ASTM D 1655 specification. Accordingly, the following fuels are made:

- Jet A – Fuel of the kerosene type with maximal freezing temperature of -40°C
- Jet A-1 – Similar to Jet A with freezing temperature of -47°C

Table 1.4: Classification of petroleum fuels (class F).

Category	Characterization
G	Gas. Gas of petroleum origin consisting basically of methane and ethane
L	Liquefied Gas. Liquefied gas of petroleum origin consisting basically of propane, propene and both butane and butene
P	Liquid Fuels. Fuels of petroleum origin excluding liquefied petroleum gases. They include gasoline, jet fuels, gas oil and diesel fuel. Heavy distillates can contain small quantities of residues
R	Residual Fuels. Petroleum fuels containing residual fractions of a distillation process
C	Petroleum Coke. Hard fuels of petroleum origin produced in cracking processes, and consisting basically of only carbon

The requirements for British jet fuels for civil aircraft are established by the specification D, Eng. RD (DERD) 2494. Previously, it was developed for military aircraft. This is fuel of the kerosene type with a freezing temperature of -47°C.

The list of control operations (AFQRJOS) is a basis for the international requirements for jet fuels.

Fuel for turbines that are made and delivered according to AFQRJOS should correspond to the most rigid requirements of the following specifications:

- The standard of the British Defense Ministry DEF STAN 91-91/ Editions 2 (DERD 2494) from May, 1996
- The standard specification ASTM D 1655
- The managing material of the international association of air transport (IATA) from December, 1994

The jet fuel that corresponds to the requirements of AFQR-JOS in the documents is marked as Jet A-1.

In the CIS (former USSR), jet fuels are made with regard to GOST T 10227. The main marks are TS-1 and RT.

Fuel classification as the normative document is absent in Russian practice. Concrete requirements are reflected in the documentation for the various kinds of fuel (e.g. GOST and OST), and also in various specifications.

1.4 LUBRICATING OILS AND LUBRICANTS

Lubricants are of multiple importance in the forging process. They serve as the cooling agent of the forging parts, support the filing of parts and prevent the fusion of work piece and tool. The main functions of lubricating oils and lubricants are:

- Reducing friction. Putting lubricating oil between moving metal surfaces decreases friction. This results in energy conservation and significantly less wear.
- Reducing wear. Lubricating oils reduce mechanical wear (caused by abrasion and erosion) and chemical wear (corrosion caused by combustion acids and water). Wear is influenced by various factors which can be divided into the fields of tool, rough part, forming machine, lubricant and actual contact area. Possible ways to reduce wear lie in the increase of annealing durability as well as in the hardness of the surface, for example through nitriding, and application of mostly ceramic wear resistant layers on the tool surface. Also, massive wear reduction can be achieved by lowering the tool temperature.
- Cooling. Lubricating oil cools the mechanical components by helping to remove heat and decreasing the amount of heat produced (by reducing friction).
- Ensuring leak tightness. Lubricants ensure leak tightness between the mechanical components such as the cylinder and the piston, improving compression and producing better engine performance.
- Removing impurities. Lubricants keep engine components clean by preventing the formation of deposits and suspended impurities (dust, combustion residue, engine wear particles) that then build up on the oil filter.

Properties differ from one lubricant to another. However, the one component that lubricants have in common is that they are composed of a main ingredient called the lubricating base. The lubricating base may be petroleum derived or synthetic. The lubricating base accounts for 75 to 85% of the lubricant. There are three different classes of lubricating bases that are used in modern oil industry:

- Mineral bases are manufactured from petroleum that has undergone a variety of complex separation processes. They are the most commonly used for both automotive and industrial applications.
- Synthetic bases or synthetics are products created by the chemical reaction of several ingredients. Two main classes are used for lubricants: esters and synthetic hydrocarbons (in particular polyalphaolefins manufactured from ethylene). These products have excellent physical properties and exceptional thermal stability.
- Semi-synthetic oils are obtained from mixing both types of bases (generally 70 to 80% of mineral oil and 20 to 30% of synthetic oil).

The development of modern lubricants and their correct use are especially important as these have substantial economic consequence on the use of these devices. Optimally chosen lubricant brings saving of energy needed for machine work, conserves the parts of these devices, reduces wear and tear, reduces the maintenance time and shortens the overhaul intervals, which can take away millions of dollars from industry. The characteristics of lubricants are improved by special substances called additives. In the current industrialized society, the use of lubricants without these enhanced characteristics is no longer conceivable. Lubricating oil additives are synthetic active substances, which are added to the basic oils in order to give the products the desired characteristics or properties. This is why, additives are the next main compounds of the oils (15 to 25%).

The following main classes of additives are available:

- Viscosity index improvers. These make the oil a sufficiently low viscous fluid when cold (to facilitate starting) by lowering the pour point to between –45 and –45°C (depending on the oil), and to viscous fluid when hot (to prevent the contact between moving mechanical components). This class of additives are polymers which are introduced into a lubricating base to produce a relative greater increase in viscosity when hot than when cold.
- Antiwear additives. These reinforce the antiwear action of the lubricant. The main family of antiwear additives are alkyl-zinc dithiophosphates and numerous phosphorus derivatives.
- Antioxidants. These eliminate or slow down lubricant oxidation, increasing the time between oil changes through improved resistance to high temperatures. Dithiophosphates that are used as antiwear additives are also excellent antioxidants. Other chemical families such as substituted phenols and aromatic amines are also used.
- Detergents. These prevent the formation of deposits and varnish on the hottest areas of the engines (such as ring grooves). Calcium or magnesium metal salts from main chemical families such as alkylaryl-sulfonate, alkylphenate and alkylsalicylate are used as detergents.
- Dispersing agents. These maintain in suspension all the solid impurities formed during engine operation. The solid impurities include unburned residue, gums, sludge, diesel soot, deposits cleaned by detergents. Usually polar compounds from the alkenylsuccinimides, succinic esters or their derivatives are used as dispersing additives.
- Alkalinity additives. Those that neutralize combustion acid residue from fuels are made primarily for diesel engines. The main compounds of this class of additives are phenols and salicylates (which are natural alkaline), and sulfonates. In the case of sulfonates, their neutralizing property can be reinforced by adding basic salts such as carbonates or hydroxides.
- Corrosion inhibitors. These prevent the corrosion of ferrous metals under the combined effects of water, atmospheric oxygen and a number of oxides formed during combustion. Mainly alkaline or alkaline-earth sulfonates, neutral or alkaline (Na, Mg, Ca) salts, fatty acids or amines, alkenylsuccinic acids and their derivatives are used as corrosion inhibitors.
- Antifreeze agents. These enable lubricating oils to retain good fluidity at low temperatures (from -15 to -45°C). The main classes of this type of additive are

polymethacrylates, maleate-styrene copolymers, naphthalene waxes and vinylacetate-fumarate polyesters.

- Foaming inhibitors. Oil foaming can be caused by the presence of other additives. For example, detergent additives act in the same way as soap in water; they clean the engine, but tend to foam. Foaming can also be caused by lubricating system design for which the major aim is to facilitate air-oil mixing. Such designs can cause turbulence during lubricant flow as well as formation of bubbles in the oil. Very small amounts of silicon oil or alkyl acrylates can be used as an antifoaming additive.
- Extreme-pressure additives. These reduce friction torques and, as a result, conserve energy and protect the surfaces from heavy loads. The most common classes of this additives are organometallic derivatives of molybdenum and some compounds derived from fatty acids, phosphorus-sulfur molecules, borates, etc.

It can be seen that the main functions of additives are to improve the lubricating properties of the lubricant and to make the lubricant as stable against adverse effects such as oxidation, corrosion, etc. It is important to know the mechanisms of the various effects that each additive is trying to inhibit in order to make the right choice of the additive or mixture of various additives to select. The most important function of all lubricants is to lubricate the surfaces in order to prevent friction between them.

As shown in Figure 1.8, there are four different kinds of friction.

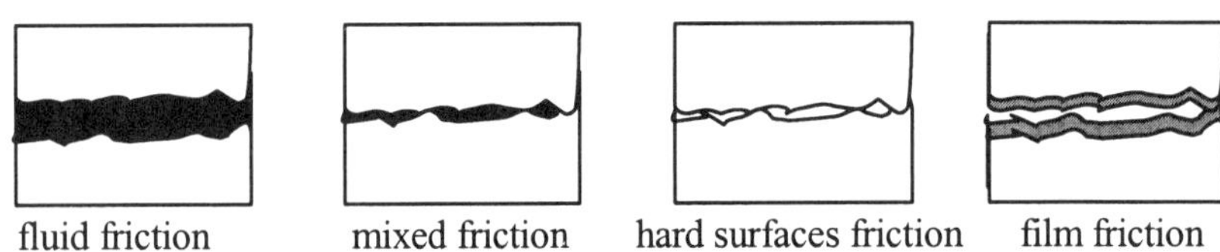

Fig.1.8: Four kinds of friction.

In fluid friction, there is no contact between two surfaces. It is obvious that wear and tear is minimal in this type of friction. However, there is still very high need for lubricants for fluid friction, even though it is economically unjustifiable to use this type of friction in big industrial devices. The most popular types of friction in industrial devices are mixed and film friction. Mixed friction is present as a combination of fluid friction and hard surface friction. The tear by mixed friction is obviously higher than by fluid friction, because there is contact between surfaces at some points. The optimal for the industry is the so-called film friction. In this kind of friction, a thin film of surface-active substance is formed. This film protects the surfaces from direct contact and high wear and tear.

It is possible to make the film for film friction artificially. Such a modification in the lubricant and the friction surfaces is carried out by addition of special additives called anti-wear (AW) and extreme pressure (EP) additives. The name EP additives is derived from the need to endure extremely high pressure at the points of contact between two surfaces. The most popular AW and EP additives are made from zinc dialkyldithiophosphate. This zinc complex with general chemical structure $Zn[(RO)_2PS_2]_2$ belongs to the group of film forming surface active substances. However, zinc dialkyldithiophosphate does not form the protecting film for film friction, but the products of their decomposition during friction do form the protecting film. The final film mostly contains hydrocarbons. However, there are inorganic compounds such as iron sulfides, iron oxides, zinc oxides, amorphous polyphosphate and so on. An example of film formation on the friction surface is shown in Figure 1.9.

One more important requirement for lubricating oils, which can be improved by adding additives, is the ability to keep their properties for a long time. One of the most important additives used to improve this ability of oils is antioxidants.

Oxidation is a radical chain reaction promoted by alkyl and peroxide radicals. This means that it is necessary to add radical acceptors to the oil to prevent its oxidation. It is possible to classify most of the modern antioxidants in six groups:

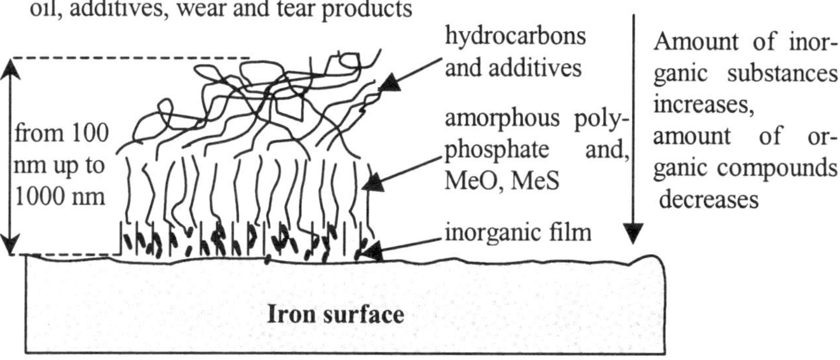

Fig.1.9: Formation of the film in film friction.

1. Antioxidants, which prevent oxidation by reaction with peroxide radicals. Such additives often contain aromatic compounds in their structure with relatively weak O-H and N-H bonds. Examples of such compounds are phenols, naphthols, aromatic amines, aminephenols, and diamines. This type of additive has strong reducing properties and reacts quickly with peroxide radicals.
2. Antioxidants which prevent oxidation by reaction with alkyl radicals. The additives react quickly with alkyl radicals. Examples of such additives are hynones, iminehynones, methylhynones, stable nitrogen oxide radicals, and molecular iodine. Alkyl radicals react quickly with oxy-

gen. This is why antioxidants of this type are effective under conditions of oils with low concentration of oxygen.

3. Antioxidants which destroy hydroperoxides. This type of additive reacts quickly with hydroperoxides without free radical formation. The following substances are good examples of such additives: sulphide, phosphite, arsenate and so on, and also thiophosphate and carbonates of various metals, and various complexes.
4. Antioxidants which deactivate the surface of metals. Some compounds of metals with variable valency react with hydroperoxides with the formation of free radicals, which strongly promotes oxidation. This type of catalytic oxidation can be prevented by adding a "complex former", which forms with a metal complex that is rendered inactive to hydroperoxide. Examples include diamines, hydroacids and other compounds that form strong complexes with metals.
5. Antioxidants with repeated action. During the oxidation of some classes of substances such as alcohols and aliphatic amines, peroxide radicals are formed, which have both oxidizing and reducing functions. In such systems, additives that prevent oxidation are regenerated again in the reducing reaction. This involves a catalytic breakage of the chain in the radical chain reaction. The number of chain breakages depends on the reaction ratio of additive regeneration and its consumption in the reversible reaction. Examples are aromatic amines, nitrogen oxide compounds, compounds of metals with variable valency, etc.
6. Antioxidants with combined function. Some additives prevent oxidation by simultaneous reactions with many compounds that promote oxidation. For example, it can react with both alkyl and with peroxide radicals (anthracene, methylhynon), or can destroy hydroperoxides and break off the reaction chain with $RO_2\cdot$ radicals (carbomate and thiophosphate of metals). Such compounds are referred to as antioxidants with combined function. The parallel reactions can occur with the same group in one additive. For example, $R\cdot$ and $RO_2\cdot$ radicals react with the double bond of methylhynon. Frequently, two or more functional groups exist in one molecule, each of which is responsible for an appropriate reaction. For example, phenolsulphide reacts with hydroperoxide with the sulfide group and with $RO_2\cdot$ with the phenol group. In a different type of reaction, it can take part in the initial reaction with antioxidant and products of its transformation. Often the mixtures of additives have combined function. For example, adding phenol and sulphide to the oil first prevents oxidation by breaking the reaction chain with $RO_2\cdot$ radical, and second reduces the reaction rate for branching reactions by destroying the hydroperoxide. Synergy may be obtained if two antioxidants strengthen the action of each other. Also, it is possible for the action of two or more antioxidants to be additive. However, if

the action of a mixture is less than the sum of the actions of each antioxidants, antagonism results.

In the last type of antioxidants we learn that an important property of antioxidants is synergy. It is to be understood that by synergy one can mix antioxidants each with a weak action against oxidation and obtain a strong antioxidant action as a combination. That may sound strange but this effect really occurs during the preparing of many industrial additives. This is why it is especially important to understand the nature of synergetic effects in order to be able to use the maximum possible potential of additives. How to obtain synergistic effects can be classified in three groups:

1. One antioxidant breaks the reaction chain and the second reduces the ratio of initiation by destroying ROOH groups or by deactivating the catalyst that promotes the formation of the ROOH.
2. Two initial substances (antioxidants or not) react with each other and form an effective antioxidant.
3. The transformation products of antioxidants strengthen the antioxidant function.

The main properties of lubricants and lubricating oils are determined by carrying out the following specific tests:

- Oxidation tests. The combined effects of ambient oxygen and temperature affect lubricant aging; this phenomenon is known as oxidation. In laboratory oxidation tests, temperatures are selected in accordance with those encountered during equipment operation.
- Corrosion tests. One of the lubricant's main functions is protection against corrosion. Accordingly, the reactions between the lubricant and various metal surfaces need to be determined and, where appropriate, modified.
- Chemical compatibility tests. These tests are used to determine the behavior of various materials when they are in contact with oil. For example, specific tests are conducted to ensure the compatibility of a lubricant with seals.
- Storage stability tests. Some lubricant combinations are formed by the association of products that are not fully miscible in oil. In this case, the final stability that represents changes during storage needs to be verified.
- Dispersion tests. The purpose of these tests is to determine the ability of an oil to maintain in suspension solids that are likely to form during operation. For example, combustion engines produce carbon residues, some of which are found in the oil. The latter must be able to maintain the soot in suspension and prevent deposits that could result in fouling or even clogging.
- Shear tests. These tests determine how well lubricants withstand mechanical loads that can cause the molecules of some components to break up.
- Pressure tests. There is currently no device available that can directly measure the ability of an oil film to withstand pressure. Various properties (film

strength, high pressure, extreme pressure) can only be assessed using bench tests in the laboratory to determine their effects on other oil properties. The same types of tests are carried out for greases as for oils, using special devices: four-ball machine, Timken, etc.

- Laboratory tests are supplemented by bench tests on machines very similar to real ones and which operate under controlled conditions.
- Engine tests are designed to determine the behavior of oils on either gasoline or diesel engines. Each test is performed in such a way as to highlight one or more lubricant properties. No single engine test is available that enables all properties to be tested simultaneously.
- Bench tests are carried out either on standard automotive multi-cylinder engines or on single-cylinder engines.
- Gear tests check the internal cohesion and pressure withstanding properties of lubricants on real gears. Various methods are used depending on the type of gear and test conditions.

1.4.1 International Standards for Lubricating Oils

1.4.1.1 Industrial oils

The most fully developed standards system of classifications of industrial oils is the series MS ISO 6749 “Lubricant materials, industrial oils and related products (Class I). Classifications of groups”.

The classification includes 18 groups of products into which this class is divided according to the area of application (Table. 1.5).

The use of the international classification according to lubricant material application allows us to group lubricant materials that are made in different countries in the classes of the application area. This considerably facilitates the task of selecting the lubricant materials for the specific equipment.

Interchangeability of lubricant materials produced in different countries and intended for specific equipment can be established by the complex estimation of their functional properties, by the laboratory devices and installations, by stands with model and natural friction units, by the real engines, in real machines, and mechanisms at the operating conditions.

The principles and criteria for estimating these properties and also equipment used to perform tests of all kinds for these oils in different countries are basically identical. However, concrete laboratory, stand, motor and operational methods can differ appreciably. Besides, there are differing viewpoints among experts even from the same country regarding the classification of the methods, i.e. this classification is substantially conditional.

Table 1.5: Classification of lubricant materials, industrial oils and related products. Division into groups according to area of application.

Group	Application
A	Open systems
B	Formless systems
C	Gear wheels
D	Compressors (including cooling)
E	Engines of internal combustion
F	Spindles, bearings and couplings connected to spindles
G	Gear wheels
H	Hydraulic systems
M	Metal treatment
N	Electrical isolation
P	Pneumatic tools
Q	Heat transfer
R	Temporary protection against corrosion
T	Turbine
X	Applications requiring greasing
Y	Other applications
Z	Cylinders for steam machines

1.4.1.2 Motor oils

The classifications of motor oils according to SAE and API have been accepted worldwide.

Motor oils are classified according to their viscosity and operational properties. Besides, motor oils are subdivided in terms of energy conservation and ability to reduce fuel charge in comparison with a reference oil.

The classification according to viscosity is based on the standard SAE J 300 "Classes of viscosity of motor oils" in which the designation of the oils is given according to their viscosity at 100 °C and at a temperature < 0 °C (Table 1.6).

The class of the oil according to SAE characterizes only the viscosity of the oil and does not give information about its operational properties.

The area of application is defined by the operational properties of the oil.

The most widespread classification by operational properties and areas of application of motor oils is the API classification (Table 1.7).

Universal oils have a double designation (SF/CD, CE/SG).

During the operation of automobiles, the classification of oil according to SAE is recommended whereas the classification according to API is applied with respect to operational properties.

Table 1.6: Classification of motor oils by viscosity — SAE J 300 (April, 1997)

Viscosity class	Viscosity at low temperature		Viscosity at low temperature		
	pumpability, Mpa·s, at the temperature (°C) ASTM D 5293	pumpability, Mpa·s, at the temperature (°C) ASTM D 4684	Kinematic viscosity, mm^2/s, at 100°C		Kinematic viscosity, mm^2/s, at 150°C
			min	max	
OW	3250 at –30	6000 at –40	3.8	-	-
5W	3500 at –25	6000 at –35	3.8	-	-
10W	3500 at –20	6000 at –30	4.1	-	-
15W	3500 at –15	6000 at –25	5.6	-	-
20W	4500 at –10	6000 at –20	5.6	-	-
25W	6000 at –5	6000 at –15	3.3	-	-
20	-	-	5.6	<9.3	2.6
30	-	-	9.3	<12.5	2.9
40	-	-	12.5	<16.3	2.9
40 .	-	-	12.5	<16.3	3.7
50	-	-	16.3	<21.9	3.7
60	-	-	21.9	<26.1	3.7

The CCMC (ACEA) Classification of motor oils has been developed for European engines. The CCMC classification is developed separately for gasoline and diesel engines, similar to what pertains for API (Table 1.8).

Each class of CCMC oils is comparable with the corresponding class of API oils and is defined by a set of complex laboratory and motor tests, according to ASTM, CBC, IP or DIN standards.

Table 1.7: API classification for motor oils by its operational properties and areas of application.

Class	Application area
Category S (Oils for gasoline engines)	
SA	Engines working at light conditions
SB	Engines working at moderate conditions
SC	Engines working with high loadings (models released before 1964)
SD	Engines working at heavy conditions (models released before 1968)
SE	Engines working at heavy conditions (models released before 1972)
SF	Engines working at heavy conditions reformed gasoline
SG	Engines released since 1989
SH	Engines released since 1994
SJ	Engines released since 1997
Category C (Oils for diesel engines)	
CA	Engines working at moderate loadings with low sulfuric fuel
CB	Engines working at high loadings with sulfuric fuel
CC	Engines working in heavy conditions
CD	Engines working in heavy conditions with high sulfuric fuel
CD-II	The same for old diesel engines
CE	Engines working in heavy conditions
CF-4	Engines released since 1990
CF-2	Improved characteristics of CD-II
CG-4	Engines released since 1994

Table 1.8: CCMC classification for motor oils by its operational properties and areas of application.

Class	Area of application
Oils for gasoline engines	
G1	Engines, working in usual conditions
G2	Engines of modern cars working in hard conditions
G3	Engines of modern cars, with high viscous properties
G4	Engines of modern cars, used for driving with high-speed
G5	Engines of sports high-speed automobiles
Oils for diesel engines	
D1	Engines working under usual conditions
D2	Engines working under heavy conditions
D3	Engines working under especially heavy conditions
PD-1	Diesel engines of the cars
D4	Engines working in heavy conditions
D5	Engines working in especially heavy conditions
PD-2	Engines with turbo-pressurization for cars

Since January of 1996, the ACEA classification has practically replaced the CCMC classification. The ACEA classification was developed as an integration of standards of European organizations: CEC, ATC and ATIEL. It was adopted in 1998.

The ACEA classification is applicable to different oil products for different gasoline engines (classes A1-96, A2-96, A3-96), diesel engines (classes B1-96, B2-96, B3-96) and diesel trucks (classes El-96, E2-96, E3-96). After the changes and additions, new classes A1-98 and A3-98 for gasoline engines, Bl-98, B2-98, B3-98 and B4-98 for diesel engines of cars and E4-98 for diesel engines of trucks have been included.

A new system known as the European monitoring system of the qualities of motor oils (EELQMS) has been developed to cover the requirements for European motor oils. EELQMS includes the requirements of registration of tests, data collecting and change of compositions. All tests should be registered in the European registration center (an independent organization, possessing experience in the control of the order of the performance of tests) before they are performed.

The car firms in the USA and Japan cooperating with ILSAC have formulated requirements for motor oils for gasoline engines of cars. In ILSAC classification, there are two classes which are designated as GF-1 and GF-2. They are close to API classes SH and SJ, respectively.

1.4.1.3 Transmission oils

The requirements for quality of transmission oils intended for various tasks are numerous. This is why there is the need for classification.

Similar to motor oils, there are two systems of classification: SAE with regard to viscosity and API with regard to operational properties.

According to SAE classification, tractor transmission oils are divided into 6 classes (Table 1.9) with regard to viscosity.

A hyphenated viscosity class implies that the oil is suitable for all seasons for the specific climatic zone. For example, the oil with SAE 75W-90 classification has the low temperature properties of the viscosity class SAE 75W (i.e. dynamic viscosity of the oil does not exceed 150 Pa*s at the temperature of -40°C). At a positive temperature, the kinematic viscosity corresponds to the viscosity of the class SAE-90, (i.e. at 100°C the kinematic viscosity is 13.5-24 mm^2/s).

The distinctions in designs of transmission units and conditions of their operation create distinctions in the requirements for operational properties of the oil.

The development of API classification was based on this principle. As such, the oils are divided according to the type and loading degree of the gears where the oil is applied (Table 1.10).

Table 1.9: The SAE classification for transmission oils with regard to viscosity.

Viscosity class	The maximum temperature of the achievement of the dynamic viscosity of 150 Pa*s, °C	Kinematic viscosity	
		Min	Max
	winter oils		
75W	-40	4.1	
80W	-26	7.1	
85W	-12	22.1	
	summer oils		
90		13.5	24
140		24	41
250		41	

The SAE and API classifications give only the general characteristic of oils without consideration of all parameters of quality. The complete requirements for physical, chemical and operational properties of the oils and their allowable limiting values are given in the specifications. In the countries of Western Europe and the USA, transmission oils are produced with two types of specifications:

- Specifications of firms making automobiles
- Military specifications for oils for supply to the armed forces of both the USA and NATO are made

Both specifications include the requirements for physical, chemical and operational properties of the oils as well as a definition of the number and methods for their tests.

Table 1.10: API classification of transmission oils with regard to area of application.

Class	Application area
GL1	Cylindrical and spiral-conic gears, working at low speeds and loading
Gl2	Cylindrical gears, working at low speeds and loading
Gl3	Usual transmissions with spiral-conic gears, working in hard conditions
GL4	Transmissions, working at normal speeds and high loading
GL5	Transmissions, working at high and low speeds and high loading
Gl6	Transmissions, working at extreme high speeds and normal loading
MT1	Mechanical transmissions

According to API classification, industrial specifications are applied to all kinds of oils from GL-1 up to GL-6. On the other hand, military specifications are only applied to GL-4 and GL-5.

These specifications provide the check for all parameters of quality requirements that are specified.

In the USA and the countries of Western Europe, the widest application has been achieved by the military specifications from the USA:

- MIL-L-2105 (1959 is replaced by MIL-L-2105B)
- MIL-L-2105B (1973 is replaced by MIL-L - 2105C)
- MIL-L-2105C (1987 is replaced by MIL-L-2105D)
- MIL-L-2105D (since 1987)

In USA in 1995, a new military specification MIL-PRP-2105E that unites the MIL-L-2105D and API MT-1 requirements was developed.

The military specifications give the descriptions of almost all requirements for lubricant oils that are available at gas stations in the USA and in a majority of other countries.

However, the requirements of some manufacturers are higher than guaranteed by these specifications. Therefore, manufacturers of automobiles have their own specifications for transmission oil for the first filing. These oils need these additional requirements to ensure specific characteristics such as cleanliness of details, serviceability of synchronizers, etc.

Nevertheless, for the first filing, oils are frequently recommended according to the MIL-L-2105D specification.

1.4.1.4 Hydraulic oils

In the framework of international standardization, hydraulic oils are classified by the ISO 3448 standards for viscosity classes and ISO 6074 for operational properties. According to ISO 6074, liquids from mineral raw material used in hydraulic systems are incorporated in group H, which in turn is subdivided into four categories depending on the structure of the oils and the application area:

- HH — mineral oil without additives
- HL — mineral oil with antioxidizing and anticorrosive additives
- HM — oil of the HL type with antideterioration additives
- HV — oil of the HM type with improved viscous-temperature properties

2

Modern Characterization and Analysis Techniques for Crude Oil

2.1 CHROMATOGRAPHIC METHODS

Chromatography is probably the oldest and most important analytical method in crude oil chemistry as well as general analytical chemistry. The first record of the use of the principle of chromatography can be traced back to ancient times when the ancient philosopher, Aristotle, wrote about the use of the adsorptive action of special substances for the purification of seawater. However, the first scientific use of chromatography was at the end of the nineteenth century when in 1895 the German scientist Ferdinand F. Runge described for the first time a chemical separation method based on the chromatography principle. The method used by Runge can be classified or called paper chromatography or a very simple form of thin film chromatography. At the time, the German scientist did not realize the importance of the method, but merely described it as a type of scientific game. At the end of the nineteenth century, R. T. Day, for the first time, attempted to separate crude oil by filtering it through a lime filter. Unfortunately, the scientist did not recognize the main principle on which this method was based but described it as a new filtration technique. The first inventor of chromatography to describe its main principles was the Russian scientist M. S. Zwet, who used the chromatography method in 1903. Zwet was a botanist and used chromatography for the study of chlorophyll. A. J. P. Martin and R. L. Synge received the 1955 Nobel Prize in Chemistry for the development of a new chromatographic analytical method. These scientists were the first to separate amino acids by the chromatographic method. Perhaps this year could be marked as the birth year of modern chromatography.

The main idea behind chromatography is the separation of a mixture of two or more compounds by using two auxiliary phases; one of which is the static phase and the other the mobile phase. The physical state of the static or stationary phase can be solid or liquid whereas the mobile phase can be liquid or gaseous. All chromatographic methods can be classified into four main classes based on the combination of the physical state of the mobile and static phases:

1. Mobile phase is liquid and static phase is solid. The main principle of the separation by this method is adsorption on the static phase.
2. Mobile phase is liquid and static phase is liquid. The main principle of the separation by this method is absorption on the static phase or distribution of the compounds analyzed between the two phases.
3. Mobile phase is gas and static phase is solid. The main principle of the separation by such method is adsorption on the static phase.
4. Mobile phase is gas and static phase is liquid. The main principle of the separation by such method is absorption on the static phase or distribution of the analyzing compounds between the two phases.

2.1.1 Gas Chromatography

In gas chromatography, the mobile phase is gas and the static phase is usually a liquid, or a solid in special cases. The main application of gas chromatography is in the analysis of complicated mixtures of many components. The high popularity of gas chromatography is due to its requirement of small sample amounts for analysis and the possibility to analyze compounds in trace amounts. The biggest disadvantage of this method is that all the compounds analyzed must be in the gaseous phase or can be evaporated into the gaseous phase.

In gas-liquid chromatography, the stationary phase is a liquid which is covered on an indifferent (inert) solid substance called the carrier. The carrier is either a porous adsorbent or the wall of a capillary. This analytical method corresponds to the principle of classical distribution chromatography, in which a liquid is used as the mobile phase. In gas adsorption chromatography, solid granulates with an active adsorptive surface, such as activated coal or silica gel, are used as the stationary phase.

Gas chromatography has been well known as a separation method for a long time. After they had invented the principle of distribution chromatography, A. J. P. Martin and R. L. Synge suggested the use of a flowing gas as mobile phase. The reason is that gas makes it possible to carry out distribution chromatography with high flow velocity, because of its low viscosity as opposed to the viscosities encountered in the liquid phase.

The rapid development of the method in the last few years has substantially contributed to the relatively simple and extremely sensitive measuring methods of gas chromatography. This makes it easy to operate continuously with connections between the separating column and the detector. Thus, separation, identification (qualitative) and quantitative detection can be made practically in one processing step. A temperature of analysis of 350°C has already been achieved for special purposes. This means that substances with a boiling point up to approximately 400°C can be analyzed by gas chromatography.

One of the advantages of gas chromatography that needs to be mentioned is that the analysis can be carried out quickly. With the correct choice of analysis conditions, it often needs only a few minutes, or in rare cases about one hour, to complete a gas chromatography analysis. However, the time needed for gas chromatography strongly depends on the complexity of the mixture being analyzed. It was shown in chapter one that crude oil is a very complicated mixture of organic substances. This is why it is hardly possible to analyze crude oil directly by this method. It becomes necessary to separate crude oil into different fractions before analysis is performed by the gas chromatography technique. This separation can be undertaken by using different methods, including distillation, solution analysis, etc. Usually such preparation of crude oil fractions for chromatographic analysis makes the results of analysis more accurate and saves a lot of time during both the analysis and the evaluation of the analysis results.

With the use of modern instruments for gas chromatography, the amount of material needed for analysis is very little. For gas analysis, the gas sample needed is usually under 0.5 cm^3, and for liquid mixtures it is under 0.5 g. There are special gas chromatographs that can handle a sample amount of under 0.01 g.

However, the biggest advantage of the gas chromatography and every other chromatographic method is that the compounds in the mixture that is being analyzed (i.e. sample) are not destroyed during the analysis.

Before starting separation for any type of analysis, it is important to understand what type of compounds (components) can be analyzed by the method. Just as in any analytical method, gas chromatography has specific requirements for the properties of components that can be analyzed. These requirements are as follows:

1. The components in the sample analyzed must be gaseous or it is possible to evaporate the sample at a temperature under 350°C.
2. In the case of analysis at a high temperature, it is required that the components be thermally stable at the analysis temperature.
3. It is still possible to analyze a sample mixture which contains thermally unstable compounds. However, in this case, the reaction mechanisms of the thermal destruction of these compounds must be defined as exact as possible; in this way, the native compounds of the sample mixture is analyzed by analyzing their cracking products.
4. The separation of the sample components must be possible on the column or columns installed in the gas chromatograph. It is important to point out that every column used in a gas chromatograph is earmarked only for a specific class of compounds. As such, the column can be used to analyze only the designated class or classes of compounds. However, using most modern instruments, it possible to install more than one column to improve the analyzing capabilities of the gas chromatograph.

It is obvious from the stated requirements that there are a lot of materials that can be analyzed by gas chromatography. However, the requirement that the material be gaseous or vapor at the column temperature provides a restriction in the crude oil compounds that can be analyzed. Actually, with the requirement for the boiling temperature to be under 350°C, it is only possible to analyze the light fractions from petroleum such as gasoline, kerosene and diesel. However, these fractions are very important for their industrial use as fuels. This makes gas chromatography one of the most popular analysis techniques in crude oil chemistry.

A simplified scheme of a gas chromatograph is presented in Figure 2.1.

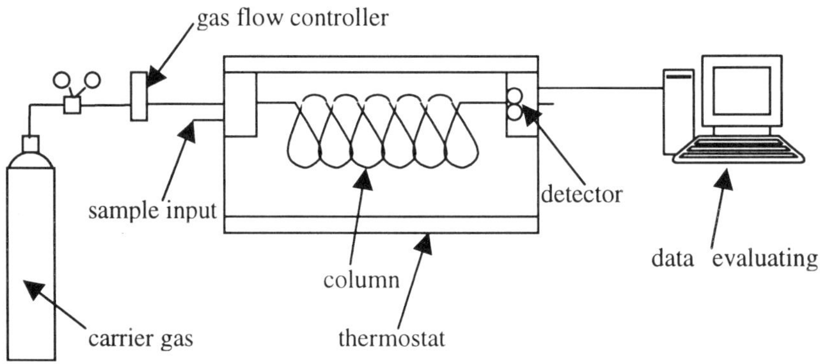

Fig. 2.1: Scheme of a gas chromatograph.

The detector and column are the most important parts of a gas chromatograph. Usually, separation by gas chromatography is carried out at higher than room temperature. As such, thermostatic heating of the column is necessary.

There are many types of columns used for gas chromatography and they are made from different materials. Usually, these are glass, plastics and metals such as copper or steel. All these materials have advantages and disadvantages for different types of analysis. For example, glass columns are thermally and chemically very stable, and normally these are relatively cheap columns in comparison to columns made from other materials. The disadvantage of the glass column is that it is very fragile, and thus it is an easily breakable material that is relatively difficult to install. Plastic columns are not easily breakable. They are very flexible but chemically stable. However, the disadvantage of such columns is that they are not thermally stable. As such, they can be used only for analysis at a low temperature. The metal columns are now the most popular columns for gas chromatography. They are not breakable, and are thermally and chemically stable. They only have the disadvantage that they may not be used for the analysis of acid rich or corrosive mixtures, because the metal can be oxidized and this can falsify the results of the measurement. However, metal columns are now the most robust and stable, and therefore the most popular types of columns. However, they are also the most expensive columns. The length of columns used now varies from 1.5 m up to 4.5 m whereas the diameter can vary from 2 mm up to 10 mm.

The most important part of the column is the filling or fill material or packing. Based on the type of filling, the columns can be classified into packed columns, adsorption columns, distribution columns and capillary columns.

The packed columns are the simplest and at the same time most modest columns for gas chromatography. The configuration of these columns depends on the gas chromatograph in which it is used. The most recommended configuration of packed columns is the U-form, because this makes it simple to clean the column or change the filling or packing. However, the biggest disadvantage of packed columns is that they are very sensitive to shaking and the packing can be easily broken. This type of column is, however, recommended for experimental work in gas chromatography.

An adsorption column is usually filled with adsorptive active substances, which could be, for example, oxidized alum, activated coal, silica gel or various types of molecular sieves. The idea in separation using the adsorption column is the separation of a mixture of compounds into a number of fractions according to the adsorption ability of the fill material for the compounds. It is important at this point to understand what adsorption is. Adsorption is the deposition of a gaseous or liquid substance on the surface of the adsorbent whereby the substances adhere to the surface of the adsorbent. The binding or adsorption power or the affinity of the adsorbent for different compounds has to be different in order for separation to take place. This allows for the separation of the compounds according to their different adsorption powers. Adsorption columns are especially effective for analysis at room temperature. This is why the use of this type of column is recommended for analyzing gas mixtures. The type of filling must be chosen depending on the nature of the gas compounds; a polar adsorbent must be used for a gas mixture with a large fraction of polar compounds, whereas non-polar adsorbents must be used for non-polar gases such as hydrocarbons. Since natural gas and most gases formed during crude oil treatment are hydrocarbons, it is reasonable to use non-polar adsorbents as filling for analyzing the gases formed during petroleum treatment. However, many gas mixtures have complicated compositions containing hydrocarbons as well as many types of polar gases such as hydrogen sulfur or carbon monoxide. This is why it may be necessary and it is indeed possible to have a series of two or more columns in one instrument for most modern gas chromatographs. This enables the analysis of complicated mixtures of gases in one step.

The analysis of evaporated liquid mixtures using gas chromatography is possible as well. For this purpose, filling poisons such as a non-volatile liquid are used. The purpose of this poisoning is to block the most active adsorptive centers of the filling. In the absence of filling poisoning, high molecular compounds in the mixture would be adsorbed so strongly on the active centers of the filling that the transport of the mixture through the column is practically impossible. This leads to the formation of zones called tailing zones. Good separation and reproducible analysis in this case is impossible. A second method to analyze liquid samples using this type of column is to increase the temperature of the column. However, this method is useful only for mixtures of compounds with relatively low boiling temperatures or which can be easily desorbed from the filling at high temperatures.

In the event that the filling of the column must be changed or filled for the first time, it is important to ensure that the filling is as homogenized as possible. In order to achieve homogeneity, the packing material is fractionated by sieving into narrow particle size ranges. Good filling requires only a narrow particle-size range fraction with the appropriate average particle size. The column must be repeatedly shaken during the filling of packing material in the column to prevent the formation of voids in the column. The column should be blown through by high-pressure air after the column is completely filled with packing. It is done for improved homogenization of the packing. After the column is packed in this way, it is plugged at both ends and it can be used for analysis after conditioning.

The other type of column used in gas chromatography is the distribution column. The idea in this column is the separation of compounds of the mixture according to their solubility in the filling of the column. This type of columns is recommended especially for the analysis of evaporated mixtures of liquid compounds.

The filling of the distribution column contains a material called carrier, which is coated with a liquid absorbent. Zeolites or glass spheres are usually used as carriers whereas non-volatile liquids such as silicone oil or poly-ethylene glycol are used as the absorbents in distribution columns.

The filling must be custom prepared for every mixture if the appropriate filling for the particular mixture is not available in the market. The filling for distribution columns must be as homogenized as possible. This is why sieve fractionating is necessary before the column is filled and absorbent introduced on the carrier. Fractionation is carried out in the same manner as in the case of filling for the adsorption column. After the fraction with the appropriate particle size range is obtained, the liquid absorbent is adsorbed on the surface of the carrier. The oils or high molecular weight liquids usually used for this purpose are highly viscous liquids, and it is very difficult to uniformly cover the surface of the carrier with such viscous liquids. In this case, the absorbent is solved in a small amount of solvent before covering the carrier. The carrier is then soaked in the solution and intensively mixed. The excess solvent is removed from the prepared filling by evaporation at a high temperature. The packing material prepared in such a way is introduced into the column in the same manner as that in the adsorption column.

The last type of column for gas chromatography is the capillary column or Golay-column. This type of column is especially interesting because they present the highest resolution in comparison to all other types of column. Resolution is a very important requirement of every column. It shows the ability of the packing in the column to separate the compounds in the sample mixture as completely as possible. The difference between a packed column and a capillary column is that the latter has no packing. Instead, it is open on both ends of a column capillary. The fact that the capillary is open on both ends allows the column to work with very low gas flow resistance. This is why this type of column usually has a length of from 20 to 100 meters. The diameter of capillary columns is usually between 0.1 and 1 mm.

The majority of the capillary columns available in the market are made from glass, steel or quartz. The glass columns are often covered with a thin film of aluminum in order to make them less fragile and less breakable. Glass columns are rarely used nowadays. They have been replaced with columns made from fused silica. The biggest advantage of such columns is that it is possible to make columns with very thin walls, which are not as easily breakable as the glass columns.

The most important feature of the capillary column, which makes it differ from other types of columns, is that they have no carrier in their configuration. There are two types of capillary columns: thin film columns and thin layer columns.

The inner wall of the thin film column is covered with a thin film of the liquid phase. The thickness of this film can vary from 1 to 3 μm. One example of the inner arrangement of the thin film column is shown in Figure 2.2.

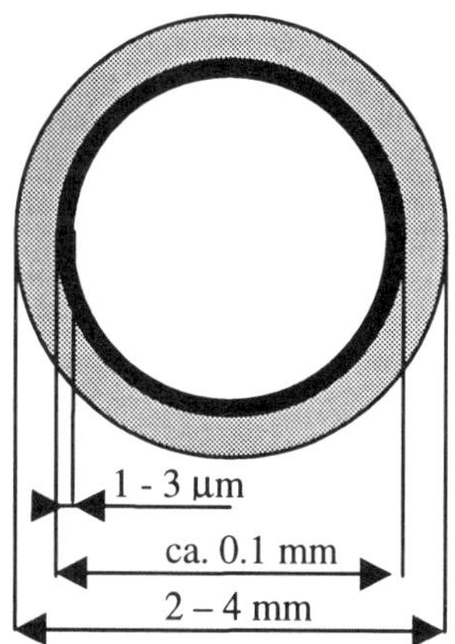

Fig. 2.2: Inner arrangement of the thin film column.

The thin layer column, also called the analytical column, is ideal for all types of quantitative and qualitative analysis, because they provide the best possible resolution in comparison to other gas chromatographic columns. A thin layer of the carrier is impregnated into the inner wall of the thin layer column. A thin film of the absorbent is then placed on the impregnated carrier and not on the inner wall of the column, as in the case of the thin film column. The carrier material is usually oxidized aluminum or silica gel. During analysis with thin layer columns, it is important to ensure that all the connections used in instrument, from the injection port to the detector, have the same size as the inner diameter of the column. Any differences in size can result in the formation of zones where the separated compounds can be mixed again, leading to the formation of wide peaks on the chromatogram and bad resolution of the column. The inner arrangement of the thin layer column is shown in Figure 2.3.

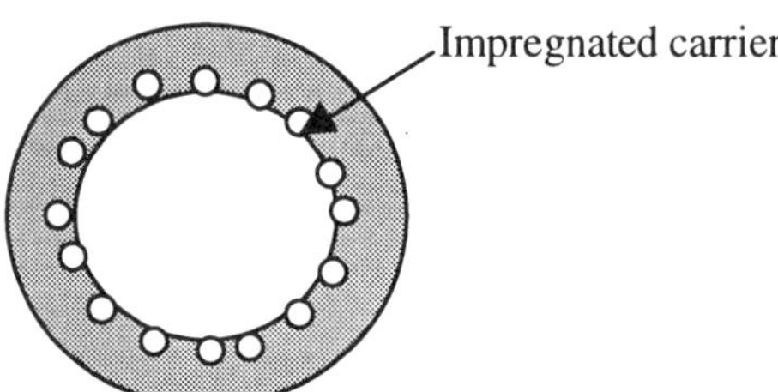

Fig. 2.3: Inner arrangement of the thin layer column.

The next important part of the gas chromatograph is the detector. The detector is the part of the gas chromatograph which detects the compounds leaving or exiting the column. It is a very important part of the gas chromatograph in the sense that even with the best column, if the detector is not able to exactly detect the compound as it exits the column, then the analysis is incomplete. There are many methods to detect the compounds present in the sample mixture. For example, in the case when the carrier gas is hydrogen, the analyzed sample can be burned as it exits from the column. The majority of compounds have unique colors of their burning flames, allowing for the detection of the compounds according their flame colors. This is just one of the simplest examples of the possibilities for detection.

Even though there are many types of detectors used in gas chromatographs, it is possible to classify them as integral and deferential detectors. The latter class of detectors is currently the most well developed class of detectors. If only the carrier gas flows through the column, modern deferential detectors will show a direct zero base line. When analyzing real sample mixtures by gas chromatographs with such detectors, the detector detects the compound immediately as it exits from the column and sends a signal to the evaluating computer. After the particular component or compound has exited the column completely, the detector sends a zero signal to the evaluating computer. These types of detectors are known as exact detectors.

The signal sent from the detector is usually too weak to be directly analyzed by the evaluating computer. This is why this signal is normally amplified. However, it must be ensured that the signal coming from the detector is strong enough to be measured but too strong to be off scale. Often, it is not possible to measure all the compounds of the sample mixture at the same level of amplification of the signal. This is the reason why attenuators are often installed in gas chromatographs. There are two types of attenuators: input and output attenuators. The input attenuator is installed before the amplifier, and it can attenuate the signal by 10^1 up to 10^5. The output attenuator is installed right after the amplifier, and this can reduce the signal sent to the evaluating computer by a factor of from ½ up to 1/4096.

There is one further classification of detectors. Detectors can analyze the molecules of the mixture with and without destroying of the molecule of the component. These two classes of detectors are called destructive and non-destructive detectors, respectively. The latter type of detector is very sensitive to the concentrations of the different compounds in the mixture. Also, the volume of the measurement cell is important for this type of detector.

Currently, there are over a hundred different types of detectors used in gas chromatographs. In this chapter, only the most popular and probably the most important detectors will be described.

An important property of each detector is the sensitivity. Sensitivity shows the ratio between the output signal of the detector and the amount of substance that is passing through the detector. This characteristic of the detector corresponds to the gradient of the amount/signal curve. One example of such a curve is shown in Figure 2.4.

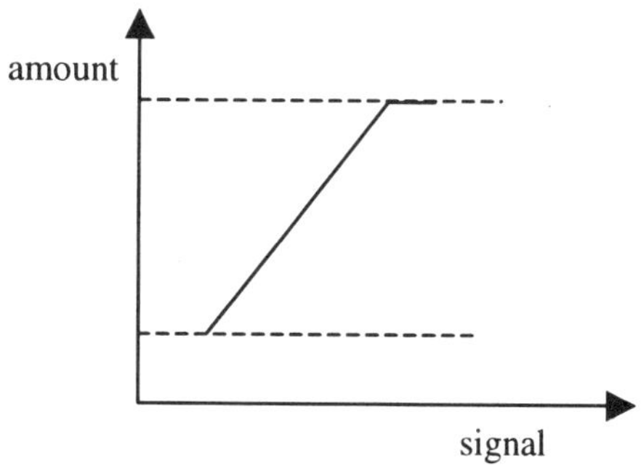

Fig. 2.4: Amount/signal curve.

This curve is defined by the minimal detectable level amount of the substance that enters the detector. Usually, this level is accepted as two or three times higher than the magnitude of the normal noise of the detector. The area where the dependence of the detector signal on the amount of the component or substance analyzed is reproducible is defined as the dynamic range.

This makes dynamic range an important characteristic of the detector. It is usually necessary to compare different detectors. The dynamic range shows the amounts of the substance analyzed needed for reproducible analysis. For some types of detectors, this area can be relatively small, for example for flame photometric detectors. A relatively wide dynamic range is typical for flame ionization detectors. Special attention must be given to the dynamic range in the choice of an appropriate detector for quantitative analysis, because this characteristic of the detector determines the minimum and maximum concentration of the different components in the mixtures, which is being analyzed by the gas chromatograph. Examples of approximate areas (dynamic ranges) for eight most popular detectors are shown in Figure 2.5.

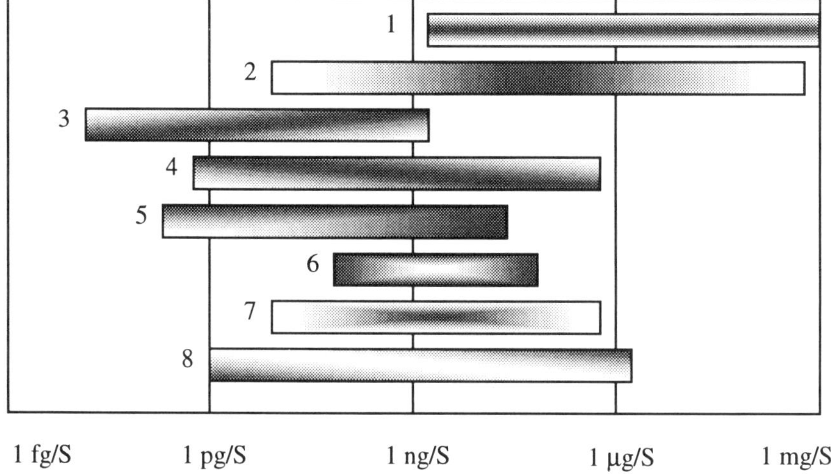

Fig. 2.5: Dynamic ranges of different detectors:

1 – thermal conductivity detector
2 – flame ionization detector
3 – electron trap detector
4 – nitrogen / phosphor thermo ionic detector for nitrogen compounds
5 – nitrogen / phosphor thermo ionic detector for phosphor compounds
6 – flame photometric detector for sulfur compounds
7 – flame photometric detector for phosphor compounds
8 – photo ionization detector
S – attitude of the signal

The last important property of the detectors is the selectivity. This characteristic of the detector shows which compounds in the sample can be analyzed by gas chromatograph with each detector. There are universal detectors which can be used for almost every compound in any sample mixture. For example, by connecting a gas chromatography to a mass spectrometer, where the mass spectrometer plays the role of a detector for the gas chromatograph, even the most complicated mixture of compounds can be analyzed. The disadvantage of this hybrid system is the strictness of the analysis by such detectors such that a full library of compounds is required at the evaluating computer for this detector. Also, experience is required of the scientist making the analysis. Universal detectors are the most used detectors in modern chromatography. If a detector is peculiar for a particular class of compounds, then it is referred to as a selective detector. Usually, such detectors are more exact than universal detectors, but the disadvantage of such detectors is that only mixtures of particular classes of compounds can be analyzed.

The most popular detector is the flame ionization detector. Using this type of detector, the component analyzed that is eluting from the end of the column is mixed with hydrogen and burned in synthetic air. It is necessary to use synthetic air to achieve a clean burning of the hydrogen and the component analyzed. Ion formation during the clean burning of hydrogen is hardly possible whereas ions are formed during the burning of all organic substances according to the following reaction:

$$CH + O \rightarrow CHO^+ + e^-$$

Electrons formed in this reaction are dropped onto the collector electrode. These electrons form the electrical signal of the detector. It is important to ensure that synthetic air contains only oxygen and nitrogen and that the collector electrode is very clean from the burning residue or other pollutants. Otherwise, this can lead to a high deviation of the measured values from the real values. One example of the flame ionization detector is shown in Figure 2.6.

Actually, the flame ionization detector is restricted to only the analysis of organic substances. As such, gases like CO_2, SO_2, NH_3 and CO cannot be analyzed by gas chromatographs with this detector type. However, analyzing inorganic gases in the petroleum chemistry becomes necessary only in some special cases such as ecological analyses of the gas formed during crude oil treatment.

The Janak detector is probably the simplest detector that exists for gas chromatography. With this detector, carbon dioxide must be used as the carrier gas. Both the carrier gas and the sample analyzed go through a vessel containing concentrated $Ca(OH)_2$ solution after exiting the column. In this way, the carbon dioxide is separated from the gas analyzed. All the gases not dissolved in the $Ca(OH)_2$ solution are collected in a special collector. The volume of the compounds is automatically measured in this collector. It is obvious that this collector cannot be used for analyzing gases that are soluble in $Ca(OH)_2$ solution.

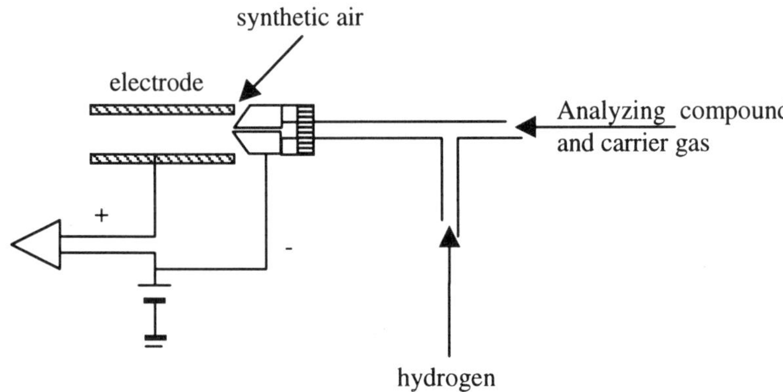

Fig. 2.6: Example of the flame ionization detector.

The next and probably most important detector in gas chromatography is the thermal conductivity detector. This detector contains four cells, which are installed in a thermostat. Two of the cells are the measurement or analysis cells. The analyzed gas and carrier gas flow through these cells. The remaining two cells are called the reference cells. Only the pure carrier gas flows through both these cells. Inside of all the four cells are platinum or wolfram electrodes. All the electrodes are heated up using exactly the same electrical stream. As such, the differences in temperature of the electrodes are only due to the difference in the thermal conductivities of the gases flowing through them. If a difference in temperature in the electrodes in the measurement and comparison cells is detected, the detector sends a signal to the evaluating computer.

The temperature difference is measured by measuring the electrical conductivity of the electrodes in the cells. All these electrodes are connected in a special bridge circuit. This circuit is in equilibrium as long as the temperature of all the electrodes is the same. The electrical conductivity depends strongly on the temperature of the electrodes. If a different gas in the carrier gas flows through the measurement cells, then this circuit is immediately out of equilibrium.

The different electrical conductivities of the electrodes lead to a change of voltage. This change is sent as the measured detector signal. One example of the cells and the bridge circuit are shown in Figure 2.7.

It is obvious from the functioning principles of the detector that the thermal conductivity detector is suitable for all types of chemical compounds. This is why its use is recommended for analyzing complicated mixtures such as crude oil fractions or products.

Another type of detector that is used relatively often for gas chromatography is called the ion trap detector. In some cases, this detector is called a mass selective detector. Gas chromatographs with this detector are called gas chromatography / mass spectroscopy hybrid units. This detector is very popular especially for the ecologists.

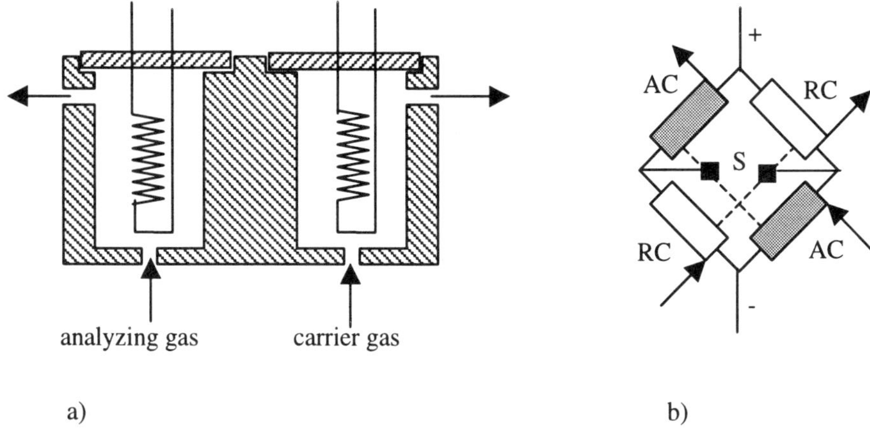

Fig. 2.7: a) Scheme of two cells of thermal conductivity detector; b) bridge circuit

The mass spectrometer serves for the exact determination of the masses of atoms and molecules as well as for the registration of the mass spectra from particle mixtures to mass and relative proportion. The first ion trap detector was developed in 1919 by F.W. Aston.

An example of the ion trap detector is shown in Figure 2.8.

The ion trap detector is operated all the time under vacuum. The measurement cell consists of two rotating symmetric pole caps and a ring electrode. The molecules of the sample are usually analyzed in the measurement cell. They are ionized by an electron beam that is only switched on for a brief period of time. The ions in the space between the pole caps and the ring electrode are captured or caught by a sturdy high-frequency field. By varying the high-frequency field, the ions leave the space between the pole caps and the ring electrode in the order of their mass to electrical charge values, and arrive at the amplifier.

It is obvious from the working principle of the ion trap detector that this is equally a universal detector as is the thermal conductivity detector. The analysis of the mixtures of compounds with this type of detector is possible not only by retention time of the compound in the column, but also by the composition of the ions of this compound. Usually, libraries of the compounds for the ion trap detector are much more extensive than the libraries for the thermal conductivity detector. However, the more extensive possibilities for knowing which compound exited the column makes this detector an excellent detector for the specialist, but introduces many difficulties for beginners.

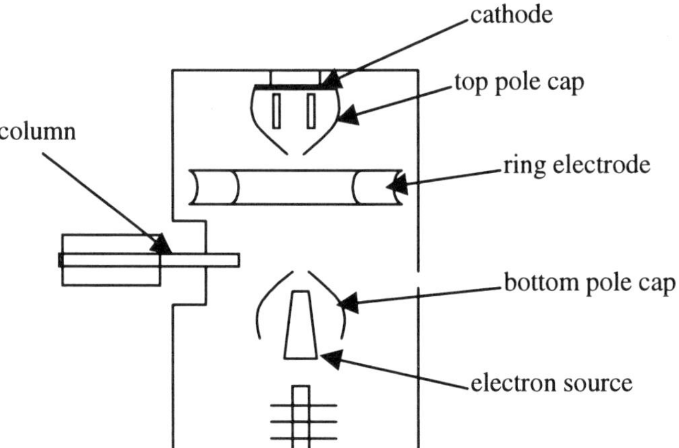

Fig. 2.8: Scheme of the ion trap detector.

The next detector to mention in this book is called the electron capture detector. The electron capture detector is a detector which possesses a special sensitivity for substances that can catch slow electrons. These include halogen organic compounds. The electron capture detector is much more sensitive for this compound class than, for example, the flame ionization detector. One example of such a detector is shown in Figure 2.9.

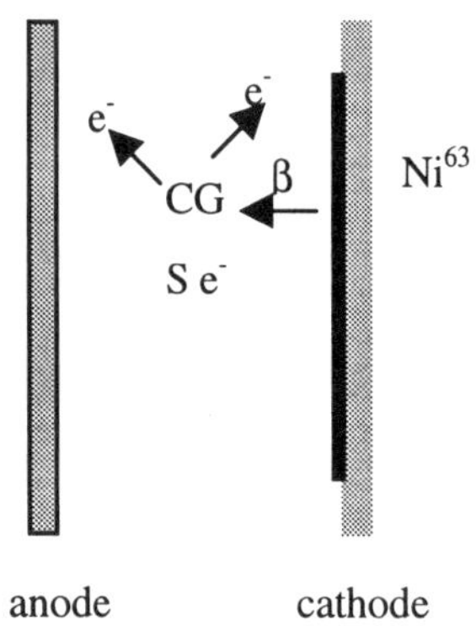

Fig. 2.9: Scheme of the electron capture detector:

e^- - slow electron

β - β - particles (fast electrons)

S e^- - analyzing substance molecule with caught slow electron

The electron capture detector has a layer of Ni^{63} as a β-radiation source. The radioactivity of the radiation source amounts from 10 up to 15 mC (370 to 550 MBq).

First, the carrier gas, which usually contains 90% argon and 10% methane for this detector type, is ionized by the β-radiation. Slow electrons are formed in this way, and are caught by the collecting electrode (anode). As long as no compounds arrive into the detector room which can react with the slow electrons, the electrical current stream remains constant. This gives rise to the zero stream. A recombination of the slow electrons with the positive feed gas ions is not possible, since the accumulation energy is equal to the dissociation energy.

If molecules of the component analyzed come into the detector together with the carrier gas, and these molecules can catch slow electrons, then recombination of the electrons with these molecules can take place. The molecules catch slow electrons. Fewer electrons arrive at the collecting electrode. This change of electron stream forms the signal stream. Negatively charged ions are formed from molecules of the component analyzed; they can transfer their electrons to the positive carrier gas ions. The high sensitivity of this detector can be decreased considerably by contaminating the electrodes. This is why cleanliness of the electrodes must be constantly checked.

The next and last type of detector discussed in this chapter is called the thermo ionic detector. This detector is also called a phosphorus / nitrogen detector. Actually, this type of detector is a modified form of the flame ionization detector, which has its special field of application with organic nitrogen and phosphorus compounds. The thermo ionic detector is structured in the same way as a flame ionization detector with only a small rubidium containing glass bead at a platinum wire as source of alkali located between the burner nozzle and the collecting electrode. The alkali bead always has a negative potential. One example of the detector constructed for the detection of phosphorus compounds is presented in Figure 2.10.

For the analysis of phosphorus compounds, the flame burns in the same way as with the flame ionization detector. A nozzle is installed in such a way that electrons formed by burning the organic compounds can flow out from the burner but not to the collecting electrode. If the nozzle is not grounded as is indicated in Figure 2.10, electrons which formed during the burning of the hydrocarbons could pass up to the negatively charged alkali bead. In such a situation, the reactions cannot occur at the alkali bead. By grounding, the electrons at the nozzle can flow off. The reactions take place at the alkali bead, and this makes the detector especially sensitive for phosphorus compounds.

The electrons formed in the reactions on the alkali bead go to the positively charged collecting electrode and form the signal stream.

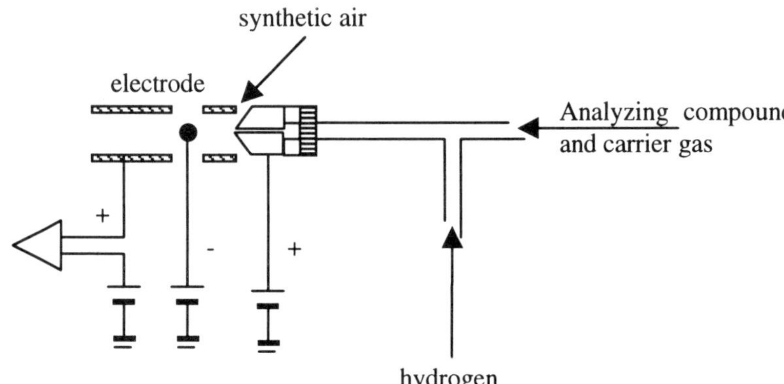

Fig. 2.10: Scheme of the thermo ionic detector for the analysis of phosphor compounds.

The alkali bead consists of rubidium glass, which has semiconductor properties. If it is cold, it is an insulator, and in the hot state, it has some conductivity. If there are alkali ions (Me^+) in the hot alkali bead, then they react with electrons into alkali atoms, which surround the alkali bead as a cloud. The alkali atoms can become thermally innervated and react under light rays into innervated atoms Me^*. Radicals (R·), which are formed from the burned component analyzed, react with innervated alkali atoms by electron transfer from the alkali atom to substance radical. The innervated alkali atoms change their electrical charge and form positively charged alkali ions, whereas the sample radicals react to form negatively charged ions R^-. The positive alkali ions are caught by the negatively charged alkali bead again, and react immediately to alkali atoms again. Also, they are ready for further reactions with the sample radicals again. The anions (negative charged ions) react to form a neutral final product. Electrons are captured or caught by the collecting electrode and the detector signal is thereby formed.

The main reactions taking place in the detector with phosphorus compounds can be shown in the five reactions given below. All these reactions are written under the assumption that only reactions with the phosphorus oxides such as PO or PO_2 occur in the detector room, since the organic substance is completely burned in the burner.

$O{=}P + Me^* \rightarrow O{=}P^- + Me^+$
$O{=}P{=}O + Me^* \rightarrow O{=}P{=}O^- + Me^+$
$O{=}P + OH^* \rightarrow HPO_2 + e^-$
$O{=}P{=}O + OH^* \rightarrow HPO_3 + e^-$
$HPO_3 + H_2O \rightarrow HPO_4$

Nitrogen has an odd number of electrons on the outer shell just like phosphorus. However, the nitrogen oxide radicals formed similarly to the phosphorus oxide radicals are cracked very quickly in the hot analyzing chamber in comparison to phosphorus oxides. Therefore, the thermo ionic detector can be reconstructed in such a way that it can be more sensitive for organic nitrogen compounds instead of the phosphorus compounds. A reducing flame condition must be created for analyzing nitrogen compounds such that instead of nitrogen oxide radicals, cyanide or cyanate radicals are formed. These react at the alkali bead in a manner similar to phosphorus oxide radicals.

During the analysis of organic nitrogen compounds, one reduces the hydrogen and air supply to the extent that the flame at the burner nozzle is turned out. At the same time, the bead is electrically heated. The hydrogen is ignited on the glowing alkali bead and burns in the form of cold plasma around the alkali bead. An example of the thermo ionic detector reconstructed for organic nitrogen compounds analysis is shown in Figure 2.11.

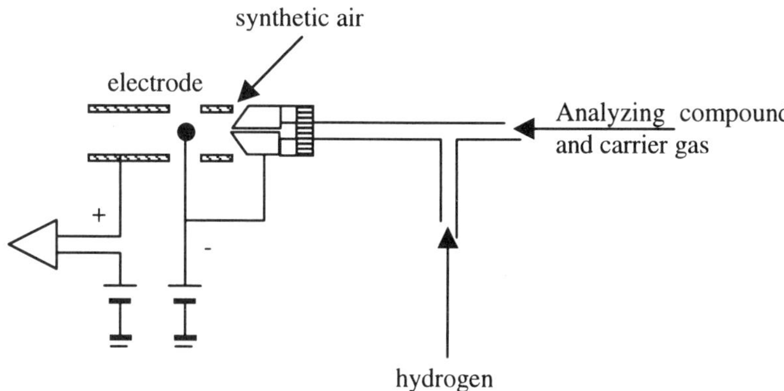

Fig. 2.11: Scheme of the thermo ionic detector for the analysis of nitrogen compounds.

The reaction mechanism of the reactions running in the analysis room of the detector can be presented by the following reactions:

$$\begin{array}{c}|\\ -\mathrm{C}-\\ |\end{array}\!\!\mathrm{N}< \rightarrow \cdot\mathrm{C{=}N} \qquad \text{(cracking)}$$

$$\cdot \mathrm{C{=}N} + \mathrm{Me}^{*} \rightarrow {}^{-}\mathrm{C{=}N} + \mathrm{Me}^{+}$$
$${}^{-}\mathrm{C{=}N} + \cdot\mathrm{H} \rightarrow \mathrm{HCN} + e^{-}$$
$${}^{-}\mathrm{C{=}N} + \cdot\mathrm{OH} \rightarrow \mathrm{HCN0} + e^{-}$$

This type of detector is very specific and can be used in petroleum chemistry only in specific cases for the analysis of nitrogen or, very rarely, phosphorus compounds. Some of such cases are shown in the last chapter of this book. These are cases where crude oil is treated or coprocessed with other feeds that contain hetero-atoms. An intensive study of this area of crude oil chemistry just started some decades ago. This is why an intensive study of the chemical mechanism for such systems is essential.

The gas chromatograph can be used for meaningful analysis only after the appropriate column and detector have been selected. However, there is still an important aspect of gas chromatographic analysis that needs to be mentioned. This involves the introduction of the sample into the column. Sample introduction is achieved by using special injectors. There are five techniques to introduce sample into the column.

The first technique is called the filled needle technique. In this technique, the needle of the injector is fully filled. This is the simplest method but its biggest disadvantage is that its reproducibility is very bad especially for volatile substances. Since most crude oil products (such as gasoline) are volatile substances, this method is not recommended for their analysis.

The next method is called the empty needle technique. In this method, the sample is sucked out from the needle. The reproducibility of this method is higher than that by the full needle technique.

Another method is the air bubble technique. In this technique, a little amount of air is sucked in the injector. The reproducibility of this method is higher than that of the first two methods.

The next two methods are called the solvent methods. In the first solvent method, the solvent is sucked into the injector directly after the first air bubble is sucked in. The reproducibility of this method is the best in comparison to all methods shown before.

The second solvent method is called the sandwich method. In this technique the sample is sucked into the injector followed by air bubble and then the solvent. After the solvent, the sample is sucked in one more time. The reproducibility of this method is the best. However, its disadvantage is that a high amount of the sample is introduced in the column. It is not every column that can function with high sample loading or large amount of the sample.

All these methods are presented in Figure 2.12 for a better understanding of the techniques they represent.

In addition to the availability of different methods of sample introduction into the column in gas chromatography, there are two different types of injectors that can be used to input the sample. These are the splitless injector and the split injector.

Splitless injectors are used in cases where the column can work well with high amounts of sample. In this case, the sample is fully injected into the column and is pushed through the column by carrier gas.

However, for capillary columns, the amount of sample is very important. This is why there is the second type of injectors: split injectors. With these injectors, it is possible to input just a fraction of the evaporated sample in the column. The other part of the evaporated sample can be pushed directly over a special split into the atmosphere. This type of injector is used only for capillary columns.

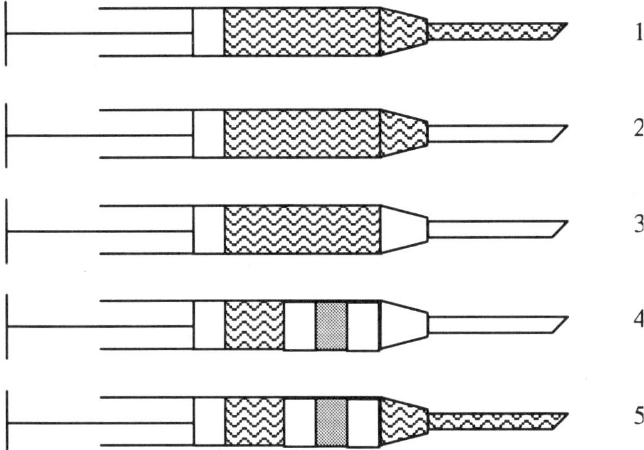

Fig. 2.12: Main techniques for the input of the sample in the column:
1 – full needle technique
2 – empty needle technique
3 – air bubble technique
4 – solvent technique
5 – sandwich technique

When all the preparation work has been completed and sample is introduced in the column, analysis can begin. The analysis result consists of a chromatogram with many peaks that correspond to the compounds present in the sample mixture. At the beginning of this chapter, it was mentioned that the main idea of chromatography is the differential adsorption or absorption of the components in a sample mixture followed by desorption. However, what is observed in the chromatogram are wide peaks instead of thin lines that should represent the compounds of the sample. Why do we have wide peaks? One of the possible theories used to explain this occurrence is the theory of resistances. The main idea of this theory is presented in Figure 2.13.

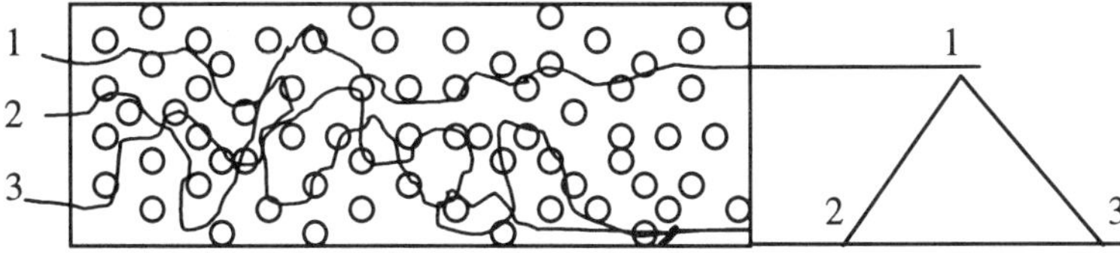

Fig. 2.13: Peak formation.

It is seen from Figure 2.13 that three molecules of the same component introduced in the column entrance at the same time pass to the end of the column using different routes. The length differences are specific to the filling or packing of the column. The peak for the same substance can be wider or thinner depending only on the type of column filling. The longer it takes the molecules to pass through the column, the shorter and wider the peak is for this component.

Despite that chromatography is a relatively exact method, it has some drawbacks such as the presence of anomalies, just like any known analytical method. The anomalies that arise from chromatography are very difficult to classify. However, an attempt will be made in the following discussion to highlight the main types of anomalies in chromatographic analysis.

Spikes can arise due to physical or chemical contamination of the column or detector. For flame ionization detectors, these contaminants can come from the air. However, the simplest possible reason for the presence of spikes is the mechanical vibration of the instrument during analysis. Typical appearance of spikes is shown in Figure 2.14.

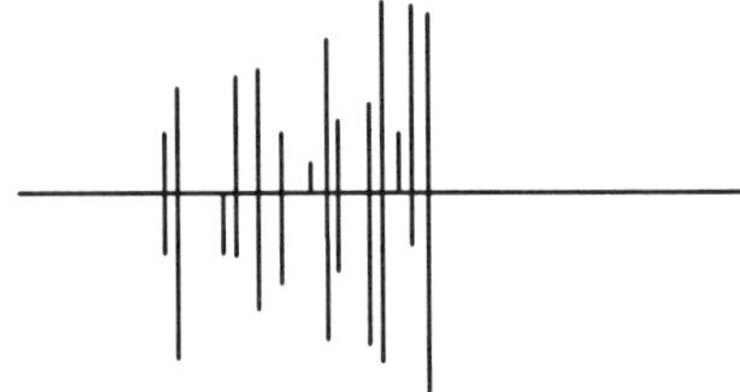

Fig. 2.14: Example of the spikes.

Noise is represented by all fluctuations that arise during the measurement. There are many reasons for noise. However, the most important ones are the wrong choice of gas flow rate or leaky connections between different parts of the instrument. One more problem that arises from the wrong choice of gas flow rate or bad control of gas flow is peak migration. Evaluation of the results of chromatograms with peak migration is practically impossible. Another reason for peak migration is poor temperature control in the column.

Some chromatograms may contain what is referred to as ghost peaks. These correspond to the components that are actually not present in the sample. The most common reason for ghost peaks is that the column was contaminated before the analysis started. This shows the importance of a cleanliness check before the analysis.

The last interesting problem arises from the wrong choice of column for analysis of the sample. It is called shared peak. An example of a shared peak is shown in figure 2.15.

There are a lot more anomalies that can arise in the use of chromatographic methods. It is not possible to discuss all of them in this short chapter. This chapter has been restricted to providing the reader highlights of the scientific and technical fundamentals of gas chromatography. However, a list of recommended literature has been provided at the end of this chapter for readers who are interested in pursuing the subject further.

Fig. 2.15: An example of shared peaks.

It was shown earlier that there are many reasons for the occurrence of anomalies during gas chromatographic analysis. This is why everything needed for analysis must be chosen very accurately. After the gas chromatograph for analyzing a specific sample is selected, the next step is to select the carrier gas. Every gas can theoretically be used as a carrier gas. However, there are some important characteristics that every carrier gas must have. The carrier gas must be chemically and physically inert to the filling or packing of the column. Also, it must not react with the compounds in the sample mixture. The last property of a good carrier gas is its level of purity. It is very important for the analysis that the carrier gas introduces no impurities into the gas chromatograph. Even very small amounts of foreign substances in the carrier gas can falsify the results very seriously. Presently, three gases are the most popular carrier gases used in gas chromatography. These are hydrogen, nitrogen and helium.

The use of hydrogen as a carrier gas has only one disadvantage. The high tendency of hydrogen to diffuse and permeate through the smallest splits in the connections between parts of the gas chromatograph makes hydrogen relatively dangerous. If the leakage in the system is inside the thermostat, it can result in explosion because of the presence of a dangerous mixture of hydrogen and air. This is why the use of hydrogen as the carrier gas must be related to ensuring that there are no leaks between connections of parts in the gas chromatograph. Consequently, a very thorough leak check is a must before the start of every experiment.

The next possible carrier gas is nitrogen. This is probably the cheapest and least dangerous carrier gas. However, it must be noted that nitrogen is also used as an inert atmosphere in many experiments in crude oil chemistry. This means that gas coming to the analysis contains nitrogen. The analysis of nitrogen in the gas mixture would be impossible if the carrier gas is the same as one of the components of the sample mixture.

Nowadays, helium is the most popular carrier gas used in gas chromatography. The use of helium as inert atmosphere in thermal or catalytic treatment experiments is, in most cases, not rational because of the high cost of this gas in comparison to nitrogen. The formation of helium from crude oil or crude oil products is absolutely impossible too. This is why helium can be recommended as the optimal carrier gas for analyzing crude oil fractions or gases formed during the thermal or catalytic treatment of crude oil.

One important factor influencing the resolution of the column is the velocity of the gas flow. This value must be chosen optimally for every column or column system used in gas chromatography. Too high a velocity of the gas in the column can lead to very bad separation of the compounds in the column. This means that it is possible for many components of the mixture to exit the column at the same time, thus making the evaluation of the results cumbersome and prone with error or practically impossible. Too low a velocity of the gas in the column leads to the formation of adsorption–desorption equilibrium, resulting in a poor resolution of the column as well. The gas flow in any modern gas chromatograph is usually controlled by a special controller called the mass flow controller. These devices are very exact. However, they need to be calibrated. There are three most popular devices to measure the gas flow and to calibrate the mass flow controller.

Soap bubble measurement is carried out in the device shown in Figure 2.16.

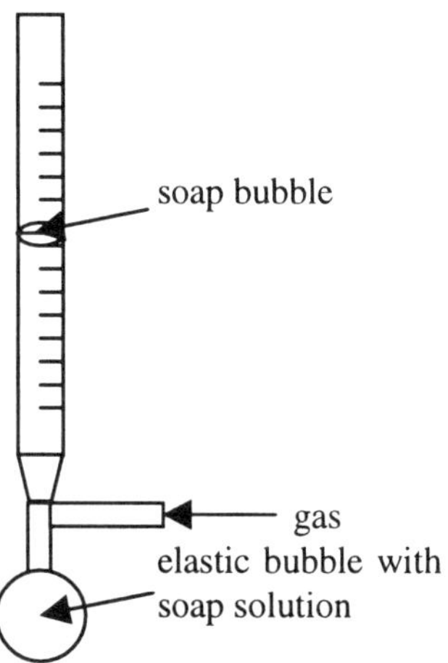

Fig. 2.16: Gas flow measurement device using the soap bubble principle.

The principle of this gas flow measurement device is simple. A push on the gummy bubble leads to soap bubble rising in the measurement cylinder. The gas flow pressures up the soap bubble. The time it takes for the bubble to pass through a defined volume of the measurement cylinder is measured by a timer. The principle of soap bubble gas flow measurement is the direct measurement of the average gas flow velocity. The advantage of this device is that a very accurate average value for the gas flow velocity can be calculated from the measurements. This device is the most accurate for the calibration of gas flow controllers. And this method is strongly recommended to measure the gas flow in the case when there are no full automatic calibration devices for the gas flow controllers.

The next method uses the principle of floatation of a heavy sphere in the gas flow. The scheme of the device using such a principle is shown in Figure 2.17. The flowing gas makes the sphere float. The higher the velocity of the gas flow, the higher the sphere is pushed. This device must be accurately calibrated before use. The biggest disadvantage of this device is that the gas flow velocity defined by this method at every moment is instantaneous velocity. All mass flow controllers, including the most modern ones, have certain errors, and gas flow can be permanently changed within the limits of these errors. It is difficult to determine gas flow velocity with controllers with relatively large errors. However, this device can be recommended for calibrating modern mass flow controllers with low errors.

Fig. 2.17: Gas flow measurement device using principle of heavy sphere floatation.

The last device for measuring gas flow and for calibrating mass flow controllers that is discussed in this chapter is called the capillary gas flow measurer. The principle of this device is used in many modern automatic calibration devices for mass flow controllers. The scheme of such a device is shown in Figure 2.18.

The main idea of this method is that the pressure needed to force the gas through the capillary is directly proportional to the gas velocity. The pressure can be measured by the method shown in Figure 2.18, by measuring the height of the liquid, which can be mercury or water or, for special cases, highly viscous liquids. However, this measurement is also possible by using other pressure measuring devices as well.

After all the preparations and experiment are over comes to the time to evaluate the experimental data. There are many methods to evaluate the gas chromatogram. Usually, every gas chromatograph is calibrated for each specific method.

The most popular and simplest method is called the peak height method. The method is applicable only if the chromatographic column is not overloaded. In this case, a linear dependence between peak height and the quantity of the respective component in the sample exists for constant analysis conditions. If the maximum possible load of the column is exceeded, the separation ability of the column decreases. As a result, a linear dependence between peak height and the quantity of the respective component is not presented any more. Deviations from linearity can occur with column that is not overloaded as well, if the compounds of the mixture have very different physicochemical properties.

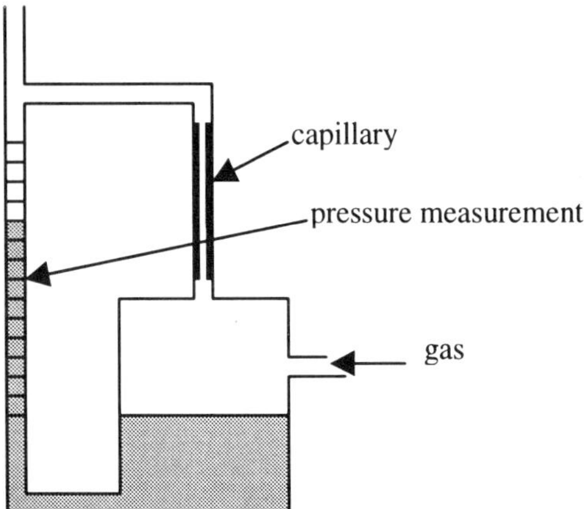

Fig. 2.18: Gas flow measurement device using principle of the capillary pressure.

This method of evaluation of gas chromatograms is suitable especially for recurring analyses with an internal standard. For example, the gas formed by thermal treatment of crude oil can be permanently analyzed by gas chromatography. However, it must be ensured that the analysis parameters are kept constant for all the analyses. In this case, all influences that affect the height of the peak are excluded.

The next evaluation method is based on the multiplication of the peak height by general retention time. This evaluating method is preferred for use with chromatographs that have detectors that are sensitive to a change in the carrier gas flow velocity. With these detectors, the magnitude of the signal increases with increasing gas flow velocity, in which case the total retention time of the compound is reduced. On the other hand, the result of the multiplication of the peak height and total retention time of the compounds remains approximately constant.

Temperature changes, which may occur if the temperature-programmed mode of the analysis is used, cannot be corrected by this method. This is why this evaluation method is applicable only for isothermally executed chromatography. The main advantages of the method are as follows:

- The analysis values obtained are very similar to the real concentrations of the compounds even in chromatograms with numerous peaks.

- Bad separated peaks can be evaluated, if the peak height and the total retention time are measurable.
- Thin peaks, which can arise frequently with capillary column chromatography, are easy to evaluate.
- With wide peaks, the peak height can be relatively accurately assessed.

The error obtained with this evaluation method is due mainly to the fact that the total retention time of the individual components depends linearly on the peak height only approximately. The amount of the component in the sample is often judged as too small for thin peaks and too high for wide peaks. The linear dependence is well suitable for homologous classes of compounds in the sample mixture analyzed.

The next evaluating method is based on the multiplication of the peak height by the width of the peak at half its height. With this evaluation method, the total area of a peak, which corresponds to a Gauss distribution curve, is determined as an area of an equal-leg triangle. The height of the triangle is equal to the maximum of the detector signal. The area of the triangle is the result of the multiplication of the triangle height with its width at the half height.

This method used frequently in comparison to all the other methods discussed previously because it is less sensitive to the influence of temperature, change of pressure and carrier gas flow.

However, this method can be used only for the evaluation of chromatograms with symmetric peaks. The advantage of the method is that the chromatogram can have symmetric peaks even if, like in some cases, the column of the gas chromatograph is overloaded.

The last method of evaluation of a gas chromatogram is called the mix method. With this method the compounds analyzed must be known. The pure compound that is to be evaluated is mixed with the sample mixture. Analysis of the sample mixture is performed twice in this method. One is in the absence of the pure compound whereas the second is a mixture with the pure compound. Evaluating the concentration of the component proceeds with a comparison of the peaks from both chromatograms. The advantage of this method is that this method is very accurate. On the other hand, the disadvantage is that the evaluation of the pure compounds is needed, and also there is the need of at least two chromatograms, which makes this method longer and more expensive than every other method discussed before.

More information about gas chromatography can be found in references 54-59 at the end of this chapter.

2.1.2 High Performance Liquid Chromatography

High performance liquid chromatography is also called high-pressure liquid chromatography. This method of chromatography was developed in the 1960s from column chromatography. The main difference between this method and gas chromatography is that non-evaporating samples can be analyzed by this method. This difference is also the biggest advantage of high performance liquid chromatography. The reason is that some crude oils contain over 80% of heavy fractions that cannot be evaporated at temperatures below 350°C. Such materials cannot be analyzed by gas chromatography but are readily analyzed by high performance liquid chromatography. The samples being analyzed by high performance liquid chromatography must have the important property that they are soluble in the solvent selected as the mobile phase for the chromatographic analysis.

It was mentioned earlier that high-pressure liquid chromatography was developed based on column chromatography. This method is very rarely used nowadays and only for special cases. However, this method is good to demonstrate the functioning principle of modern high-performance liquid chromatographs. The scheme of a column chromatograph is presented in Figure 2.19.

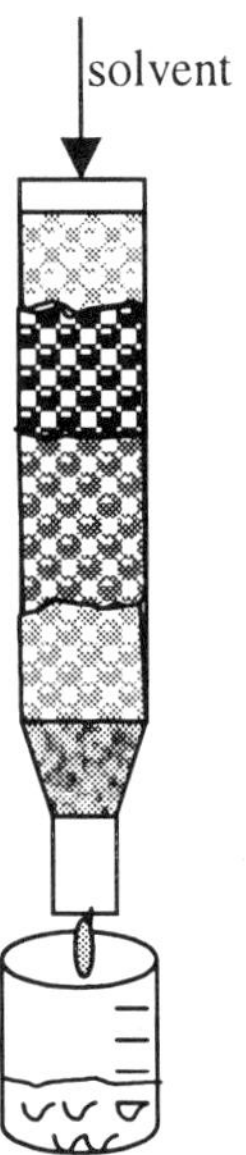

Fig. 2.19: Scheme of the classical column chromatography.

The setup shown in Figure 2.19 is the simplest example of liquid chromatography. The column diameter for such chromatographs usually varies from 1 to 5 cm. The column is filled with fine adsorbent particles. The top of the column is open for the entire duration of the experiment. The sample dissolved in a small amount of the solvent is introduced first in the column. After this, the solvent is introduced slowly but continuously into the column and separated compounds or fractions of compounds exit the column at the bottom. In this method, the solvent flows only under the influence of gravitational force. The disadvantage of this method is that a single experiment can take hours for the separation and the column packing must be prepared specifically for each experiment. The compounds exiting the column are often detected by optical methods. In the example shown in Figure 2.19, the sample was separated into three compounds or three fractions. These fractions can be detected by the difference in color. The need of a long time for every experiment is the reason why column chromatography was developed into high performance chromatography. However, the old classical column chromatography can still be found in many research petroleum laboratories even now. This method is used for the separation of heavy oil fractions into fractions of compound groups, as for example paraffinic, naphthenic and aromatic compounds. Fractionating by this method can be carried out to obtain more fractions of petroleum than was shown in this example. However, the analysis of individual compounds of crude oil is impossible by this method. But some individual compounds of heavy petroleum fractions can be analyzed by high performance liquid chromatography.

The main principles in the high-performance liquid chromatography technique are the same as for the gas chromatography technique. However, the mobile phase in the high-pressure liquid chromatography is a liquid solvent, which is forced through the column under high pressure. This is why no gas tank but a solvent reservoir is used in liquid chromatography. The new accessories needed in liquid chromatography are high-pressure pumps. Depending on the chromatograph, these pumps can achieve pressures up to 500 bar. One example of a high-pressure liquid chromatograph is shown in Figure 2.20.

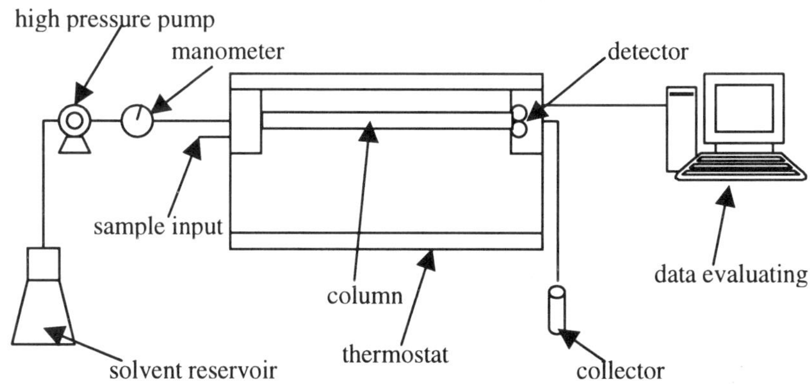

Fig. 2.20: Scheme of the high performance chromatograph.

Some notable differences between the high-pressure liquid chromatograph and the gas chromatograph can be seen clearly in Figure 2.20. For example, the columns used for liquid chromatography are shorter than the columns used for gas chromatography. The packing material used for the columns in liquid chromatography is usually very fine: 3–50 μm. The biggest advantage of high-pressure liquid chromatography in comparison to the classical column liquid chromatography is that a very precise analysis can be performed in a very short time.

As in the case for gas chromatography, the column and detector are the heart of the high performance liquid chromatograph. There are many types of columns used for liquid chromatography. These are classified as guard, derivatizing, capillary, fast, and preparatory columns.

The guard columns are placed anterior to the main separating column. This serves to protect the separation column from impurities or contaminants, which could be present in the sample. Usually, these columns have a length of 5 cm, and the diameter of the guard column must be the same as that of the separating column. There are many types of guard columns designed for the different purposes. For example, there are guard columns to filter the particulates from the sample or separate ions that can influence the analysis. The use of these columns by petroleum chemists is important, because the composition of petroleum is so complicated that it is very possible for the sample to contain insoluble particulates, which can damage the separation column and make it unusable. These columns must be changed frequently in order to optimize their protective function. The size of the packing varies with the type of protection needed.

Derivatizing columns serve to chemically change the initial compounds. The use of these columns can give accurate data for the sample in the case where the analysis of the initial mixture resulted in doubtful data.

Capillary columns are used in the current high performance chromatographs. The functioning principle of this type of column is the same as was described for gas chromatography. There are three types of capillary columns used in liquid chromatography: open tubular, partially packed, and tightly packed. The advantage of these columns is that they allow us to work with very small amounts of sample.

Fast columns are designed specifically to decrease the time needed for analysis. Usually, these columns are shorter than the usual separating columns; the packing material of such columns consists of a very fine material with particle size of about 3 μm. The advantages of using these columns are decreased analysis time and increased sensitivity of the column.

Preparatory columns serve to prepare the bulk sample for laboratory preparatory applications or further analysis. A preparatory column usually has a large diameter designed to facilitate large volume injections into the chromatograph.

There are more types of columns used in liquid chromatography designed for special cases or for specific chromatographs. However, the most important part of the column is the parking of the column. The high-pressure liquid chromatography can be classified as adsorption, distribution, gel permeation, affinity and ion pair chromatography according to the type of fillings or packing.

The principle of adsorption chromatography is the same as for gas chromatography, which was discussed in some detail in a previous section. Another description is not very essential at this point. The most popular filling or packing for adsorption liquid chromatography is silica gel. There is only one difference as compared to the case for gas chromatography. This concerns the particle size of the packing material. Usually, high performance liquid chromatography makes use of packing material with the smaller particles sizes.

The main principle of distribution liquid chromatography is the same as for gas chromatography as well. However, the stationary phase for liquid chromatography is in most cases a non-polar liquid. Non-polar stationary phase is called reverse-phase and the method used for chromatography with such a stationary phase is called reverse phase technique.

Gel permeation chromatography is distinct from every other type of chromatography. There is no physical interaction between the sample and the stationary phase in this method. The stationary phase in gel permeation chromatography is usually a highly porous material with particle sizes in the range from 6 μm to 10 μm. Separation proceeds according the sizes of the components in the sample. This means that the sample is separated into fractions with regard to the molecule size or molecular weight of the components.

The main principle of the gel permeation chromatography is as simple as shown in Figure 2.21.

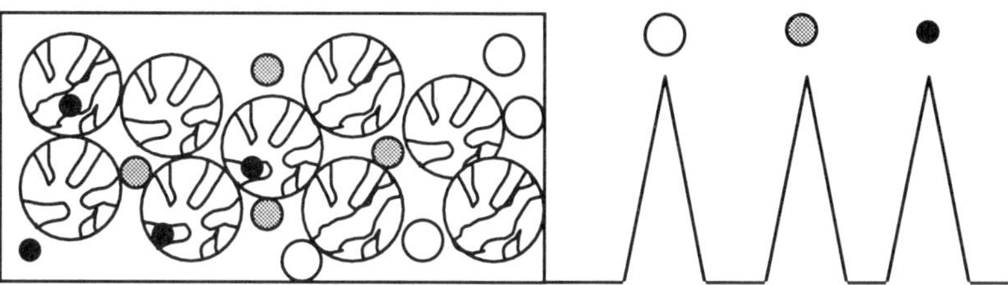

Fig. 2.21: Principle of gel permeation chromatography.

During gel permeation chromatography, the solution flows through the chromatographic column filled with porous packing material. The velocity distribution of the solution and its components depends on the pathway through which the components or particles of the solution pass. The particles that move around the packing particles are the fastest because this is the shortest possible route. Components or particles that flow through the pores of the porous packing material are slower since they also have a longer distance to cover (through the pores).

Small molecules shown in the example in Figure 2.21 as black circles are distributed in the whole volume of the column and their average moving velocity is very slow. Bigger molecules, gray circles, do not go through the smallest pores in the same way as the lightest compounds of the sample. Instead, they are only able to pass through the bigger pores. As such, the flow velocity of this set of molecules is higher than that for small molecules. The biggest molecules, white circles, cannot pass through the pores of the column packing and move through the column around the particles with the fastest possible velocity. As a result, the biggest molecules exit the column first, and the smallest ones come out last.

The gel permeation chromatographic column is characterized by the minimum and maximum sizes or molecular weights of molecules that can be analyzed by that column. This is because all the molecules with molecular weight higher than maximum analyzable molecular weight cannot pass through the pore and all move with the same velocity. The result is that they cannot be separated. The same problem arises with the molecules smaller than the smallest analyzable molecule. They all pass through the smallest pores of the packing material with the same velocity.

The affinity chromatography is a chromatographic method which uses specific interactions between the sample molecules and the stationary phase. Affinity means the tendency of the molecules to react with each other. The first component in this case is the one bounded to the stationary phase. The other component is the component in the sample. This is adsorbed from the solution onto the ligand. The components, which have no affinity for the ligand, are transported by the mobile phase through the column. In order to desorb the analyzed component molecules from the stationary phase, a solvent with greater affinity for the ligand in comparison with the stationary phase is introduced into the column. In some cases, it is sufficient to change the pH value of the solution to desorb the components being analyzed from the stationary phase.

The last type of column used in liquid chromatography is the column used for ion pair chromatography. In this method an ionic solvent, which has the opposite charge to the sample compound, is used as the mobile phase. The use of such a mobile phase leads to the formation of an ion pair, which is the material eventually analyzed in this method. Samples which cannot be analyzed by adsorption or distribution chromatography because of the high electrical charge of ions, should be analyzed by this method.

An important part of the high performance liquid chromatograph is the detector. The main tasks of the detector for this type of chromatography are the same as for gas chromatography. However, the detectors used in liquid chromatography are different from those used in gas chromatography, because of the necessity to perform analysis with liquid samples.

One of the most popular detectors in high-pressure chromatography is the ultra violet (UV) detector. Compounds that can absorb ultra violet light, are detected using this detector. The UV detectors have a relatively large dynamic range. The great advantage of UV detectors is that they are only minimally sensitive to temperature changes during analysis. An example of the UV detector is shown in Figure 2.22.

After leaving the column, the mobile phase flows through a quartz cell, which acts as the analysis cell. The analysis cell and a reference cell (which contains the mobile phase) are irradiated with the same source of UV light. A low-pressure mercury lamp or deuterium lamp is usually used as the UV light source. The UV light intensity is measured with photoresistors. Before starting the real analysis, it is necessary to carry out an electronic calibration of the photoresistors of the analysis and reference cell. The pure mobile phase flows thorough both cells during the calibration. The mobile phase should be optically permeable with the wavelength generated by the UV light source. During the real analysis, a component of the sample mixture eluting from the column into the detector absorbs the UV light, and changes the resistance of the photoresistor. The difference in resistances between the photoresistors of the analysis and reference cell produces a detector signal, which is sent to the evaluating computer.

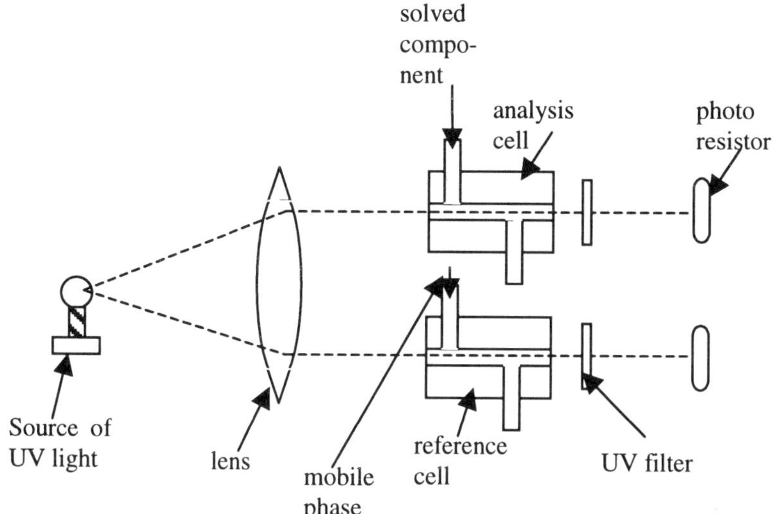

Fig. 2.22: UV detector.

Photodiode array detectors are a class of UV detectors. The main difference of this type of detectors from the classical UV detector is that photodiode array detectors scan the entire spectra from UV all the way to visible light. This detector type is very beneficial in the sense that with well-designed software, it is possible to select the best wavelength for every component in the sample analyzed. As such, the analysis can be carried out more accurately.

The next detector often used in liquid chromatography is the refractive index detector. This detector uses the property of the sample molecules to bend or refract light. The disadvantage of these detectors in comparison to UV detectors is that refractive index detectors are less sensitive than the UV detectors. Refractive index depends strongly on the temperature of the sample. This is why the refractive index detector must be well thermally insulated. A scheme of the analysis cell of the refractive index detector is shown in Figure 2.23.

The refractive index detector contains two cells: one analysis cell and one reference cell just as UV detectors. Light from the lamp goes through the analysis and reference cell at the same time. The pure mobile phase flows thorough the reference cell and the sample dissolved in the mobile phase flows through the analysis cell. If the light in the analysis cell is bent differently compared to the light in the reference cell, the detector produces a signal, which is sent to the evaluating computer.

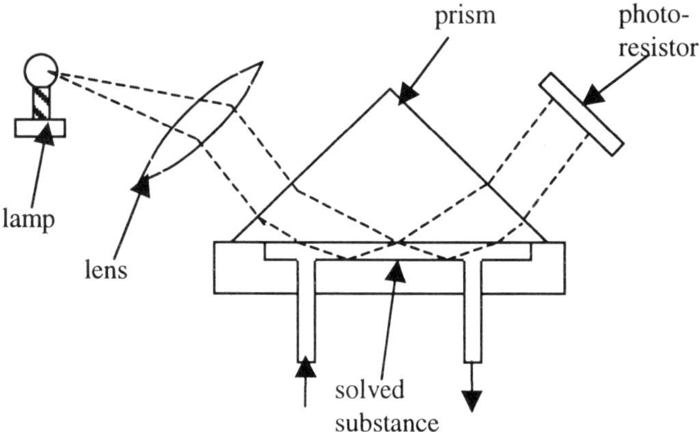

Fig. 2.23: Analysis cell of refractive index detector.

The most modern but not especially popular refractive index detectors are those that use a laser as the light source. The use of lasers brings many advantages. First of all, the resolution of a laser refractive index detector is much higher than by using classical detectors. However, the price for such detectors is relatively high and it is used only for special analytical cases.

The next type of detector used in liquid chromatography is fluorescence detectors. These detectors use the ability of components in the sample analyzed to absorb and re-emit light at defined wavelengths. Each compound has a characteristic fluorescence. Ultra violet lamps are usually used as the light source in such detectors. The technique of fluorescence detectors is different as compared to the usual detectors. They contain only one flow cell. The light goes through this cell. The photoresistor measures the intensity of the light passing through the cell and the monochromator measures the re-emitted light, i.e. fluorescence.

Electrochemical detectors are a very interesting example of detectors used in liquid chromatography. These detectors can detect only compounds that are able to react by oxidation or reduction reactions. Usually, this technique functions by measuring the gain or loss of electrons from the flowing sample as it passes between two electrodes maintained at an electrical potential difference.

The last two popular detectors in liquid chromatography are nuclear magnetic resonance and mass spectrometers. The mass spectrometer as a detector was described in the section in this chapter on gas chromatography. On the other hand, the principles of the nuclear magnetic resonance are described in chapter 5 of this book because of its important in asphaltene chemistry.

It was indicated that most detectors used in high-pressure liquid chromatography are optical detectors that use ultra violet, visible or laser light sources. It stands to reason that the solvents that can be used as the mobile phase for liquid chromatography must have special properties. The most important properties of the mobile phase for high performance liquid chromatography are:

- Transparency for ultra violet light.
- The boiling point of the mobile phase must be between 40°C and 100°C. The limit of 100° is important only for distillation of the solvent after the analysis. The limit of 40°C is important for the analysis, because a lighter solvent can by evaporated in the pump leading to the formation of vapor bubbles in the pumps.
- Miscibility is important in case a mixture of many solvents is used as the mobile phase for the elution.
- Polarity indication of which compounds can be used in a given mobile phase.
- Mobile phase must be free of all solved gases. Nitrogen and oxygen have good solubilities in most solvents used in liquid chromatography. There are three methods to degas the mobile phase: to degas the mobile phase by helium stream, to degas the mobile phase by treatment of the solvent with ultrasound and to degas the mobile phase by distillation.

In degassing the solvent with a helium stream, helium gas is bubbled continuously through the solvent until all the solved gases are removed. In this method, it is important that the mobile phase does not come in contact with air during the degassing. This is the simplest and most popular method.

In degassing with ultrasound, a glass with solvent is placed in the bath with the source of ultrasound. During the treatment with ultrasound, bubbles of the gas removed can be seen issuing from the solvent. The ultrasound treatment must be carried out until no more gas bubbles are issuing from the solvent.

The last method, the distillation method, is the most effective but also the most complicated in executing. In this method, the solvent is evaporated in an inert atmosphere and then condensed again. It is important to ensure that no gas is dissolved in the solvent during the condensation. For this reason, the distillation unit must be evacuated before degassing and then filled with a gas that is insoluble in the mobile phase.

Elution with the selected and degassed solvent as mobile phase can be done in two different methods. The first method is simplest, but cannot be used for all samples especially those containing complicated mixtures of compounds in crude oil fractions. As such, this method is only rarely used. It is called isocratic elution. In this method, the same composition of mobile phase is used during the entire analysis time. The main property of samples that can be analyzed by this method is that the components in the sample have approximately the same polarity. The second elution method is called gradient elution. This method can be used for complicated mixtures of compounds which have different polarities and which cannot be analyzed by only one solvent. However, this method demands the use of special pumps to support the stepwise mixing of the various solvents.

From the last shown elution method, it is clear that the pump is a very important accessory for high-pressure liquid chromatography. There are three main types of pumps used in modern chromatographs:

- membrane pumps
- reciprocating piston pumps
- syringe type pumps
- constant pressure pumps

In the membrane pump, a disk brings the piston to a reciprocating movement. Special hydraulic oil passes the impulse from the piston onto the membrane. The functioning principle of the membrane pump is presented in Figure 2.24.

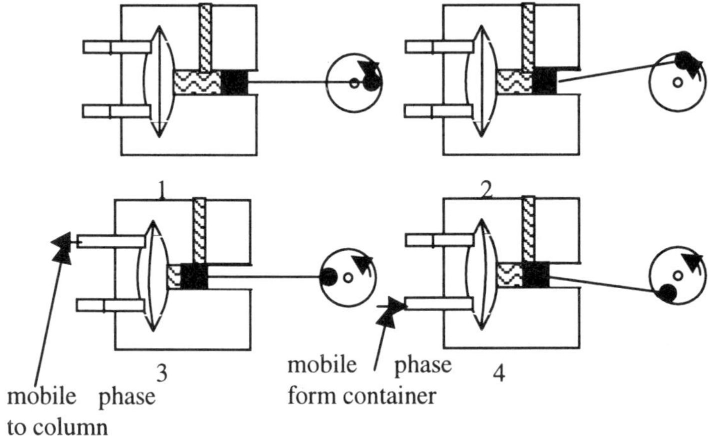

Fig. 2.24: Functioning principle of the membrane pump.

On the back stroke, the separation column valve is closed, and the piston pulls in solvent from the mobile phase container. On the forward stroke, the pump pushes the solvent out to the column from the hydraulic chamber. A wide range of flow rates can be attained by altering the stroke frequency. As shown, the functioning principle of the membrane pump is relatively simple. However, there is the disadvantage that the pumping of the mobile phase is only done discontinuously or intermittently. A schematic showing the pulsating pumping of the solvent by a membrane pump is illustrated in Figure 2.25.

The problem with discontinuous pumping can easily be solved by using two pumps functioning at the same time such that if one pump sucks in the solvent another pump presses the solvent to the column (out of phase with each other). This type of pump system is significantly smoother because one pump is filling while the other is in the delivery cycle. Such connections of two membrane pumps are relatively often used in liquid chromatography.

Fig. 2.25: Example of discontinuous pumping.

The working principle of the reciprocating piston pump is similar to that of the membrane pump. The difference is that the reciprocating piston pump has no membrane. The solvent is pumped directly by the piston. This is why the piston of such pumps must be made from special chemically and mechanically stable materials. A reciprocating piston pump consists of a small motor driven piston, which moves rapidly back and forth in a hydraulic chamber that may vary from 35-400 μl in volume. The reciprocating piston pumps have the same disadvantage as the membrane pumps and the same principle is used to prevent intermittent pumping. Dual and triple head pumps are used to make the pumping continuous, and this arrangement consists of identical piston-chamber units which operate at 180 or 120 degrees out of phase.

Syringe type pumps are usually used for chromatography with capillary columns because this pump can deliver a finite volume of mobile phase before it is refilled. These pumps have a volume of hydraulic chamber between 250 to 500 ml.

The mobile phase is pressed by gas continuously at constant pressure into the column by constant pressure pumps. The big advantages of this type of pump are really the constant pressure and the continuous pumping of the mobile phase.

Pumps are the last important part of high performance liquid chromatography. In principle, if the column, detector and pump are appropriately selected and installed in the chromatograph, analysis can commence. The chromatograms of the high-pressure liquid chromatography are similar to the chromatograms of gas chromatography and can be evaluated by using of the same techniques, which were described in detail in the section on gas chromatography.

More about high performance liquid chromatography can be found in references 60-62 at the end of this chapter.

2.1.3 Thin Layer Chromatography

Thin layer chromatography, often called film-development chromatography, is the simplest, quickest to perform, and cheapest type of chromatography, in comparison to gas chromatography and liquid chromatography. However, this method can only be used for qualitative analysis. This type of chromatography is very popular with crude oil geologists, because it can be used conveniently for field investigations. The method does not need any special or complicated accessories. Thin layer chromatography is also used in the laboratories by crude oil chemists. Usually, this is only for quick screening research of many samples. This method is recommended for field analysis of organic substances in ground samples during crude oil prospecting.

Thin layer chromatography differs from all other types of chromatography discussed before by the simplicity of the technique used. Thin layer chromatograph has a stationary phase and a mobile phase just like every other chromatographic method. However, the stationary phase for thin layer chromatography is not located in the column as in gas or liquid chromatography. Instead, it is fixed on a glass, aluminum or plastic plate as a thin layer.

Thin layer chromatography is very similar to paper chromatography. Thin layer chromatography has a wide variety of possibilities depending on the choice of the stationary phase. Adsorption, distribution and ion chromatography can be carried out in thin layer chromatography.

The preparation of the experiment for thin layer chromatography begins with the preparation of the analysis chamber. The analysis chamber for this chromatographic method is a glass filled with a small amount of the mobile phase, and a filter paper is introduced. The analysis chamber is prepared by dipping the filter paper in the mobile phase for circa half hour. This time is needed for the chamber to reach vapor formation in equilibrium with the mobile phase. The fact that the mobile phase vapor in the analysis chamber is saturated and in equilibrium with the mobile phase liquid is especially important for executing the analysis, because it has a great influence on the velocity of the mobile phase during the analysis. An example of the analysis chamber that has been prepared correctly is shown in Figure 2.26.

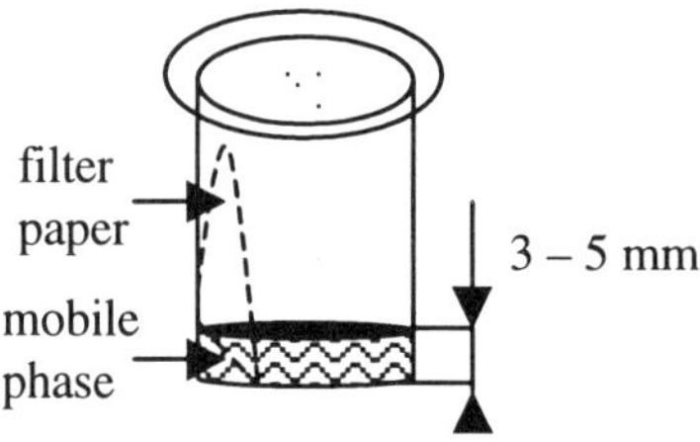

Fig. 2.26: Analysis chamber for thin layer chromatography.

The humidity that comes with the mobile phase filter paper ensures the evenness of the vapor pressure in the analysis chamber and acts as an accelerator for the achievement of saturated vapor with the mobile phase. The filter paper can also act as an indicator; if the solvent front reaches the upper limit of the filter paper, it is usually postulated that the analysis chamber is ready for the commencement of the analysis.

It is important to select the appropriate mobile phase for the analysis (i.e. for the separation of components in the sample). The right choice of the mobile phase has a great influence on the resolution during analysis. The most popular solvents used as mobile phase for thin layer chromatography are n-hexane, tetrachloro methane, benzene, dichloro methane, trichloro methane, acetic acid ethyl ether, dioxane, acetone, i-butanol, i-propanol, ethanol and methanol. Despite a good choice of solvent for the mobile phase, it is often required to use a mixture of solvents, because no single solvent has all the properties needed. There is a simple method used to select the most appropriate solvent for thin layer chromatography. It is called the spot test. This method is presented in Figure 2.27.

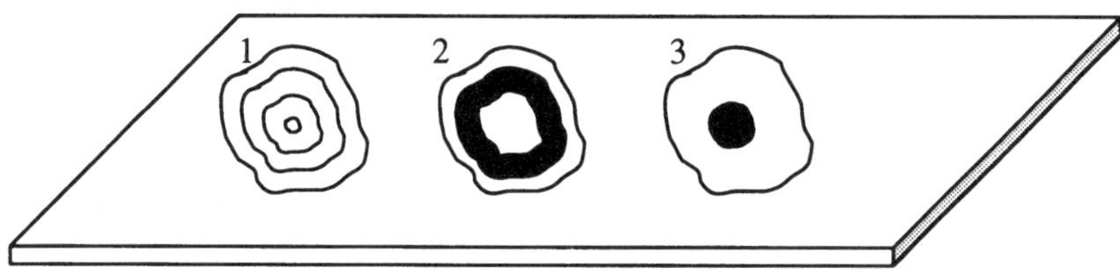

Fig. 2.27: Spot test.

In the spot test, the sample is spotted on the chosen thin layer plate and a drop of the mobile phase tested is dropped in the center of the sample spot. The solved sample begins distribution over the thin layer. The best solvent is chosen based on evaluation of the spot formed after the radial distribution of the sample has stopped. Solvent 1 in Figure 2.27 is the most appropriate solvent and has the best resolution in comparison to solvents used on spots 2 and 3. Solvent 2 moves the compounds of the sample too fast and the spot is clearly overdeveloped. In contrast to the second solvent, the solvent used on spot 3 moves the compounds of the sample too slowly. The spot is therefore clearly underdeveloped.

As the next step, the sample must be prepared for the analysis. The sample is usually dissolved in a volatile solvent. The concentration of the sample in the solvent can vary form 1 to 10% or higher for special cases. The solved sample is introduced with a pipette on a spot (i.e. spotted) at a point approximately 15 mm from the bottom part of the thin layer. The right choice of the material for the thin layer is important for the analysis. The most popular thin layer material is silica gel. It can be covered with various highly viscous liquids to enable distribution chromatography. The principle in the choice of thin film material is the same as the choice of the packing material for columns used in gas or high-pressure liquid chromatography. However, the thin layer can be impregnated for special purposes with various solids, which can react with the sample. Phosphates and borates are popular for use as impregnated solids.

Analysis can start after the right mobile phase and thin film plate have been chosen. Before the thin layer plate is placed in the analyzing chamber, the excess solvent used in solving the sample must be removed slowly to avoid touching the thin layer plate. The analysis chamber is then placed on a sturdy table to prevent vibrations during analysis. This is important for accurate analysis results. Then, the thin layer plate is placed in the analysis card in the analysis chamber. The analysis chamber with the thin film plate is shown in Figure 2.28.

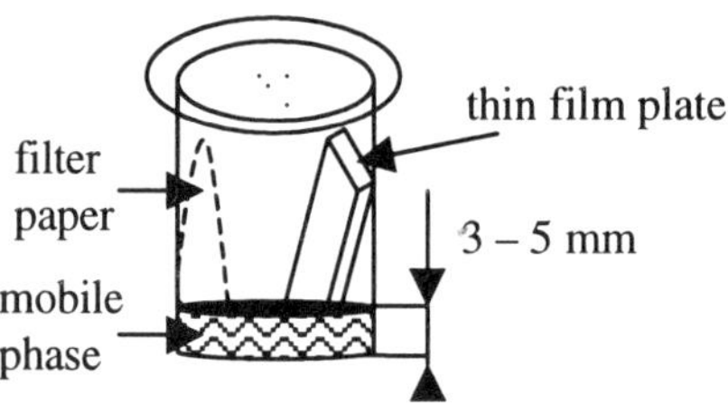

Fig. 2.28: Working analysis chamber.

The development of the sample begins due to the capillary action of the mobile phase on the thin layer plate, after the thin film plate has been placed in the analysis chamber. Usually, the time needed for the analysis varies from twenty to sixty minutes. However, the analysis proceeds until the solvent reaches the upper limit of the thin film. In the case where appropriate resolution is achieved earlier than when the solvent front reaches the upper limit of the plate, analysis can be stopped early to save time.

Evaluation of the analysis results proceeds by measurement of the differences in the distance from the start line until the analyzed spot and front of the mobile phase, as presented in Figure 2.29.

After measuring the distances d_1 and d_2, a special factor R_f is calculated according to the formula (2.1) (see Figure 2.29). The R_f values depend on the solid absorbent, the compound polarity, and the eluting solvent polarity. The factor R_f for constant analysis conditions is a characteristic property of the substance. If the thin-film card is calibrated for permanently analyzing similar samples, the substance analyzed can be detected just by the R_f value.

$$R_f = d_2 / d_1 \tag{2.1}$$

The evaluation can also be carried out visually as shown in Figure 2.29 or by transferring the data into digital form by a scanner and evaluating the results in a computer. Fluorescence active impregnation can be used in the case of analysis of almost transparent samples. The spots of such samples are hard to see in the visible spectra of light. However such substances can be easily detected by viewing the thin film with UV light and observing the fluorescence picture.

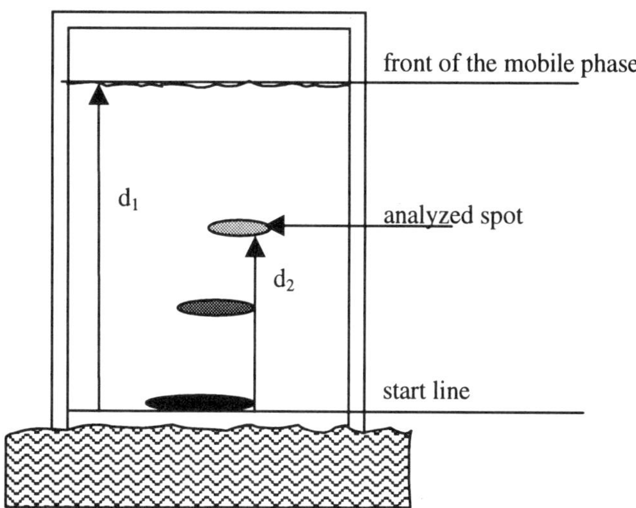

Fig. 2.29: Evaluating the results of the thin film chromatography.

The evaluation of some samples is sometimes not possible after the first development. This can be caused by a wrong choice of solvent or by special properties of the sample. However, it is possible to develop the thin film one or more additional times. There are two methods for developing the thin-film card more than one time: one-dimensional and two-dimensional development. The thin-film card must be dried so that there are no residues of the mobile phase on the thin film before the development in the second step.

After drying, the thin film is placed in the analysis chamber in the same way as was done for the first analysis. The analysis can be repeated until an appropriate resolution of the card is achieved. The R_f value for one-dimensional analysis is calculated by the formula in equation (2.2), with regard to the number of times (n) it was repeated.

$$^{n}R_f = 1 - (1 - R_f)^n \tag{2.2}$$

Two-dimensional developing of the thin film cards is needed only in special cases. In this method the card is placed in the analyzing chamber the second time but turned by an angle of 90° in comparison to the first analysis. The advantage of this method is that the spots are located diagonally on the thin film plate and are easy to evaluate. Usually in this method, two different solvents are used. However, this method is relatively complicated in comparison to the one-dimensional developing and it is recommended only for the analysis of samples which cannot be separated by only one solvent.

The qualitative evaluation of the spots discussed above is possible only in some cases. Often, a more accurate analysis of the separated substances is needed. For this purpose, the spot of the sample analyzed must be accurately removed from the plate. Then, the substance is extracted from parts of the thin film by using appropriate solvents and analyzed by spectral or other methods, which will enable the identification of the structure of the compound.

It was already mentioned that thin film chromatography is used firstly as a qualitative analysis method. There is no accurate quantitative analysis that is possible by thin film chromatography. However, there is an empirical formula, equation (2.3), which allows the approximate calculation of the weight of the substance (W) from the spot area (A). This method is very inaccurate and can be used only for gross estimation.

$$A^{1/2} \approx \log W \tag{2.3}$$

The thin film chromatography is the last chromatographic method discussed in this chapter. More about this method can be found in references 63-66 at the end of part I of this book.

2.2 SPECTROSCOPIC METHODS

2.2.1 Infrared Spectroscopy

Infrared red spectroscopy is based on the ability of the substances to absorb light of a given wavelength. Infrared spectroscopy is today one of the most important spectral analytical methods in the crude oil chemistry, because of its high information content and the variety of possibilities for sample preparation.

The direct analysis of the structure of sample components without calibration with reference substances is impossible. However, defined chemical groups in the sample absorb infrared light in defined areas of the spectra. The direct prediction of the structure of the sample or components, in this case, is possible with the use of special empirical tables for infrared spectroscopy.

In order to predict the structure of the sample analyzed, it is important to understand the principle of analysis. Infrared spectroscopy is based on the measurement of the absorbed infrared light by the sample analyzed.

When a beam of infrared light of intensity I_0 is passed through a sample, it can either be absorbed or transmitted, depending upon its frequency and the structure of the molecules. The final intensity I of the infrared light that passes through the sample can be calculated by the Lambert-Beer law (2.4) which is applicable to all types of electromagnetic radiation.

$$I = I_0 \cdot 10^{-E \cdot c \cdot d} \tag{2.4}$$

Where c is the concentration of the component analyzed in the sample and d is the thickness of the sample layer. The factor E is a specific characteristic value of the component analyzed. This coefficient however depends weakly on the concentration of the component (c) in the sample. Therefore, it is important to calibrate the spectrometer for all the possible concentrations of the component analyzed before the analysis.

The infrared light, a type of electromagnetic radiation, is energy and hence when a molecule absorbs radiation it gains energy and undergoes a quantum transition from one energy state ($E_{initial}$) to another (E_{final}). The frequency of the absorbed radiation is related to the energy of the transition by Planck's law, which is presented by equation (2.5).

$$E_{final} - E_{initial} = E = h \cdot \nu = h \cdot c / \lambda \tag{2.5}$$

Thus, if a transition exists which is related to the frequency of the incident radiation by Planck's constant ($h = 6.626 \cdot 10^{-34}$), then the radiation can be absorbed. Conversely, if the frequency (ν) does not satisfy Planck's expression, then the radiation will be transmitted. A plot of the frequency of the incident radiation against some measure of the percent radiation absorbed by the sample provides the absorption spectrum of the compound or component. The absorption spectrum is characteristic for the compound and this spectrum is often called the fingerprint of the compound. Infrared spectroscopy is based on the measurement of the absorption of electromagnetic radiation that arises from the altering of the vibration level of the component's molecule. An example of the adsorption and transmission of the infrared radiation is shown in Figure 2.30.

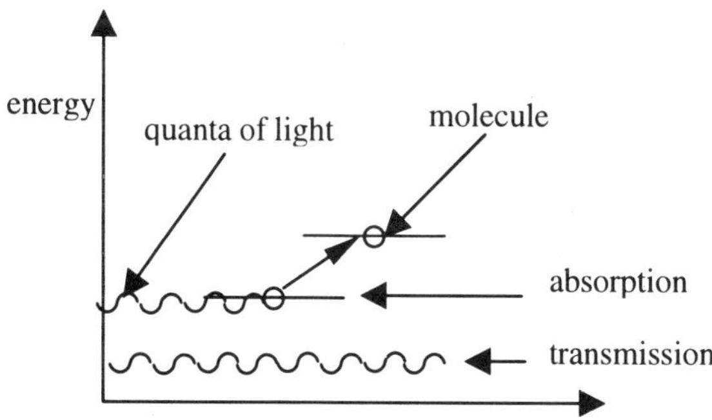

Fig. 2.30: Absorption of infrared radiation.

As stated above, the absorption of infrared radiation takes place by altering the vibration level or movement energy of the component's molecule. There are several types of motion that a molecule may undergo. First, the molecule may move through space in some arbitrary direction and with a particular velocity. This type of motion is called translational motion, and the translational kinetic energy of the molecule is associated with it. The energy of the translational motion can be calculated by equation (2.6).

$$E = mv^2/2 \tag{2.6}$$

where v is the velocity of the center of mass of the molecule
m is the mass of the molecule

The velocity with which a molecule translates may be resolved into components along each of the three axes of the Cartesian coordinate system, as shown in Figure 2.31.

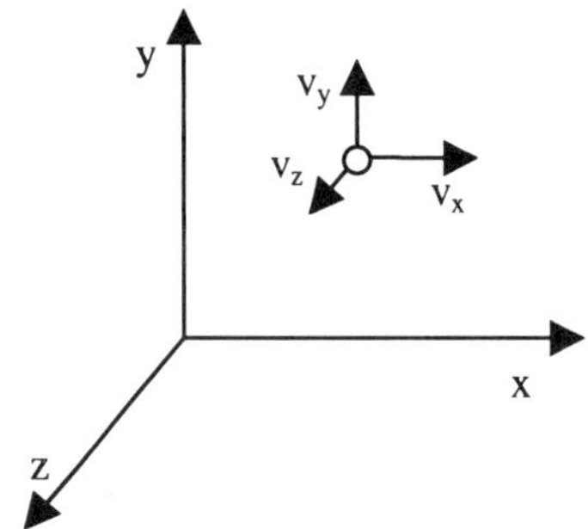

Fig. 2.31: Translational motion of a molecule.

If the molecule motion is observed as is shown in Figure 2.31, then the energy of the translational motion is calculated by equation (2.7).

$$E = mv^2/2 = mv_x^2/2 + mv_y^2/2 + mv_z^2/2 \tag{2.7}$$

where v_x is the x-component of velocity
v_y is the y-component of velocity
v_z is the z-component of velocity
m is the mass of the molecule

This equation shows that the total translational energy of the molecule can be made up of three parts, each of which represents the kinetic energy of the molecule along one of the reference directions. Any translation of the molecule may be considered to arise from the vector sum of its motions along the three axes. Thus, the kinetic energy may always be broken up into the sum of three contributions, one arising from motion along the x axis, one from motion along the y axis, and one from motion along the z axis. It means that the molecule has three translational degrees of freedom, one corresponding to each axis.

The next type of motion of the molecule is the rotational motion. The molecule can rotate about some internal axis. This axis may be resolved into components of the x-, y-, and z-axes of the coordinate system, so that any rotation of the molecule may be resolved into three mutually perpendicular components. The energy of the rotational motion can be calculated by an analogous method to the transformational motion. This is given by equation (2.8).

$$E = I_x\omega_x^2/2 + I_y\omega_y^2/2 + I_z\omega_z^2/2 \qquad (2.8)$$

where I_x is the moment of inertia about the x axis
I_y is the moment of inertia about the y axis
I_z is the moment of inertia about the z axis
ω_x is angular velocity about the x axis
ω_y is angular velocity about the y axis
ω_z is angular velocity about the z axis

The final type of the motion of molecules is called vibrational motion. This type of molecule motion is very important in infrared spectroscopy since the absorption of infrared radiation by this motion forms the fingerprint of the sample analyzed. There are many types of vibrational motions, and these are shown below. It is important to know the right number of degrees of freedom for the vibrational motion of the sample molecule. This can be calculated by using the following general equation (2.9).

$$Z = 3 \cdot N - 6 \qquad (2.9)$$

where N is number of centers of mass in the molecule

As was shown for translational and rotational motions, there are three degrees of freedom for vibrational motion for every center of mass in the molecule. The number six on the right hand side term of equation (2.9) arises from the total number of degrees of freedom for translational and rotational motion, which do not belong to vibrational motion. It should be known that for linear molecules, there are only two degrees of freedom for rotational motion. This is why for this case there is a special equation for the calculation of the degrees of freedom for vibrational motion (2.10).

$$Z = 3 \cdot N - 5 \qquad (2.10)$$

The number of the degrees of freedom for the vibrational motion (Z) calculated by the equations (2.9) and (2.10) is called degrees of freedom of normal vibrations.

For example one three-atom linear molecule has four degrees of freedom for normal vibrations (2.11):

$$Z = 3 \cdot 3 - 5 = 4 \qquad (2.11)$$

This means that four types of vibrational motion are possible for such a molecule. The first motion is the symmetric vibration of the atoms in the direction of the center of mass as shown in Figure 2.32. Since this vibration is symmetric, it does not lead to the altering of the dipole moment of the molecule. This means that electromagnetic radiation cannot be absorbed by this type of motion. Such vibrations are called infrared inactive vibrations

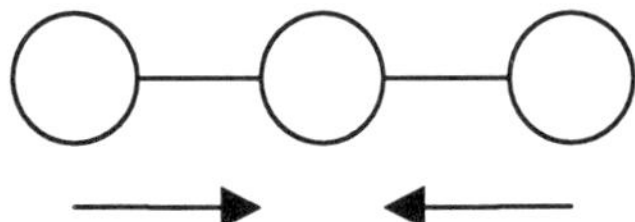

Fig. 2.32: Symmetric vibrational motion of the linear molecule.

Both end atoms move asymmetric to the central atom during the second type of vibrational movement. Such a movement leads to the altering of the dipole moment of the molecule, and this type of vibration is active for infrared analysis. An example of this movement is shown in Figure 2.33.

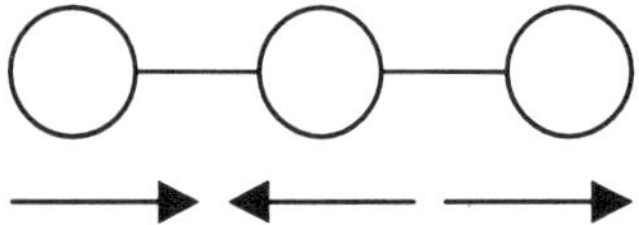

Fig. 2.33: Asymmetric vibrational motion of the linear molecule.

The next two types of vibrational movement are called deformation vibration. The end atoms move vertical to the central atom. The first type of vibrational movement is shown in Figure 2.34.

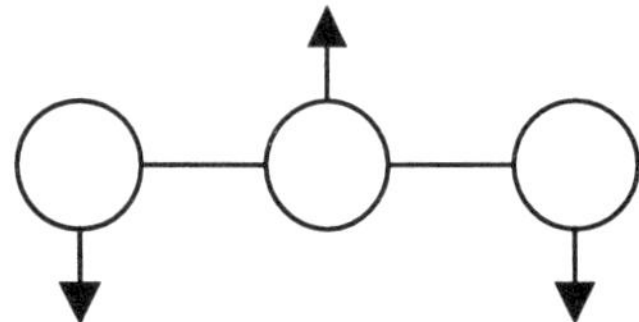

Fig. 2.34: First type of deformational vibrational motion of the linear molecule.

This type of motion leads to the altering of the dipole moment of the molecule as well as to an asymmetric vibrational motion. Thus, this vibration is active for infrared analysis.

The last type of vibration for the molecule shown is principally the same as the third type of motion. However, the vibration proceeds horizontally to the central atom. This type of motion is shown in Figure 2.35

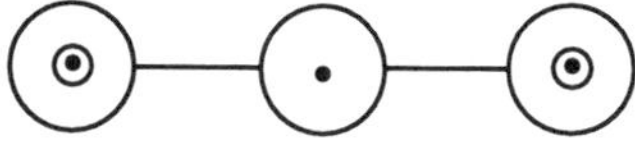

Fig. 2.35: Second type of deformational vibrational motion of the linear molecule.

This type of motion is shown as an extra motion because the motion proceeds in different coordinates in comparison to the first shown deformational motion. However, the vibration frequency and the altering of the dipole moment by these last two vibrations are the same. Such type of motions, where the vibration frequency of two or more motion types is the same, is called downgraded motion. A special value, called degradation number, is given for characterization of such motions. This value shows the number of molecules that have the same vibration frequencies. For example, the degradation number for the linear molecule shown above is two.

A three-atom nonlinear molecule is shown as the next example. Nonlinear molecules are more popular in petroleum chemistry, because most organic substance are nonlinear. The number of degrees of freedom of normal vibrations for three-atom nonlinear molecule is calculated using equation (2.12).

$$Z = 3 \cdot 3 - 6 = 3 \tag{2.12}$$

This means that three vibrational motions are possible for this type of molecule.

The end molecules vibrate symmetric to the central atom by the first possible motion. In opposition to the linear molecule this motion by a nonlinear molecule leads to the altering of the dipole moment. This means that this vibration is active for infrared analysis. An example of such a vibration is shown in Figure 2.36.

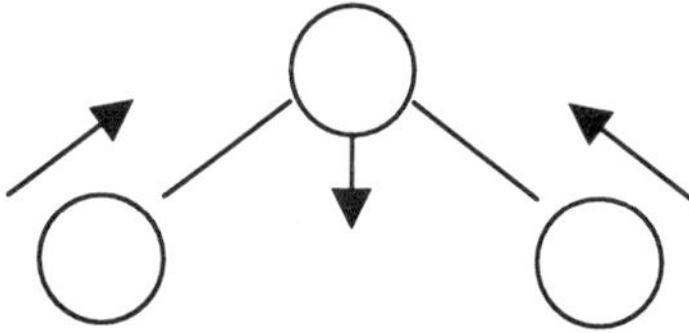

Fig. 2.36: Symmetric vibrational motion of the nonlinear molecule.

As an analogue to the linear molecule the second vibrational motion for the non linear molecule is asymmetric vibration. This type of motion is shown in Figure 2.37.

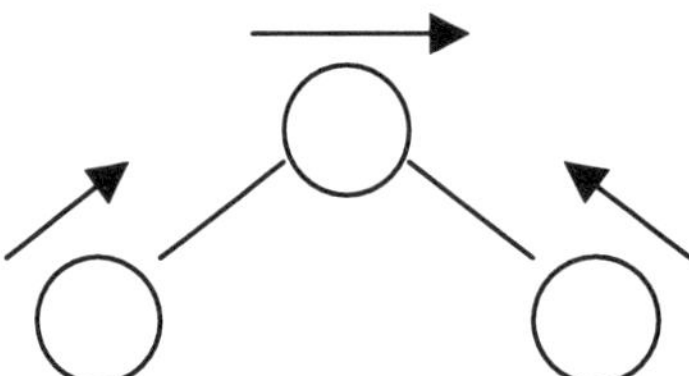

Fig. 2.37: Asymmetric vibrational motion of the nonlinear molecule.

The asymmetric vibration leads to the altering of the dipole moment as well, and as a result this motion is active for infrared analysis.

The last possible vibrational motion is the deformational motion. This type of vibration is presented in Figure 2.38.

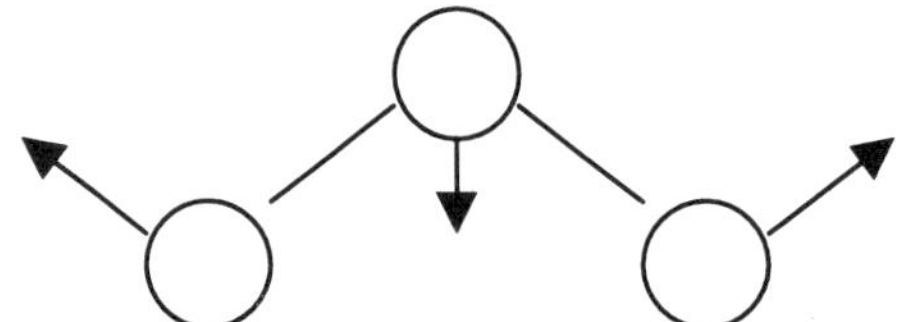

Fig. 2.38: Deformational vibrational motion of the nonlinear molecule.

From Figure 2.38, it is easy to see that for nonlinear molecule, only the type of deformational motion shown is possible, because the deformational motional in different coordinates other than the ones shown in figure 2.38 are analogues to the rotational motion of the molecule. Thus, a three-atom nonlinear molecule has only three degrees of freedom of normal vibrations whereas a three-atom linear molecule has four degrees of freedom of normal vibrations.

There are many types of vibrational motion of atoms in molecules. However the intention of this chapter is to show the fundamentals of infrared spectroscopy. In the list of references at the end of this chapter, many books and articles are recommended for a more detailed study of this type of analysis.

Each of the vibrational motions of a molecule occurs with a certain frequency that is a characteristic value for the groups in the molecule analyzed. The energy involved in a particular vibration is characterized by the amplitude of vibration, so that the higher the vibrational energy, the larger the amplitude of the motion. According to the results of quantum mechanics, only certain vibrational energies are allowed for the molecule, and thus only certain amplitudes are allowed. Associated with each of the vibrational motions of the molecule, there is a series of energy levels. The molecule may change from one energy level to a higher one by absorption of a quantum of electromagnetic radiation. This can be calculated by equation (2.5). In undergoing such an energy of transition, the molecule gains vibrational energy, and this is represented by an increase in the amplitude of vibration. The frequency of light required to cause an energy transition for a particular vibration is equal to the frequency of that vibration, so that it is possible to measure the vibrational frequencies by measuring the frequencies of light which are absorbed by the molecule. So, infrared spectroscopy deals with energy transitions between vibrational energy levels in molecules, and is therefore also called vibrational spectroscopy. An infrared spectrum is generally displayed as a plot of the energy of the infrared radiation.

Infrared spectra are measured by special instruments called infrared spectrometers. These instruments measure the differences in the intensity of the infrared light of a certain wavelength that penetrates into the sample and goes out from the sample. The most important parts of the infrared spectrometer are:

- light source, which produces an intensive infrared radiation
- monochromator
- detector

A schematic of the infrared spectrometer is presented in Figure 2.39.

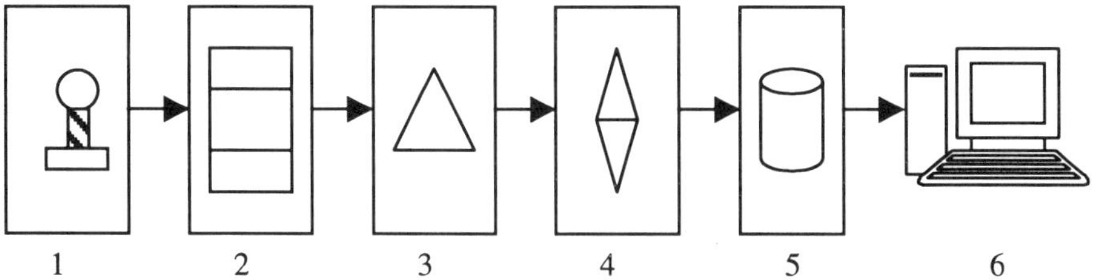

Fig. 2.39: Infrared spectrometer:
1 – infrared light source
2 – sample chamber
3 – monochromator
4 – detector
5 – amplifier
6 – evaluating computer

The most frequently used source of infrared light for infrared spectrometers is so called the Nernst stick. This stick is about two to four centimeters long and one to three millimeters thick, and is made from zirconium oxide with additions of yttrium oxide and oxides of other metals. This mixture of oxides has a negative temperature coefficient of electrical resistance. This means that its electrical conductivity increases with an increase in temperature. At room temperature, the Nernst stick is a non-conductor. Thus, an auxiliary heating is necessary for ignition of the Nernst stick. Even if the Nernst stick is red-hot, it can be heated further by electricity. The normal operating temperature of this infrared light source is approximately 1900 K.

Because of its high operating temperature and the energy distribution in the middle infrared spectrum area the Nernst stick is the most used infrared radiation source. However, it is mechanically very sensitive and can be deformed by heating. This can change the optical characteristics of the spectrophotometer.

The next source of infrared light that is relatively often used is silicon carbide stick, called the glowbar. The glowbar is usually 6-8 mm in diameter and is thicker than the Nernst stick. This gives the high mechanical firmness to the glowbar. Due to its electrical conductivity at room temperature, the glowbar can be ignited directly by electricity. The disadvantage of the glowbar is its low operating temperature (approximately 1500 K). The lower intensity of infrared radiation produced by the glowbar as compared with the Nernst stick is a result of the low operating temperature.

Ceramic infrared light sources are used in some spectrometers. A ceramic stick is heated by a metallic conductor, made from platinum or a platinum alloy, and wound around the ceramic stick. The conductor is surrounded with a sintered layer of aluminum, thorium oxide, zirconium silicate or a similar material. The heating conductors made from chrome nickel or tungsten wire are preferably suitable for short-wave spectral analysis.

The next important part of an infrared spectrometer is the sample chamber. The sample chamber is used for placing the cuvette that contains the sample or for placing any other accessory that contains the sample. The sample chamber is installed between the infrared light source and the monochromator.

It is very important to make the right choice of the cuvette material for liquid and gas samples. This material must be transparent to the infrared light. Sodium chloride is the most often used material for the cuvettes and the optics of the infrared spectrometer. Other material such as special types of glass, quartz, aluminum oxide, calcium chloride, potassium bromide and so on are also used for special purposes.

Another component part of the infrared spectrometer is the monochromator. This is the most important accessory of infrared spectrometer. The function of the monochromator is to split polychromic infrared light into many monochromic light wavelengths. However, this splitting does not go strictly into monochromic wavelengths, but the infrared radiation is split into very narrow wavelength intervals. The splitting in strictly monochromatic light, as it is for radiation of individual emitted spectral lines by vacuum mercury lamp or radiation emitted by a laser, cannot be achieved by a monochromator. The grated part of the infrared radiation is lost due to splitting of the infrared light by the monochromator. This leads to the requirement of a high sensitivity of the detectors used in infrared spectrometers.

Modern monochromators consist of a rift system, the optics and the infrared radiation splitting system, which is usually presented by prism or diffraction grid. The following two types of monochromators are most popular in modern infrared spectrometers:

- Littrow monochromator

- Ebert monochromator

The Littrow monochromator has a prism as a splitting system while the Ebert monochromator has a diffraction grid for the same purpose. The principles of the schemes of both monochromators are presented in Figures 2.40 and 2.41, respectively.

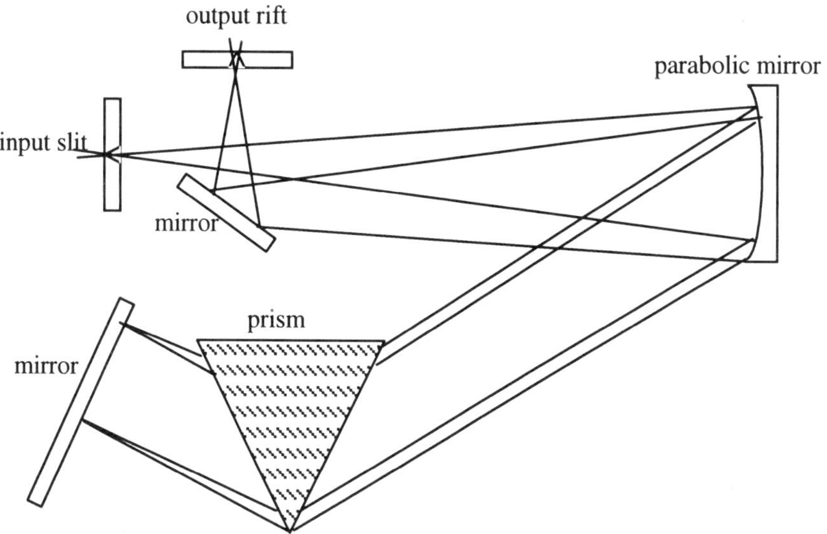

Fig. 2.40: Littrow monochromator.

The diverged infrared radiation from the input slit is directed to a parabolic mirror and returned toward the splitting system (prism or gird). Depending on the type of optical principle, the parallel reflected infrared light passes through the prism or split by the diffraction gird. It is then reflected back by a plane mirror at the same parabolic reflector for the Littrow monochromator or at a second parabolic reflector for the Ebert monochromator. After this, the monochromic infrared radiation is directed to the output slit.

The most important part of the monochromator is the infrared radiation splitting system. One of the most popular splitting systems is the prism. The function of a prism is based on having various refraction angles for radiation with various wavelengths. The function principle of a prism is shown in Figure 2.42.

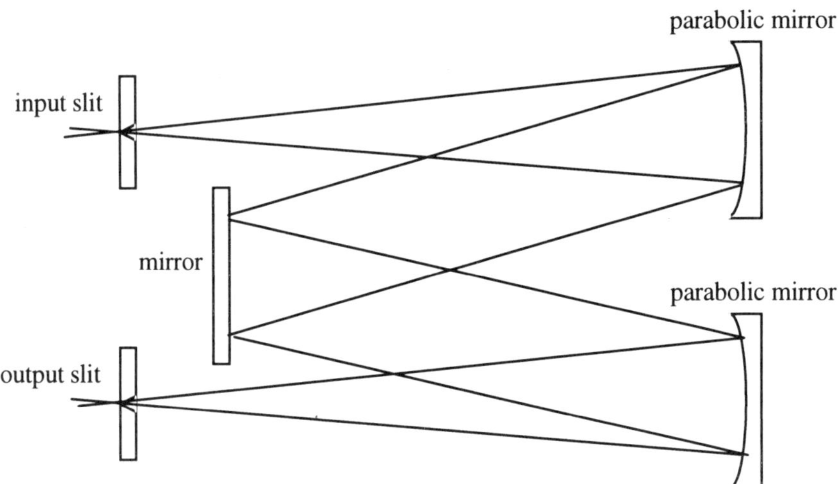

Fig. 2.41: Ebert monochromator.

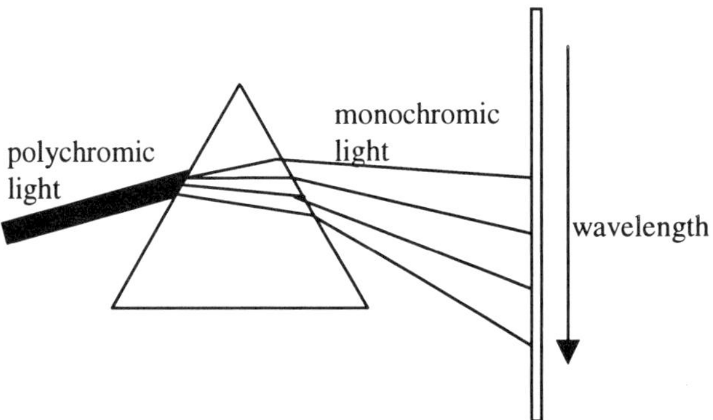

Fig. 2.42: Functioning principle of the prism.

The material of the prism is important in infrared spectroscopy, since it must be transparent to infrared light. The material most frequently used for analysis in the middle wavelength region is sodium chloride. Prism materials for the analysis of short and long wave infrared light are usually potassium bromide, cesium bromide, and cesium iodide.

The prism is made from a faultlessly grown single crystal. The most important values for the resolution ability of the prism are the base size and the refractive index. Both must be selected to be as large as possible in order to get a good resolution. The refraction angles depend upon the value of the refraction index and usually vary from 35° to 72°. The base length is limited by technical reasons. It varies depending on the type of spectrometer from 60 mm to 100 mm.

Almost all old commercial spectrometers had only prisms in the monochromators. Refraction grids are relatively popular in modern infrared spectrometers. Some of the unfavorable properties of the prism materials that relate to hygroscopy, reflection, light absorption and the need to change the prism in order to analyze in different wavelength areas could be prevented by using refraction grids. At the same time, a substantially better and constant resolution for wide wavelength areas can be achieved by using refraction grids.

Furrows scratched in an even glass or metal surface or a thin wire is applied to produce refraction grids. Infrared light is split into individual wavelengths by these furrows or wire.

The next component part in the infrared spectrometer is the detector. The most important types of detectors used in infrared spectroscopy are the thermal detectors. In this type of detector, radiation energy is first absorbed and then converted into heat energy. The actual measured value is an electrical voltage, which is produced or changed by the heating. Despite their higher sensitivity, photo electric detectors have a lower popularity due to the limits they have of the analyzable wavelength area.

The detectors used in infrared spectroscopy are generally delicate but small in order to make the thermal capacity of the detector as small as possible. This leads to a very high mechanical sensitivity of the detector. Since thermal detectors are influenced not only by the radiation energy that has to be measured, but also by changes in the temperature of environment, the measured radiation is modulated. This means that it is modulated by periodic interruption of the light beam usually with the frequency 13 Hz. Infrared radiation, in such a situation, produces an alternating voltage. This is then magnified in an alternating voltage amplifier and then sent to the evaluating computer.

The most important characteristics of a detector are its sensitivity and the relation of measuring signal to noise level.

The last component part of an infrared spectrometer other than the evaluating computer is the amplifier. The purpose of the amplifier is to amplify the signal coming from the detector to enable the computer to evaluate the signals.

The first step in carrying out an infrared spectroscopic analysis is the preparation of the sample. The types of samples analyzed by infrared spectroscopy in crude oil chemistry are solid or liquid samples.

A simple method to prepare a solid sample is called the potassium bromide pressing technique. The prerequisite for this method is that the particles of the solid sample are smaller than the wavelength, which must be measured. Larger particles lead to the unwanted Christiansen effect.

In the potassium bromide pressing technique, a mixture consisting of approximately 300 mg potassium bromide and 1 mg of sample is put into a press under vacuum. The sample/potassium bromide mixture is pressed to tablets by a pressure of about twelve tons for a duration of two minutes. The material under this high pressure has the characteristic of a cold fluid. It becomes viscous and encloses the sample particles completely.

The next method of the preparation of solid samples is called the film technique. In this method, the sample is solved in an appropriate solvent and smeared on a glass plate in the form of a thin film.

The preparation of solid sample by the film technique is the same as the preparation of liquid samples. In the two cases, there are many special requirements for the solvent used. The most important of them are as follows:

- solvent must be chemically inert to the sample analyzed
- solvent must be chemically and physically inert to the material of the cuvette
- solvent must be as transparent as possible with regard to infrared radiation

For the analysis of liquid samples, the cuvette material must chosen so that the sample is physically and chemically inert to this material.

The result of an infrared spectroscopy is the spectrogram or spectrum. A typical example of the infrared spectrogram (i.e. spectrum) is shown in Figure 2.43.

Usually, the infrared spectrum is drawn in terms of wave number / absorption or wave number / transmittance coordinates as shown in Figure 2.43. The wave number is a characteristic value of the electromagnetic radiation, and can be calculated by equation (2.13).

$$\upsilon = f / c \qquad (2.13)$$

where f is frequency of the light wave
c is velocity of light

The prime value characterizing the interaction of infrared radiation with sample analyzed is called transmittance (Tr). It is defined as relation between intensity of the infrared radiation at the input in the sample and the intensity of the output radiation from the sample. Transmittance is calculated by equation (2.14):

$$Tr = I / I_0 \qquad (2.14)$$

where I_0 is input intensity of infrared radiation
I is output intensity of infrared radiation

The concentration of a component in the sample analyzed can be calculated by the Lambert–Beer law, which is shown in equation (2.4). In order to evaluate quantitatively an infrared spectrum, equation (2.4) should be rewritten as equation (2.15).

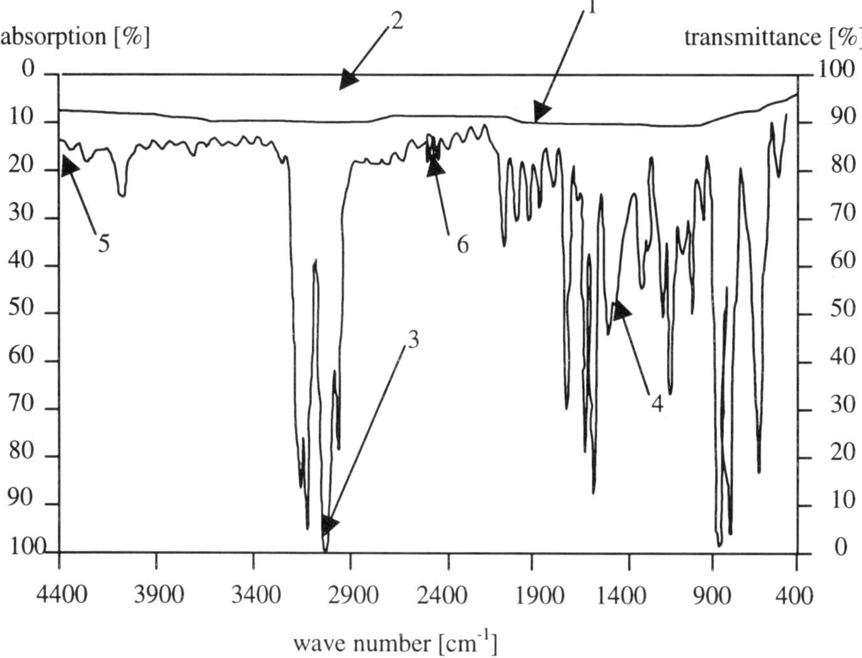

Fig. 2.43: Infrared spectrum:
1 – basis line
2 – background absorption
3 – absorption peak
4 – shoulder
5 – analysis start point
6 – spikes

$$\log (I / I_0) = -E \cdot C \cdot d \tag{2.15}$$

where E is extinction coefficient, which is characteristic for substance
C is concentration of the substance in the sample
d is thickness of sample layer in sample chamber

It can be seen that the concentration of the substance in the sample can be calculated by comparing the input and output intensity of the infrared radiation. However, the input intensity is actually not a constant value. It varies depending on the frequency or wave number of infrared radiation. This means that the initial intensity of infrared radiation must be measured for every frequency before analyzing the sample. However, this problem is solved in modern infrared spectrometers by using a double beam technique. In this method, the beam of infrared radiation from the radiation source is split into two beams by a system of reflectors. One beam passes through the sample and is then analyzed. The second beam is directed directly to the monochromator and then analyzed.

Every infrared spectrum obtained from analysis has six elements as shown in Figure 2.43. First is the baseline. The baseline shows the transmittance line of the spectrometer if there is no sample in the sample chamber. Usually, this line is not registered during analysis as it is a characteristic for the instrument and not of the sample. Number 2 in figure 2.43 represents the background absorption area. This area shows the absorbed infrared radiation in the absence of the sample in the sample chamber. A peak of a typical chemical group in the sample is shown with in number 3. This peak is characteristic of every chemical group or substance, and they give the information for quantitative analysis of the sample. The peaks of some chemical groups or substances present in the infrared spectrum may not be well defined. This may lead to the formation of a shoulder, which is shown as number 4 in Figure 2.43. The shoulder represents poorly resolved peaks. However, shoulders are not the only cause of inaccuracies in infrared spectroscopy. Spikes are shown in Figure 2.43 in number 6. Spikes are caused by undesired fluctuations in the voltage. They lead to the formation of false signals from the detector or amplifier.

Quantitative evaluation of the infrared spectrum is done with regard to the empiric tables or digital libraries. So this can be estimated quickly. For example, if sample contains aromatic groups, information can be collected in the infrared spectrum by looking for the peaks typical of the aromatic carbon.

The most important samples for analysis by infrared spectroscopy for crude oil chemists are organic substances. For organic molecules, the infrared spectrum can be divided into three important regions. First is the absorption of infrared radiation within the wave number range of 4000 and 1300 cm^{-1} which is caused by functional groups and different bond types. Second is the absorption between 1300 and 909 cm^{-1} that is typical for more complex interactions in the molecules. And last is the absorption between 909 and 650 cm^{-1}, which is usually associated with the presence of aromatic compounds in the sample.

Infrared spectroscopy is used for the analysis of almost all the fractions and products of crude oil. However, in the last century, a very interesting purpose of the infrared spectroscopy has been developed. It is the dynamic monitoring of the changes in the structure of lubricating oils as it undergoes degradation. Many processes such as oxidation or polycondensation in oils can be studied by infrared spectroscopy.

More detailed information on infrared spectroscopy can be found in references 67-70 at the end of this chapter.

2.2.2 Raman Spectroscopy

Raman spectroscopy gives results similar to those from infrared spectroscopy. This is why Raman spectroscopy is often used together with infrared spectroscopy in order to receive additional information about the sample analyzed. The motions of the molecule involved in the analysis of the sample in Raman spectroscopy are similar to those by infrared spectroscopy. These include rotational and vibrational motions. However, the physical causes of the resulting spectrum are different.

The Raman effect used in Raman spectroscopy arises from the interactions of monochromatic radiation with the shell atom. In contrast to infrared spectroscopy, these interactions are independent of the wavelength of the light used for the analysis.

For Raman spectroscopy, an intensive monochromatic laser radiation is directed towards the sample. A major part (99.99 %) of the laser light passes through the sample while a very small part is strewn from the substance in all directions. This is the elastic scattering of the light quanta at the molecules otherwise called Raleigh scattering. This has the same frequency as the laser used. A much smaller part of the used radiation (approximately 10^{-6} %) is scattered non-elastically. This is called Raman scattering. This light scattering contains information about the sample molecule. It is also called Raman fingerprint of the substance. The scattering process in Raman spectroscopy can be explained by the scheme shown in Figure 2.44.

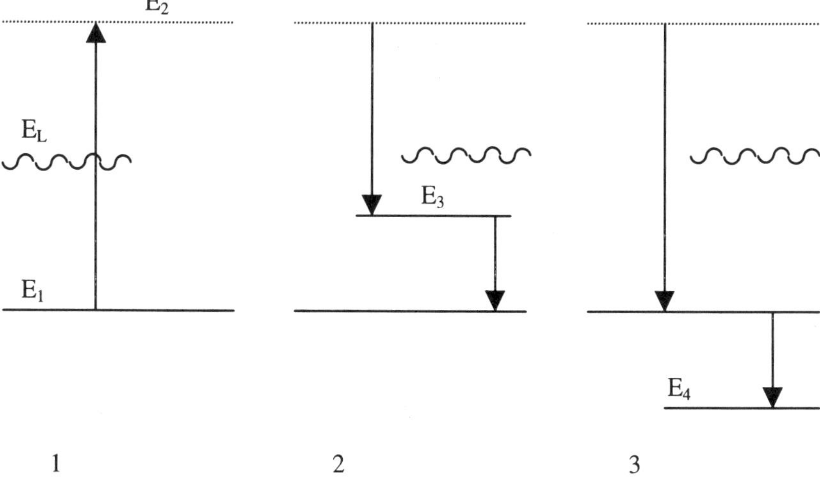

Fig. 2.44: Light dispersion process in Raman spectroscopy.

In step one shown in Figure 2.44, the molecule is brought by laser light from the base energy level E_1 to the higher energy level E_2. The higher energy level is an unstable state for the molecule. The molecule migrates quickly from the higher energy level to the end energy level. This can proceed in two different ways. The first way is given as picture 2 in Figure 2.44 which indicates that light quanta is being emitted. The energy of the resulting light quanta can be calculated by equation (2.16). This type of scattering is called stokes scattering.

$$E_L = E_1 - E_3 \tag{2.16}$$

The second possibility of the light quanta emission is shown as picture 3 in Figure 2.44. In this case, part of the vibrational energy of the molecule is transmitted to the emitted light quanta. Radiation with a shorter wavelength is emitted in this case. The energy of the light quanta emitted in this case can be calculated by equation (2.17). This type of light scattering is called anti-stokes scattering.

$$E_L = E_1 + E_3 \tag{2.17}$$

The results of Raman spectroscopy are usually given as a difference in the wave numbers between the used laser and the scattered light, whereby the wave number for the laser (in presentation of the results) is set to zero, and this point on the spectrogram is called the Raman shift. In this case, stokes scattering is marked by a plus sign and anti-stokes scattering is by a negative sign. The results of the Raman spectroscopy of the molecule presented in Figure 2.44 could look as shown in Figure 2.45.

Stokes scattering is usually more intensive than anti-stokes scattering if the analysis is carried out at room temperature. However, polar groups as –O-H or –S –H are strongly active for infrared spectroscopy whereas non-polar groups and compounds such as –C=C- are strongly active for Raman spectroscopy. There is a simple rule for linear symmetric molecules; infrared active motions cannot be Raman active at the same time. Hence, Raman spectroscopy complements infrared spectroscopy in giving additional information for clarification of the molecular structure. Accordingly, both the methods used on a sample can exactly clarify whether a molecule has linear or nonlinear structure.

Raman spectrometers are used to carry out Raman spectroscopic analysis. The scheme of a Raman spectrometer defers from that of an infrared spectrometer. First of all is the sequence of light source and detector. These are not arranged in a parallel format as was done in infrared spectrometers. Instead the light source is built perpendicular to the detector. The scheme of the Raman spectrometer is shown in Figure 2.46.

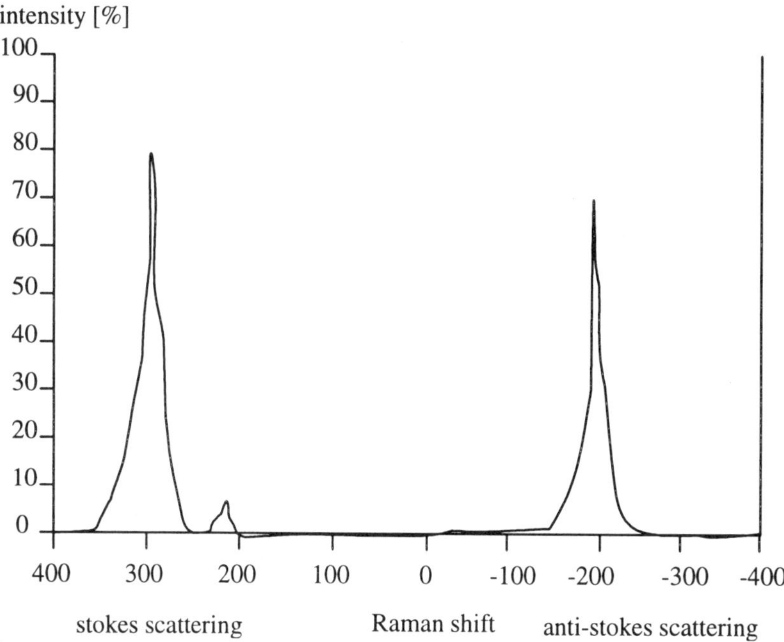

Fig. 2.45: Raman spectrogram.

Most of the component parts used in Raman spectroscopy such as the monochromator and sample chamber have the same functioning principle as in infrared spectrometers. All these were described in detail in section 2.2.1.

Samples of all physical states can be analyzed by Raman spectroscopy. However, analysis of gas samples is difficult because of the low density of the sample.

In order to implement Raman spectroscopy, a reference sample is firstly installed in the sample chamber in moveable cuvette. Sulfur is usually used as the reference sample because of its high Raman activity. The reference sample is moved in the sample chamber until an optimum position is found. The optimum position is marked by maximum scattered radiation recorded by the detector. After the optimum position has been found, the reference sample is replaced with actual sample. It is important to note that the geometrical characteristics of the actual sample must be the same as those of the reference sample. Calibration with regard to wavelength of the light used for the analysis depends on the actual sample and Raman spectrometer used. Different methods can be found in handbooks for a given Raman spectrometer.

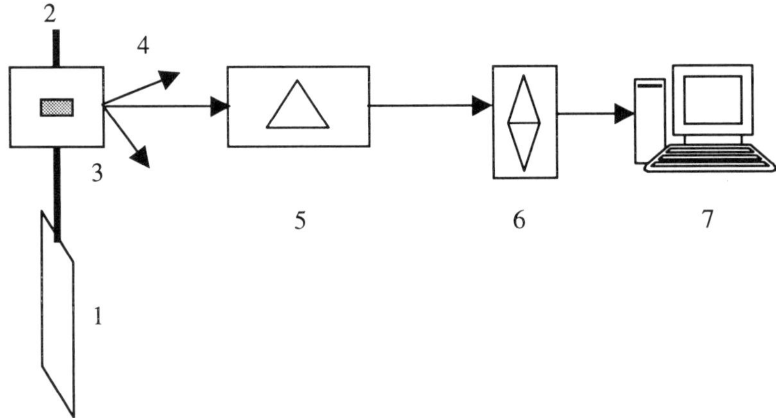

Fig. 2.46: Raman spectrometer:
1 – laser, light source
2 – transmitted light
3 – sample chamber
4 – scattered light
5 – monochromator
6 – detector
7 – evaluating computer

The Raman spectroscopy can be used for analyzing temperature sensitive and color samples. However, following problems could arise:

- heating up of the sample by laser radiation
- fluorescence of the sample

All these problems can be prevented as much as possible by using modern lasers as light source for Raman spectroscopy, for example, Neodymium-YAG laser. The intensity of the light can often be around 0.5 Watts. This laser device is installed in most modern Raman spectrometers.

Qualitative evaluation of the Raman spectrograms proceeds in the same way as was shown for infrared spectroscopy. On the other hand, quantitative evaluation in Raman spectroscopy is executed by using calibration curves or tables that are usually supplied with a given spectrometer in the form of software for the evaluating computer.

There are many possibilities for the use of Raman spectroscopy by crude oil chemists. Many references, shown at the end of this chapter, describe many different investigated areas of the use of Raman spectroscopy. It was shown earlier that heavy fractions constitute the majority of the problems that can be investigated by Raman spectroscopy. At the time most of these problems were solved, an interesting use of Raman spectroscopy was the determination of impurities in lubricating oil. In this analysis, Raman spectroscopy can give more interesting and more complete results than infrared spectroscopy.

More information on this topic can be found in references 71-75 at the end of this chapter.

2.2.3 Colorimetry and Photometry

Colorimetry and photometry are the next spectral optical analysis methods. Both the methods measure absorbed light as was shown for infrared spectroscopy. However, for both these analyses, light with shorter wavelength is used. Colorimetry uses light with wavelength of only the visible spectral area and photometry uses the visible light, ultraviolet, and in some case, infrared area. A comparison of the different spectral areas is shown in Figure 2.47.

Photometry is similar to infrared spectroscopy; a method for measuring the light transmitting ability of a solution in order to determine the concentration of a light absorbing material present in the solution. Crude oil chemists generally use spectrometry in three ways:

- to determine the concentration of a substance which is not undergoing a chemical change in the spectrophotometer cell
- to measure the rate of change in the concentration of a substance which is participating in a chemical reaction
- to determine the absorption spectrum of a substance

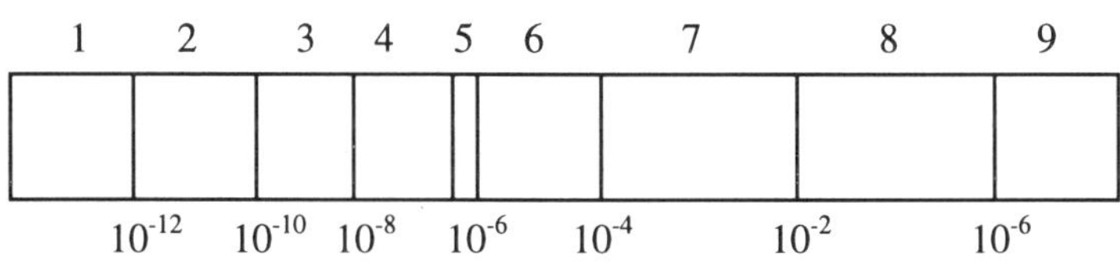

Fig. 2.47: Spectrum of electromagnetic radiation:
1 – cosmic rays
2 – gamma rays
3 – X-rays
4 – ultraviolet rays
5 – visible light
6 – infrared rays
7 – microwaves
8 – radio waves
9 – electric power

The visible region of the electromagnetic spectrum consists of electromagnetic radiation with wavelengths from approximately 400 to 700 nm. It was shown in Figure 2.47 that this region is very narrow in comparison to all other spectral areas. The short wavelength cut off is due to absorption by the lens of the eye whereas the long wavelength cut off is due to the decrease in sensitivity of the photoreceptors in the retina for longer wavelengths. Light at wavelengths longer than 700 nm can be seen if the light source is intense. Every color of the visible spectrum corresponds to a definite wavelength. The distribution of colors in the visible spectrum is shown in Figure 2.48.

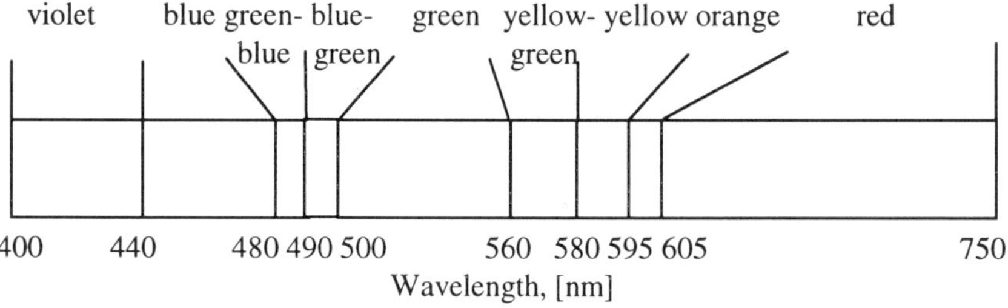

Fig. 2.48: Colors of visible spectrum.

The motions that are analyzed by colorimetry and photometry are no longer of rotational or vibrational nature as was shown for all previous spectral analyses. The motions analyzed at the visible and ultraviolet spectral regions are specific for each atom or chemical group. In order to understand the nature of these motions, it is important to understand the structure of the atom.

There are lots of models on the structure of the atom that can help us to understand the nature of matter. These models are simplified representations of real atoms. The models can be made by atomic physics to be more accurate but this makes the model more complicated. Today, it is possible to describe atoms very accurately by using quantum mechanics. However, the question that cannot be answered accurately is whether the true model has been found or not.

Which of the models is the best depends on what it is used to describe. For example, there are applications in which it is sufficient to regard the atoms as small particles.

The name "atom" derives from the Greek word "atomos" meaning indivisible or unbreakable. The use of the name "atom" was initiated by Demokrit, a Greek philosopher, in the fifth century before Christ. Demokrit supported his theories about atoms through hypotheses that he did not then prove by experiments. His theory of the atomic structure of matter could not become generally accepted against the competing concept of a continuous universe.

The first experimental verification of the existence of atoms came from chemists and was based on the clarification of the chemical nature of matter. In the nineteenth century, English chemist John Dalton formulated a law that now bears his name; according to this law, all chemical compounds are always formed by a fixed relation of the individual materials of the compound. Thus, the idea of atom

atom came about that chemical compounds could be formed by the reactions of individual atoms.

The idea of indivisibility or indestructibility of the atom was disproved by Joseph J. Thomson, when he discovered the cathode radiation in 1897. He was able to prove that this radiation consisted of charged particles that originated from atoms. This meant that atoms could be cracked with the formation of the charged particles, which are now called electrons.

In 1911, a new idea about atomic structure was described by Ernest Rutherford. In his famous scattering experiments, Ernest Rutherford discovered that atoms consisted of a major part that contained nothing and that almost all the mass of the atom was concentrated in a very small core.

Two years later, Danish scientist, Niels Bohr, described a new model of atomic structure. Nowadays, the Bohr atom model is the most popular model used by scientists. The electrons fly in the Bohr atom model on circular paths around the core. The binding energy that holds the electrons in the atom is of an electrical nature. Opposite charges attract each other, whereas like charges repel or push off each other. The binding energy between two charged particles becomes smaller with increasing distance from each other. Since the atom core is positively charged while electrons are negatively charged, the electrons are attracted to the core. The planets of our solar system have a similar attraction by the sun. As the planets are held by the sun in their circular paths, so also are the electrons held by the core in their circular paths.

In addition to the fact that the electrons fly on circular paths, Bohr also postulated that only certain orbits are permitted. Transitions between these orbits occurred immediately or spontaneously without the possibility to take any intermediate positions. It is impossible to understand these postulates with the laws of classical physics. Only quantum mechanics could explain the second Bohr postulation.

The Bohr atom model was an attempt to understand the presence of individual lines in the optical spectra of atoms. Since the atomic nucleus attracts the electrons, energy is needed to keep the electrons away from the core. Exactly as the case with the planet orbits, this energy arises from the rotating motion of the electrons around the core. The further the electron is from the core, the more is the energy stored by the electron. If an electron moves from a higher orbit to a lower orbit, then it loses energy, which is radiated in form of a light quantum. An electron can move to a higher orbit only by the supply of energy. This means by absorption of a light quantum.

The fact that only light quanta with special energy levels can cause transitions of electrons between orbits, shows that only certain orbits exist in atoms. The energy levels of the light quanta absorbed can be explained as the energy difference between different orbits.

In a more exact observation of the optical spectra of different atoms, it was detected that many of the spectral lines are additionally split up. This means that the Bohr electron orbits differ somewhat by some type of electron transition that needs only a small energy for transition. Sommerfeld postulated that there are not only circular orbits, but also elliptical orbits, which contain approximately the same energy level as circular orbits. Figure 2.49 shows the Bohr–Sommerfeld model of a hydrogen atom.

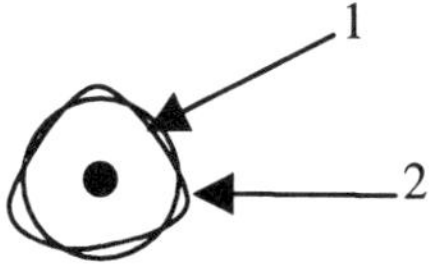

Fig. 2.49: Bohr atom model:
1 – Bohr orbit
2 – Sommerfeld orbit

The instruments which analyze a substance by virtue of the electron shells of the atoms in the visible area of the spectra are called colorimeters. These instruments measure the difference in the color intensity of a reference solution with a defined concentration of the component analyzed and the color intensity of the sample solution.

Colorimeters are constructed based on the principle of continuous ray. That means the light ray passes continuously from the light source to the detector. The main elements of every colorimeter are a light source, two cuvettes, a detector and a monitoring device. One cuvette is used for the reference solution and has the same thickness of the sample room. Usually, the thickness of the sample room and consequently the thickness of the analyzing sample can be varied in the analyzing cuvette. The principle scheme of the colorimeter is shown in Figure 2.50.

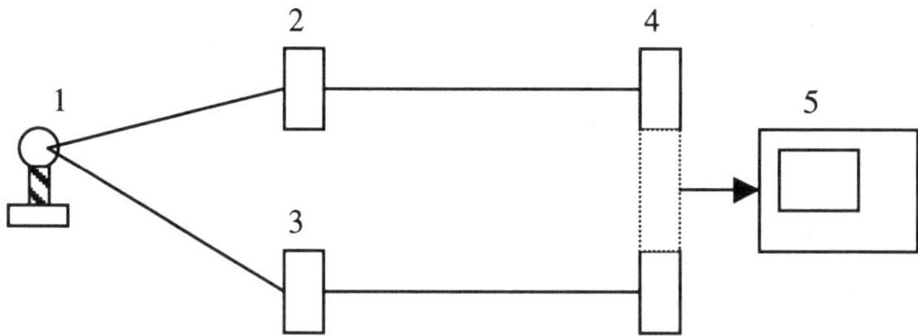

Fig. 2.50: Colorimeter:
1 – light source
2 – sample cuvette
3 – reference cuvette
4 – detector
5 – monitoring device

However, colorimeters are very rare to see in crude oil chemical laboratories, because as was indicated earlier, almost all the important functions of the colorimeter can be reproduced by modern photometers.

There are two important types of photometers: the one-ray photometer and the two-ray photometer. The construction of the one-ray photometer is very similar to the colorimeter. The scheme of the one-ray photometer is presented in Figure 2.51.

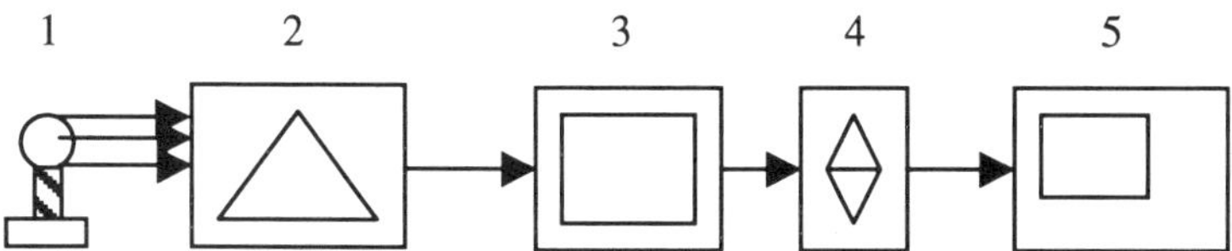

Fig. 2.51: One-ray photometer:
1 – light source
2 – monochromator
3 – sample cuvette
4 – detector
5 – monitoring device

Usually, the one-ray photometer has no dynamic change possibility for the wavelength of the light used for the analysis whereas such a possibility exists for the two-ray photometers. The main difference between the two-ray photometer and the one-ray photometer is that the two-ray photometer has two cuvettes in the sample chamber. These are the sample cuvette and the reference cuvette. The reference cuvette is filled with pure solvent while the sample cuvette is filled up with a solution of the sample analyzed in the solvent. The results of photometry are evaluated by comparison of the intensity of light that passes through the sample solution and the pure solvent. The scheme of the two-ray photometer is presented in Figure 2.52.

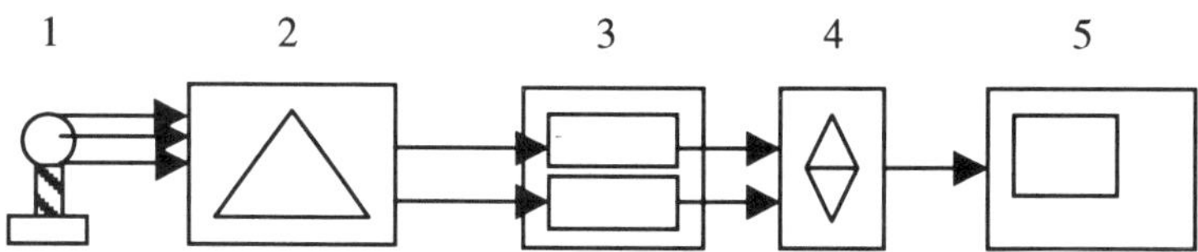

Fig. 2.52: Two-ray photometer:
1 – light source
2 – monochromator
3 – sample chamber with reference cuvette and sample cuvette
4 – detector
5 – monitoring device

All the component parts used in photometers have the same working principle as those already described in other spectrometers, for example, the infrared spectrometer. The prism and refraction grids are used as monochromators. The detector is usually made of different types of photoresistors depending on the instrument type.

Photometry and colorimetry are used by crude oil chemists to determine the content of different metals and heteroatomic compounds in crude oil and petrochemical products. Many references on photometry and colorimetry are given at the end of this chapter. Many authors have described the successful analysis of different metals in motor fuels by photometric and colorimetric methods. The composition of additives used during fuel production can be characterized by photometric and colorimetric methods because very many additives contain metals. It is not only fuels that can be characterized by photometry and colorimetry. Lubricants, which contain metals as an important component, can be successfully determined by these methods. These methods can quickly give qualitative information on heavy metals and heteroatomic compounds such as oxygen and sulfur in crude oil. More on this topic can be found in references 76 and 77 at the end of this chapter.

2.2.4 Fluorescence and Phosphorescence Spectroscopy

Fluorescence and phosphorescence spectroscopy are generally referred to as fluoremetry. Fluoremetry uses the ability of some materials to emit light at a certain wavelength. The emitted spectrum by the substance is characteristic for that specific material alone. The fluorescence effect arises if an electron is moved from an orbit with a higher energy level to an orbit with a lower energy level. Primarily, the electron in the atom under analysis is transmitted from the base orbit to the orbit with a higher energy level by a monochromic light from the light source of the fluoremeter.

The instruments used for the fluoremetry, fluoremeters, are very similar to the ones used for Raman spectroscopy. However, the light spectral region used in fluoremetry is usually the ultraviolet or the visible spectral area. The principle of the scheme of a fluoremeter is shown in Figure 2.53.

Pulsing lasers are used in most modern flourometers as the light source. Measurement with such lasers makes it possible to measure the florescence effect right after the prime transmission of the electron from the base to the higher energy level and during the transmission. The main advantage of using such a laser is the possibility to measure the time delay effect of phosphorescence. Measuring this effect is impossible by application of a continuous light source.

The most popular light sources in modern fluoremeters are gas lasers. At the present time, nitrogen, XeCl, XeF and KrF are used quite frequently for fluoremetry. The reason to use lasers as the light source for fluoremetry is the same as was given for Raman spectroscopy: lasers emit monochromic light with very high intensity in comparison to the classical light sources.

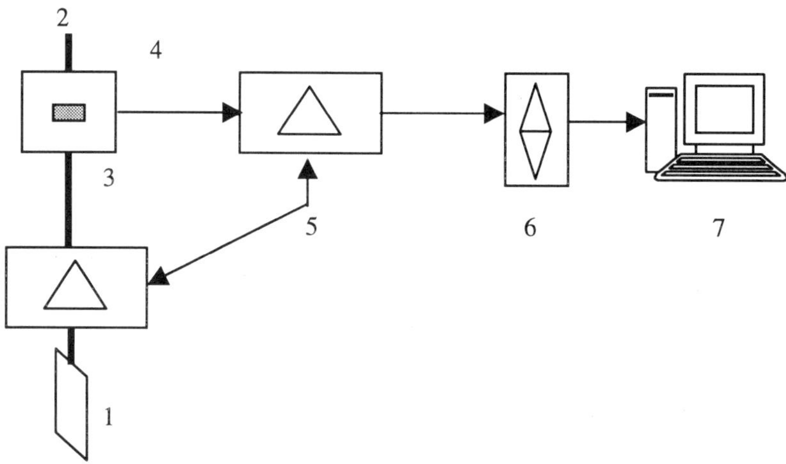

Fig. 2.53: Fluoremeter:
1 –light source
2 – transmitted light
3 – sample chamber
4 – emitted light
5 – monochromators
6 – detector
7 – evaluating computer

Almost exclusively, only diffraction grids are used as monochromators in fluoremeters. The functioning principle of this device was described in detail in the section in this chapter devoted to infrared spectroscopy.

The evaluation of the results of fluoremetry can be done both quantitatively and qualitatively. Qualitative evaluation is done in the same way as for most spectral methods. There are tables of spectra of known compounds or the spectra are saved in digital libraries, which are delivered together with the instrument. There are three methods used for quantitative evaluation:

1. direct method for the fluorescence-able samples
2. converting of non-fluorescence-able samples into fluorescence-able samples
3. fluorescence extinction of the indicator substance

The first method is the most used. The second and third methods are usually used in connection with high performance liquid or thin film chromatography.

For quantitative evaluation according to the first method, equation (2.17) is used.

$$I_F = \phi \cdot I_0 \cdot (1- 10^{E \cdot C \cdot d}) \tag{2.17}$$

where I_F – intensity of the emitted light
ϕ – light emission coefficient
I_0 – intensity of the light entering into the sample
E – specific characteristic value of the component analyzed
C – concentration of the component analyzed
d – sample thickness

If the extinction (E) according to the Lambert-Beer law (see equation 2.4), which is calculated by equation (2.18), is less than 0.01,

$$E = E \cdot C \cdot d \tag{2.18}$$

Then, equation (2.17) can be rewritten as equation (2.19):

$$I_F = 2.3 \cdot \phi \cdot I_0 \cdot E \cdot C \cdot d \tag{2.19}$$

In equation (2.19), it is seen that with a solution containing a relatively low concentration of substances absorbing the light, the concentration of the component analyzed depends linearly on the intensity of the fluorescent light. Nonlinear dependence of highly concentrated solutions having extinction coefficients over 0.01 leads to the filtering effect.

Fluoremetry belongs to the class of relative methods. This means that quantitative evaluation by such methods is possible only after calibrating the instrument by reference solutions with a known concentration. If the factors ϕ, I_0 and d are constant during calibration and measurement, then the concentration of the sample or a component in the analyzed sample can be calculated by equation (2.20).

$$C_a = C_C \cdot I_{Fa} / I_{FC} \tag{2.20}$$

where C_a – concentration of the sample or sample component
C_C – concentration of the sample or sample component in the reference solution
I_{Fa} – fluorescence intensity from the sample solution
I_{FC} – fluorescence intensity from the reference solution

Fluoremetry has two main advantages in comparison to photometry that uses the same light spectrum for analysis. These are the high sensitivity and the high selectivity of the method. High sensitivity of the method can be explained on the basis of the absence of such effects as background absorption, which are typical of absorption spectral methods. High selectivity is due to the difference in wavelengths between the prime light and the emitted light. By changing the wavelength, the florescence of single compounds in the sample mixture can be measured.

Fluoremetry is especially useful for the determination of condensed aromatic and heteroatomic compounds in crude oil and its products. The use of fluoremetry in crude oil chemistry is presently so highly developed that already there is a special classification of types of fluoremetry typical for crude oil chemists. These are the qualitative, visible and quantitative methods widely used by crude oil chemists.

More detailed descriptions of techniques and methods of fluoremetry for crude oil chemistry can be found in references 78 and 79 given at the end of this chapter.

2.2.5 Atomic Absorption Spectroscopy and Atomic Emission Spectroscopy

Atomic absorption spectroscopy and atomic emission spectroscopy are based on Kirchhoff's law. According to this law, all atoms are able to absorb light quanta with the same wavelength that they are able to emit.

A light source which emits a continuous light spectrum produces a black spot on a photographic plate of a spectrometer over the whole detectable area. If one installs a burner between the light source and a photographic plate, as it is usually used for atom absorption spectrometry, and sodium chloride solution is spattered in the flame, then one receives again a continuous spectrum of appropriate density over the whole detectable area on the photographic plate. There are two bright lines on the spectrum, presenting the characteristic wavelengths of sodium. This is a well-known feature from the solar spectrum called Frauenhofer D-line. The presence of these characteristic lines, or better: these places of smaller density, can be explained as follows. In the colder, outside zone of the sun, free atoms of sodium are present, which absorb light with the same wavelength which was emitted by hot sodium atoms present inside of the sun. In other words, sodium steam, which is not so highly heated up, that is able to emit light absorbs the light of exactly the same frequency or wavelength as light which it emits, if it is hot enough to emit light. This law can be applied to every known metal.

The spectra of atomic absorption are obtained with instruments called atomic absorption spectrometers. These instruments, as already described for other types of spectrometers, consist of the light source, monochromator and detector. However, the atomic absorption spectrometers and atomic emission spectrometers differ from all other spectral spectrometers by the absence of the sample chamber. Instead of the sample chamber, they contain a burner. A schematic of the atomic absorption spectrometer is shown in Figure 2.54.

Two types of lamps are used as source of monochromatic light. These are:

1. Gas discharge lamps. They are manufactured for the determination of sodium, potassium, mercury, cadmium and thallium atoms. They emit specific mono or polychromatic radiation for these elements.

2. Cathode lamps. These lamps are used for calcium, magnesium, silver, gold, chrome, copper, iron, manganese, nickel, zinc and many other elements.

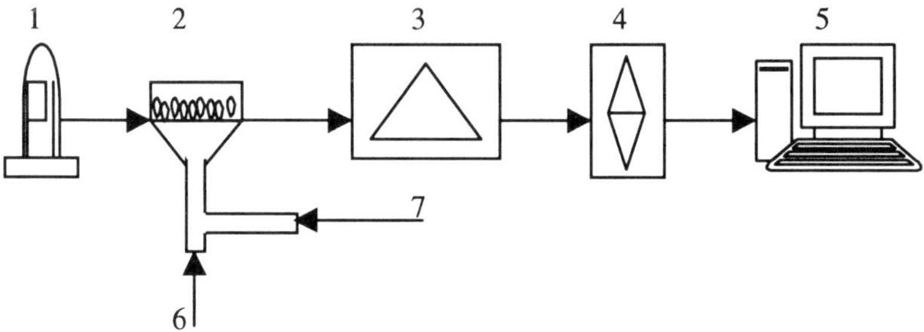

Fig. 2.54: Atomic absorption spectrometer:
1 –light source
2 – burner or atomizer
3 – monochromator
4 – detector
5 – evaluating computer
6 – gas fuel
7 – air or oxygen under pressure with solution of analyzing sample

The functioning principle of the gas discharge lamps can be explained based on Figure 2.55. A small amount of the element that is analyzable by this lamp is melted under pressure in the quartz chamber. The element emits light of a certain wavelength by a high frequency electrical stream.

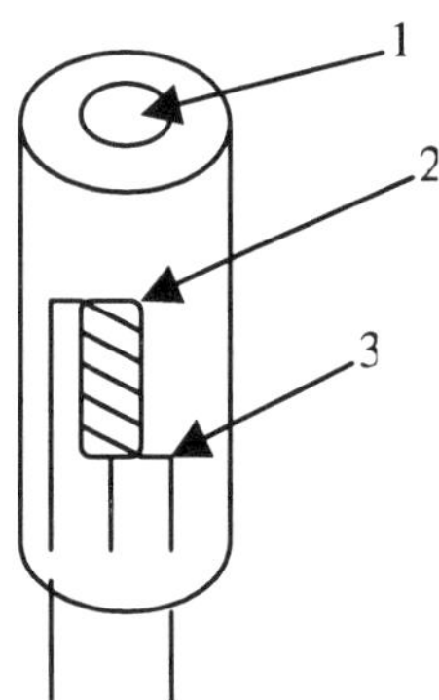

Fig. 2.55: Gas discharge lamp:
1 –quartz window
2 – quartz chamber with meted in element
3 – source of high frequency electricity

The gas discharge lamps have higher intensity of emitted light and can work for a longer time without being changed in comparison to cathode lamps.

There are many types of light sources often used in modern atomic absorption spectroscopy. The scheme of such a lamp is represented in Figure 2.56.

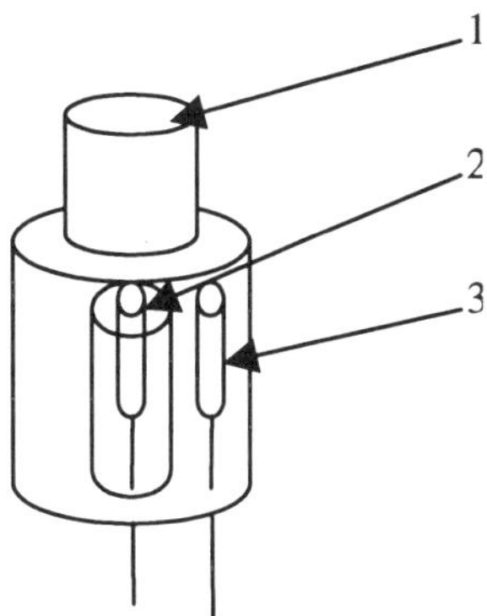

Fig. 2.56: Cathode lamp:
1 –quartz window
2 – cathode
3 – anode

The cathode lamps usually consist of one cylindrically arranged cathode made from a metallic element and an anode. The cathode is protected with a glass cylinder. The glass cylinder is usually filled with argon or neon. The anode is represented by wire made from tungsten or nickel. After creation of an electrode voltage of up to 600 V, positively charged ions are transmitted from the cathode, followed with transfer of the electrons from the base state to a higher energy level. The light for analyzing certain metallic elements is emitted by return of electrons again to the base state.

The next component part in the scheme of the atomic absorption spectrometer is the atomizer. Burners are often used as atomizers in atom absorption spectroscopy. Mixtures of air/acetylene, laughing gas/acetylene or hydrogen/argon are usually used as the fuel gas. The aim of the burner in atomic absorption spectroscopy is to evaporate the solution of the sample and to disintegrate the sample to the atomic state. Especially important for successful analysis is that the path of light through the flame of the burner be made as long as possible. Hence, the use of fissure burners.

Graphite pipe technique is the next method used to atomize the sample. The graphite pipe filled with sample is used in this method. This pipe is electrically heated up such that the sample is atomized. The biggest advantage of this method is that the atoms of the sample can be held in the pipe much longer than in the flame. It makes the results of the analysis more accurate.

Antimony, arsenic, selenium, tellurium, bismuth and tin are able to form volatile hydrides by reaction with $NaBH_4$. This property of these metals is used for the hydride atomizing technique. In this method, the metal hydrides are atomized in quartz cuvette by electrical heating.

The last method used for atomizing is the cold vapor method. Only mercury can be analyzed by this method. It is known that mercury is the only metal which has a relatively high vapor pressure even at 20°C. Mercury can be introduced in the atomizer at room temperature without prior dissolution.

The next component part shown in Figure 2.54 is the monochromator. The monochromators used in atomic absorption spectroscopy are the same as for all the other spectral analytical methods. A detailed description of the monochromators can be found in the section in this chapter devoted to infrared spectroscopy.

Detectors used in atomic absorption spectroscopy are usually photometric detectors.

Atomic absorption spectroscopy belongs to a class of relative analysis methods, meaning that direct quantitative evaluation of the results of this analysis is impossible. The most popular method of quantitative evaluation of atom absorption spectroscopy results is the standard addition methods. Certain known amounts of the element under analysis are added to the analyzing solution many times, and the intensity of the specific wavelength measured each time, thereby producing a calibration graph. An example of a calibration graph is shown in Figure 2.57.

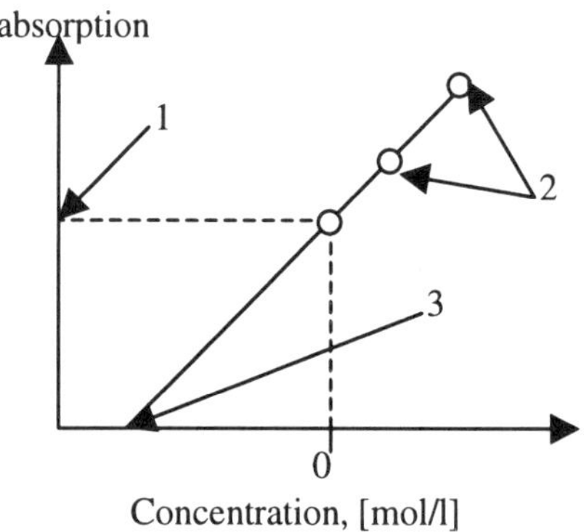

Fig. 2.57: Calibration graph for atomic absorption spectroscopy:
1 – measured absorption for the sample solution
2 – measured points of the solutions with added element under analysis
3 – concentration of the element in the sample solution

The concentration of the element in the sample solution can be found as the intersection point of the calibration line with the concentration axis at which point the absorption equals zero. The concentration of the element in the sample solution is accepted as zero for the calibration as shown in Figure 2.57.

Atomic emission spectroscopy is very similarly to atomic absorption spectroscopy. The difference between these methods can be seen from their names. Emitted light of the atom under analysis is analyzed by atom emission spectroscopy. The schematic for the atomic emission spectrometer is very similar to that for atomic absorption spectrometer. The schematic for the atomic emission spectrometer is presented in Figure 2.58.

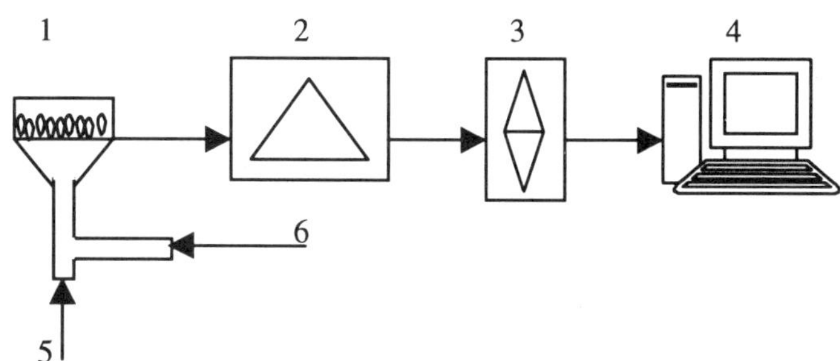

Fig. 2.58: Atomic emission spectrometer:
1 – burner or atomizer
2 – monochromator
3 – detector
4 – evaluating computer
5 – gas fuel
6 – air or oxygen under pressure with solution of analyzing sample

In Figure 2.58 it is seen that the atomic emission spectrometer has no light source. Atoms emit light by the influence of the thermal energy of the burner. However, the energy needed to enable the atoms emit the light is much higher than the energy needed to only atomize them. Hence, the much higher temperature required for the burner of the atomic emission spectrometer than that for the atomic absorption spectrometer. This is also the reason why other gas mixtures such as fuel gas should be used for atomic emission spectroscopy. Mixtures of methane/air, hydrogen/air, acetylene/air, methane/oxygen and acetylene/oxygen are usually used as a fuel gas for this analysis.

The next possible way to make atoms emit light is the flashover method. In this method, flashover is the source of thermal energy. The biggest advantage of this method is the ability to heat the sample up to 10000K. It is possible to analyze almost all known metals at this temperature.

In most modern spectrometers, plasma sources are used as source of thermal energy.

Both methods, atomic absorption spectroscopy and atomic emission spectroscopy, are used in modern crude oil laboratories, with atomic absorption spectroscopy being estimated as being the more popular, because of the following advantages:

- better precision of the method
- less spectral divergences
- wider limits of concentrations of the elements analyzed in the sample, by using the graphite pipe technique

Both methods are used for the determination of the metal content. The determination of the concentrations of heavy metals in crude oil and its products is an important topic in environmental chemistry. Heavy metals are always present in crude oil, especially in the heavy fractions such as residual fuel. The control of the concentration of heavy metals in such fuels is very important for ecological monitoring in crude oil chemistry. More about the methods, techniques and modern research results made by atomic absorption spectroscopy and atomic emission spectroscopy can be found in references 80-82 shown at the end of this chapter.

2.2.6 X-ray Fluorescence Spectroscopy

X-ray spectroscopy can be classified in the same manner as every other type of spectral analysis into absorption and emission spectroscopy. However, the most popular method of x-ray spectroscopy in crude oil chemistry is the emission spectroscopy, also called x-ray fluorescence spectroscopy. The effect used by this type of spectral analysis is the same as was described for fluorescence analysis. However, x-rays are used for this analysis instead of the ultraviolet radiation used for fluorescence analysis.

The scheme of the x-ray fluorescence spectrometer is similar to that for the fluoremetry spectrometer. This is presented in Figure 2.59.

X-ray pipes are used as the light source for x-ray fluorescence spectroscopy. There are very many types of x-ray pipes in the modern market. The functioning principle of the x-ray pipe is the same as for cathode lamps described in an earlier section of this chapter. The x-ray pipe contains an electrical heated cathode, anode and radiation output window. This window is made from beryllium because this material is transparent to x-rays. The x-rays pipes offered in the market differ because they have a different anode material, and consequently the spectral characteristics of the emitting radiation are different.

The monochromator for x-ray fluorescence spectroscopy is called the analyzing crystal. It differs from all the monochromators described earlier for all the other optical analytical instruments. The effect used in this type of monochromator is not diffraction, but interference. The wavelength of the analyzing light is changed by rotation of the analyzing crystal by certain angle.

The homogeneity of the sample is very important for successful x-ray fluorescence spectrometry. Hence, the preparation of solid samples for this analysis by melting. The samples analyzed by crude oil chemist are, in most cases, liquids or can be melted at a relatively low temperature. This is why the problem of homogeneity does not arise in analyzing crude oil or their products.

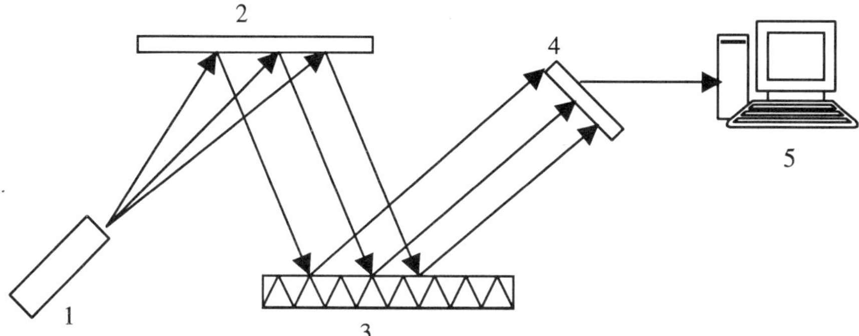

Fig. 2.59: X-ray florescence spectrometer:
1 – light source
2 – sample
3 – monochromator
4 – detector
5 – evaluating computer

Qualitative evaluation of the results of x-ray fluorescence spectroscopy is carried out in the same way as for all other spectrometry methods described in this chapter. This is done based on the table in the literature or on digital libraries supplied with the evaluating software for the specific spectrometer.

X-ray spectroscopy, just like many other spectroscopy methods, belongs to the class of relative analysis methods. This means that quantitative evaluation is possible only by comparison of the analysis results with calibration results. The following methods are used for the evaluation of x-ray fluorescence spectrometry:

- comparison with calibration curves
- standard addition method
- addition of inert substance
- statistical calculation method

The first three methods are very similar to the methods used for other spectroscopic methods. Statistical calculation methods can be used only in modern x-ray fluorescence instruments that come with the appropriate software. Different manufacturers or companies use different algorithms their instruments. The main purpose of this software is to minimize the influence of measurement errors when computing the results. A wide variety of statistical methods are available. The statistical calculation method saves a lot of experimenting time, because only the analysis of the sample is needed for every analysis. Calibration or analyses of sample with added substance is not required in this case.

The x-ray fluorescence analysis is used by crude oil chemists as an express method for the determination of the metal contents in crude oil and its products. This is very important because the metals in crude oil are poisons for cracking catalysts, and are also undesired in the fuels, since they lead to environmental pollution. All the elements with atomic number greater than eleven in the periodic table can be analyzed by x-ray fluorescence spectrometry. An important task of this method in crude oil chemistry is the determination of heteroatoms such as sulfur, oxygen, and nitrogen. Frequently, x-ray fluorescence spectrometry is used to verify the presence of certain additives in oils.

More about x-ray fluorescence spectrometry and the evaluating methods can be found in references 83 and 84 at the end of this chapter.

2.3 OTHER METHODS FOR ELUCIDATING THE STRUCTURE OF CRUDE OIL

2.3.1 Separation Methods

All the separation methods used to classify physical and chemical properties of crude oil can be classified into the following classes:

- chromatography
- spectroscopy
- adsorptive methods
- evaporating methods
- extraction
- thermal diffusion
- crystallization
- complex formation methods

The first two methods were described in detail in the first two sections of this chapter. The adsorptive methods used by crude oil chemists are actually similar to those described for the chromatographic methods. The setup used for this type of analysis is the same as was shown for column chromatography (see Figure 2.19). There are two different adsorptive separation methods:

- desorption method
- elution method

The preparation of sample for both methods is done in the same way as was shown for column chromatography. However, both methods differ in the solvents used for desorption. In the case of the desorption method, the solvents used should have a stronger absorption ability than the compounds in the sample mixtures. Solvents used for analyzing crude oil and its products include alcohols, ketones, ethers, chlorine-containing solvents, benzene, toluene and so on. It is possible to use many solvents in this method. It is very popular in analyzing heavy fractions of crude oil.

Large amounts of solvents with lower adsorption ability than the compounds of the sample mixture are used in the elution method. Pentane, hexane, heptane and petrol ether are frequently used by crude oil chemists as solvents for elution adsorptive analysis.

Evaporation methods can be classified in three types:

- distillation
- rectification
- molecular distillation

The first two methods are described in detail in chapter 5 of this book because of the high importance attached to these methods for the industrial separation of crude oil. However, it will be said at this point that both methods use the difference in the boiling temperatures of the compounds in crude oil for separating the compounds. Rectification actually involves carrying out the distillation as a multistage process. The reason to carry out such a multistage process is that distillation (i.e. in one stage) does not provide a satisfactory separation into the desired fractions. The distillation method is however frequently used, and it is a standard method in all international standards for characterization of light fractions of crude oil such as gasoline and diesel. The biggest advantage of this method in comparison to rectification is that this analysis needs less time to be carried out.

The third method is molecular distillation. This is distillation under very low pressure. Usually the pressure for this analysis varies from 0.133 Pa up to 0.013 Pa. The heavy fractions of crude oil can be separated with this method. The name molecular distillation is derived from the fact that the vapor pressure of the high molecular compounds depends almost linearly on the boiling point of these compounds under vacuum. It means that the separations by this method proceeds according to the molecular weight of the compounds in the sample. The quality of the separation depends on the evaporation velocity of the compounds in sample mixture.

Extraction methods are based on the different solubility of the compounds of sample in solvents. Usually, many different solvents or one solvent at different temperatures are used to separate the sample into fractions according their solubility in the solvent. This method is important for analyzing the heavy fractions of crude oil. This is why this method is described in detail in chapter 8 that is devoted to asphaltene chemistry.

The thermal diffusion method is used rarely in comparison to all the other analysis methods in crude oil characterization. However, this method is very useful and can be successfully used for the analysis of heavy crude oil fractions. The setup used for this analysis is really simple. It is represented in Figure 2.60.

The setup for thermal diffusion analysis contains two cylinders installed one in another as shown in Figure 2.60. The sample is introduced between these two cylinders. During analysis, the wall of one cylinder is heated up to a temperature between 100 and 180°C while the wall of the other cylinder is cooled down to a temperature between 4°C and 20°C. Compounds in the sample are in motion in a direction towards the top at the hot wall as well as in a direction towards the bottom of the cold wall. Compounds of the sample are thus separated according to their physical properties (i.e. according to the density and viscosity of the compounds in the sample).

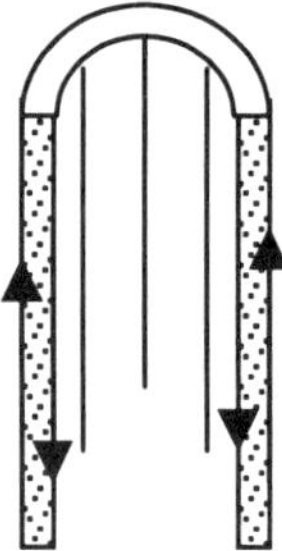

Fig. 2.60: Thermal diffusion analysis setup.

The next method is the crystallization method. This method is very frequently used by crude oil chemists for the classification of lubricating oil fractions. This method is based on difference in the crystallization temperatures of the compounds in the sample. The samples which are analyzed by this method are usually highly viscous. This makes it difficult to separate the crystallized substances from the sample. Consequently, the sample is usually dissolved in a special solvent. Examples of such solvents are liquefied propane, ethyl ether, and a mixture of acetone and toluene.

The last method that was mentioned at the beginning of this section was called the complex formation method. This method is based on formation of complexes of crude oil compounds with other substances. The most popular methods of complex formation are complex formation with $CO(NH_2)_2$ and $CS(NH_2)_2$. For $CO(NH_2)_2$, it is typically to form complexes with *n*-alkanes and their derivates with relatively long paraffinic chains with normal structure. The formed complexes are crystallized from the sample. The separation of the *n*-alkane fraction from $CO(NH_2)_2$ can proceed by adding hot water to the crystallized complex. The $CO(NH_2)_2$ is very soluble in water whereas paraffins are insoluble. This is why two layers result in this separation: the fraction of aqueous solution of $CO(NH_2)_2$ and the paraffin fraction. Because the analysis is done at room temperature where paraffins are usually solid, the paraffin plate can be easily taken off from the top of the analysis glass. The analysis with $CS(NH_2)_2$ is carried out in the same way as the analysis with $CO(NH_2)_2$. However, $CS(NH_2)_2$ forms a complex with *iso*-alkanes. By using both of these methods, a relatively exact separation of *n*-paraffin and *iso*-paraffin fractions is made possible.

2.3.2 Chemical Analysis Methods of Crude Oil Products: Determination of Unsaturated Compounds

Unsaturated compounds are not present in crude oil. However, a significant amount of these compounds can be found in the products of crude oil treatment. An example is the product of thermal cracking. Many problems are associated with unsaturated compounds. For example, these compounds tend to undergo strong oxidation reactions. This can quickly change the quality of the produced products. The oxidized compounds can react by polycondensation reactions and form insoluble residues in crude oil products. This can even be dangerous especially for motor fuels, because this residue plugs the fuel filters and makes engines not function properly.

All the chemical methods for the determination of the amount of unsaturated compounds are based on the addition reaction of halogens such as bromine or iodine to double bonds. Bromine number or iodine number is determined by this method. The bromine number or iodine number shows the amount of bromine or iodine needed to saturate all the unsaturated bonds in a hundred grams of the sample.

Determination of the bromine number or iodine number proceeds generally in two steps. First bromine, iodine or their derivates are mixed with the sample. The halogenization reaction is carried out. After this step, the mixture of the sample and indicator is titrated with sodium thiosulphate. This step shows how much of the halogen is left in the mixture. The bromine number or iodine number is determined as the difference between the halogen added and halogen remaining by equation (2.21).

$$N_h = w_h \cdot 100 / w_s \tag{2.21}$$

where N_h – bromine or iodine number
w_h – weight of bromine or iodine needed for halogenization
w_s – weight of sample

The average number of unsaturated bonds in one molecule of the product can be calculated by equation (2.22).

$$N = N_B \cdot M / 160 = N_I \cdot M / 254 \tag{2.22}$$

where N – average number of unsaturated bonds
N_B – bromine number
N_I – iodine number
W – average molecular weight of the sample

There is a special method for the determination of diolefin compounds. This method is based on the reaction with moleine anhydride as represented by equation (2.23).

$$R\text{–CH}_2\text{–CH=CH–CH=CH–CH}_2\text{–R} + \text{maleic anhydride} \rightarrow \text{adduct} \qquad (2.23)$$

The formed derivatives of moleine anhydride are solid crystalline compounds and can easily be separated from the liquid sample or solution of the sample in an appropriate solvent.

2.3.3 Structural Bulk Analysis of Heavy Crude Oil Fractions: n-d-M Method

The main idea of structural bulk analysis of heavy crude oil fractions is based on the existence of the so-called average molecule. The average molecule describes the chemical and structural properties of all molecules present in the sample mixture. It means that the result of structural bulk analysis is not the determined amount of compounds or compound groups, but it is the amount of carbon in certain structural groups, for example in aromatic groups, naphthenic groups and paraffinic groups.

The most popular and simple method for structural bulk analysis for heavy crude oil fractions is called n-d-M analysis. The main idea behind this method is the linear dependence of aromatic, naphthenic and paraffinic carbon in heavy fraction on refractive index, density and value reverse to the molecular weight. This dependence is described by equation (2.23).

$$C_L = a / M = b \cdot \Delta d + c \cdot \Delta n \qquad (2.23)$$

where C_L – amount of aromatic, naphthenic or paraffinic carbon

a, b, c – empiric constants

Δd – density difference between sample density and hypothetical paraffin density

Δn – refraction index difference between sample refraction index and hypothetical paraffin refraction index

The hypothetical paraffin is only theoretically existent n-paraffin with an endless chain length in liquid state. The density and refraction index were theoretically calculated for this paraffin:

$n_D^{20} = 1.4750$ - refraction index at 20°C

$n_D^{70} = 1.4600$ - refraction index at 70°C
$d^{20} = 1.8510$ - density at 20°C
$d^{70} = 1.8280$ - density at 20°C

The refraction index, density and average molecular weight of the sample must be measured in order to use the n-d-M method. Evaluating the results of this method begins with calculating four factors: v, X, W and Y – according to equations (2.23), (2.24), (2.25) and (2.26).

$$v = 2.51 \cdot (n_d^{20} - 1.4750) - (d^{20} - 0.8510) \quad (2.23)$$
$$X = 2.42 \cdot (n_d^{70} - 1.4600) - (d^{70} - 0.8280) \quad (2.24)$$
$$W = (d^{20} - 0.8510) - 1.11 \cdot (n_D^{20} - 1.4750) \quad (2.25)$$
$$Y = (d^{70} - 0.8280) - 1.11 \cdot (n_D^{20} - 1.4600) \quad (2.26)$$

The calculation of the amount of carbon in different structural groups begins after all these factors have been calculated. The amount of aromatic carbon is calculated first. If the factor v is positive and all the needed values were determined at the temperature 20°C, then the amount of carbon in aromatic groups can be calculated by equation (2.27).

$$C_A = 430 \cdot v + 3660 / M \quad (2.27)$$

where C_A – amount of aromatic carbon

If the factor v is negative, the amount of aromatic carbon is calculated by equation (2.28).

$$C_A = 670 \cdot v + 3660 / M \quad (2.28)$$

For the case when all the measurements are carried out at the temperature 70°C, factor X is used. If X is positive, then the amount of aromatic carbon is calculated by equation (2.29).

$$C_A = 410 \cdot v + 3660 / M \quad (2.29)$$

If factor X is negative, then equation (2.30) should be used.

$$C_A = 720 \cdot v + 3660 / M \quad (2.30)$$

The general amount of carbon in cyclic structural elements of the sample (C_G) is calculated next. The factors W and Y are used for this calculation. The factor W is used for the evaluation of the measurements made at 20°C. If the factor W is positive, then the amount of cyclic carbon can be calculated by equation (2.31).

$$C_G = 820 \cdot W - 3 \cdot S + 10000 / M \tag{2.31}$$

where S – sulfur concentration in the sample

Equation (2.32) should be used for negative factor W.

$$C_G = 1440 \cdot W - 3 \cdot S + 10600 / M \tag{2.32}$$

Factor Y is used for the evaluation of the measurements made at the temperature 70°C. If this factor is positive then the general amount of carbon in cyclic groups can be calculated by equation (2.33).

$$C_G = 775 \cdot W - 3 \cdot S + 11500 / M \tag{2.33}$$

For the case of negative factor Y, equation (2.34) should be used.

$$C_G = 1440 \cdot W - 3 \cdot S + 12100 / M \tag{2.34}$$

The amount of carbon in naphthenic groups can be calculated by equation (2.35).

$$C_N = C_G - C_A \tag{2.35}$$

Finally, the amount of carbon in paraffinic groups is calculated as a difference between 100 % and amount of cyclic carbon.

$$C_P = 100 - C_G$$

Average molecules can be drawn based on the results of the structural bulk analysis. For example, the molecule shown in Figure 2.61 has 33.3% of aromatic carbon, 13.3% of naphthenic carbon and 53.4% of paraffinic carbon.

It should be noted that the same analysis results could be obtained, for example, for a mixture of different derivates with different lengths of the paraffinic chain or different number of aromatic rings. Now it is obvious that the information obtained from the n-d-M method describes the average structure of the molecules in the sample mixture and the amount of carbon in the different groups. This is why this method is called structural bulk analysis.

The evaluation method described in this chapter was developed by Van-Ness [85]. More about the n-d-M method and similar methods for the analysis of heavy crude oil fraction can be found in many references at the end of this chapter.

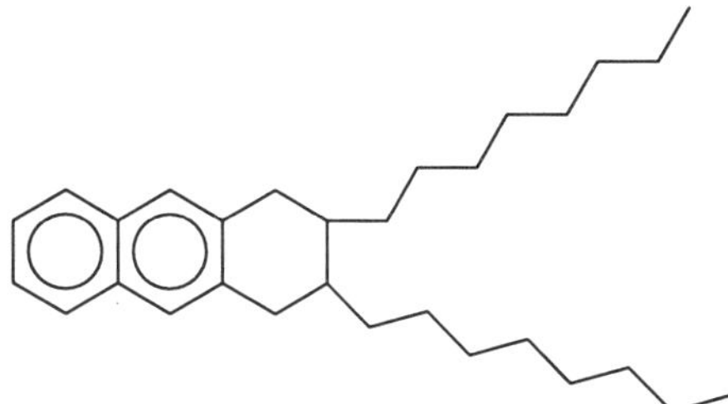

Fig. 2.61: Average molecule.

2.4 METHODS OF CHARACTERIZATION OF COLLOIDAL PROPERTIES OF CRUDE OIL AND ITS PRODUCTS

The colloidal properties of the crude oil are important especially for the chemistry of the heavy fractions of crude oil, which contain high amounts of asphaltenes and resins. The main purpose of all these methods is to determine the size of the colloidal particles in crude oil product, and consequently to estimate the colloidal stability of these products, meaning how long the product can keep the physical properties in the bulk volume in equilibrium.

All the methods used for colloidal characterization of crude oil can be classified as direct and indirect methods.

2.4.1 Direct Methods

All the spectral methods belong to the direct methods. All these methods use interference or light scattering effect. The spectrum used for this analysis is all the wavelengths from infrared until x-ray. However x-ray scattering method is the most popular method of elucidating the colloidal properties of crude oil. The technical fundamentals of such a measurement are the same as was shown for x-ray spectroscopy. However, in this measurement, it is the intensity of radiation scattered at a certain angle that is determined. Equation (2.36) is used for evaluating the measurements results.

$$I_S = I_0 \cdot e^{(-S^2 \cdot r^2 / 3)} \tag{2.36}$$

where I_S – intensity measured by angle factor S
I_0 – intensity measured at the angle 0°
S – angle factor
r – particle radius

The angle factor S depends on scattering angle and wavelength of the radiation used for the analysis. It is calculated by equation (2.37)

$$S = 4 \cdot \pi / \lambda \cdot \sin(\theta/2) \qquad (2.37)$$

where $\pi = 3.14159265$
λ – wavelength
θ – scattering angle

The next direct method used to characterize the colloidal properties of crude oil is the sedimentation method. It is obvious from the name of the method that this method is based on the sedimentation effect. There are two possibilities to carry out this method: the first is the sedimentation under the influence of gravitational force and the second sedimentation under influence of centrifugal force. The choice between these methods depends on the viscosity of the sample and the size of the particles of the disperse phase. Viscous samples or samples with relatively small particles should be analyzed by the second method.

The setup used for sedimentation analysis is shown in Figure 2.62.

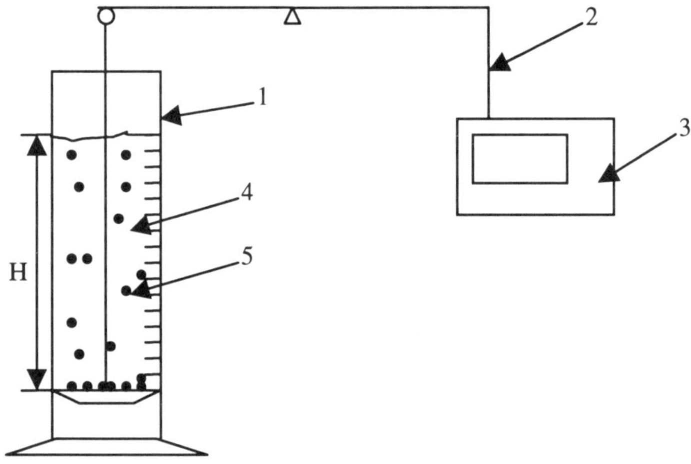

Fig. 2.62: Sedimentation analysis:
1 – measurement cylinder
2 – sedimentation scales
3 – monitoring device
4 – continuous phase of sample
5 – disperse phase of sample

The simplest evaluation method for colloidal systems with similar sizes of colloidal particles is based on equation (2.38).

$$r = \sqrt{\frac{9 \cdot \nu H}{2 \cdot (d - d_0) \cdot t_s}} \tag{2.38}$$

where ν – viscosity of the sample
H – height of the sample level (see Figure 2.62)
d – density of disperse phase of the sample
d_0 – density of continuous phase of the sample
t_s – sedimentation time
r = sedimentation rate

The sedimentation time is measured as a time from when the analysis starts until the time when the weight on the sedimentation scale does not change any further.

The sedimentation method belongs to the classical methods of characterization of the colloidal properties of disperse systems. These methods can be used for the analysis of colloidal solutions with size of colloidal particles between 1 and 100 micrometer. The analysis of solutions with smaller particles leads to relatively high errors as a result of Brownian motion.

The next direct method for the determination of colloidal characteristics of crude oil is the conductive method. This method is based on the measurement of the electrical conductivity of the sample during the time the colloidal particles pass through a calibrated microhole. The schematic of the setup used for measurement by the conductive method is shown in Figure 2.63.

There is a constant voltage between two electrodes (see Figure 2.63). The sample is sucked out continuously by a pump system shown as 5 in Figure 2.63. Thus, there is continuous stream of the sample passing through the calibrated microhole throughout the analysis time. Because of the difference in conductivity of the discontinuous phase and the continuous phases in the sample colloidal solution, the measured conductivity of the sample is changed continuously as particles of discontinuous phase pass through the calibrated microhole. The resulting electrical impulses are registered by a monitoring device. The frequency of these impulses depends on the concentration of the discontinuous phase and their intensity is influenced by the size of the particles passing through the calibrated microhole. Evaluation of the measurement results are based on calibration tables.

The biggest disadvantage of the conductive method is that only colloidal solutions with a low concentration of the discontinuous phase can be analyzed. This is due to the fact that only a few particles can pass through the calibrated hole at the same time. This may lead to a wrong record of the particle size and concentration, because these few particles are registered as one particle of a size that is greater than the size of each of the constituting particles.

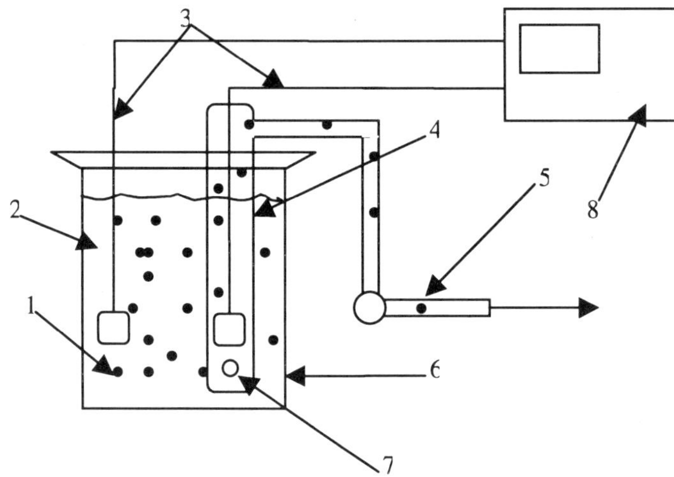

Fig. 2.63: Setup for conductivity analysis:
1 – colloidal particles
2 – continuous phase of sample
3 – electrodes
4 – glass cylinder
5 – pipe for sample output
6 – glass
7 – calibrated microhole
8 – monitoring device

The last direct method often used for determination of the colloidal properties of crude oils is gel permeation chromatography. The principles of this method were described in section 2.1.2. Normally, this method is used for analyzing the molecular weight distribution of substances. However, it is possible to use it to analyze colloidal properties as well if an appropriate solvent is used as the mobile phase. This solvent must not change the native disperse particles. Almost all the solvents that can be used in this analysis as a mobile phase change the size of native particles. This is why this analysis is usually used for estimating the particle size in the sample solution.

2.4.2 Indirect Methods

All the indirect methods that are based on the determination of the colloidal properties of crude oil measure macroscopic properties such as density, viscosity, and molecular weight. An example is that the molecular weight of asphaltene particles or asphaltene molecules is a result of the determination of the molecular weight by methods such as osmotic pressure measurement. There are very many indirect methods developed for certain oil or certain cases. The main idea in all these methods is the development of models that describe the relation between macroscopic properties and colloidal properties of the sample system.

2.5 DETERMINATION OF THE PHYSICAL PROPERTIES OF CRUDE OIL

2.5.1 Density Determination

Density is not only just one of the most important physical parameters of crude oil and its products, but it is also an important characteristic for measuring the quality for crude oil and its products. The density of a sample shows its mass in specific volume. The classical definition of density is presented in equation (2.39).

$$d = W / V \tag{2.39}$$

where d – density
W – weight of the sample
V – volume of the sample

The density that is calculated in equation 2.39 is known as absolute density. However, this density is rarely used by crude oil chemists and only in special cases. Relative density is the parameter that is usually used for the characterization of crude oil and its products. Usually, relative density is measured at a reference temperature of 20°C. Relative density is calculated by equation (2.40).

$$d_{20}^{20} = d_s / d_w \tag{2.40}$$

where d_s – density of the sample at the temperature 20°C
d_w – density of water at the temperature 20°C (0.99821 g/cm^3)

Relative density can be calculated by equation (2.40) by using the water density at 20°C. The absolute density can be compared with water density at 4°C. Then, relative density should be calculated by equation (2.41).

$$d_4^{20} = d_s/d_w \tag{2.41}$$

where d_s – density of the sample at the temperature 20°C
d_w – density of water at the temperature 4°C (1.00000 g/cm^3)

It is obvious that relative density that is calculated with equation (2.41) is equal to the absolute density that is calculated at 20°C.

There are two methods for the determination of density that are popular with crude oil chemists. The first is the hydrometer method. The hydrometer is a glass body, which is dipped into the sample. After a short equilibration time, it will float vertically at a certain level. This level results from when the mass of the hydrometer is equal to the buoyancy effect. The higher the density of the sample, the less the hydrometers will sink into the sample. The level of equilibrium shows the density on the calibrated scale. An example of density determination is shown in Figure 2.64.

The hydrometer method is the simplest and fastest one. This method only has one disadvantage, and that is that the hydrometer usually has a very small measuring range. This requires many hydrometers with different measuring ranges for the determination of the density of a sample with an unknown density.

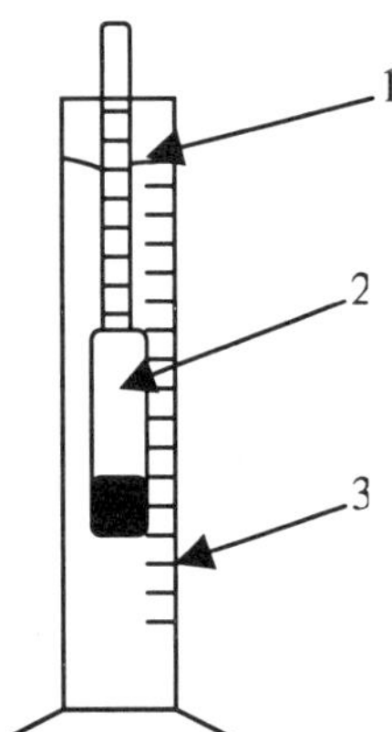

Fig. 2.64: Hydrometer density determination:
1 – level at hydrometer scale showing density of the sample
2 – hydrometer
3 – cylinder

The second method is called the pycnometer method. A pycnometer is a glass beaker of defined volume. It is shown in Figure 2.65.

Fig. 2.65: Pycnometer.

Firstly pycnometer is weighed without sample, then filled with the sample thermostatted and weighed again. The absolute density is then calculated by using equation (2.42):

$$d = (W_F - W_E) / V \tag{2.42}$$

where d – density
W_F – weight of pycnometer with the sample
W_E – weight of empty pycnometer
V – volume of the sample in pycnometer

It is important to note that density is very sensitive to temperature. Thus, it is very important to thermostat the sample before any measurement is made.

2.5.2 Viscosity Determination

Viscosity is the second most important physical parameter for crude oil. It characterizes not only one of the physical properties, but also the quality of most crude oil products such as lubricating oils and lubricants.

The term viscosity is derived from the flow behavior of a liquid. It is a measure of resistance to flow. Viscosity can also be defined as inner friction or inner resistance of the sample against flow.

There are two types of viscosity:

- dynamic viscosity and
- kinematic viscosity.

The two types of viscosity are often used by crude oil chemists for various purposes. The kinematic viscosity can be measured with the Ostwald viscometer. The Ostwald viscometer is the most popular instrument used by crude oil chemists for determination of kinematic viscosity. A schematic of the Ostwald viscometer is presented in Figure 2.66.

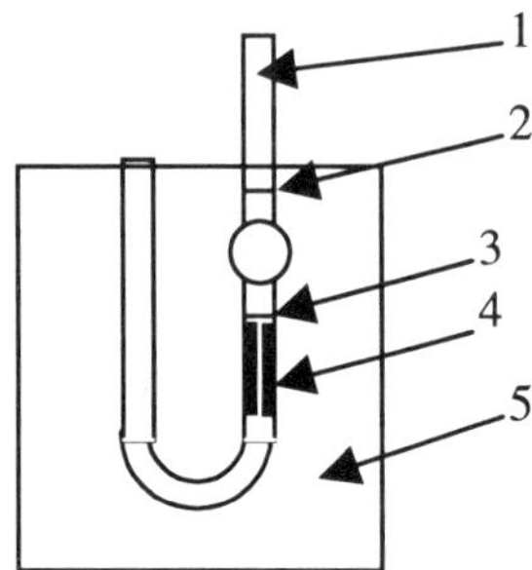

Fig. 2.66: Ostwald viscometer:
1 – viscometer
2 – sample level 1
3 – sample level 2
4 – capillary
5 – thermostat

The sample is introduced inside the viscometer for the analysis so that the top limit of the sample is located exactly on sample level 1 (see Figure 2.66). After the sample is well thermostatted in the thermostat 5 the analysis can start. The sample is released for free motion through the capillary 4 for the analysis. The analysis result is measured in terms of the time which the sample needs to flow from sample level 1 to sample level 2. The viscosity result is evaluated from equation (2.43).

$$\eta = k \cdot t \qquad (2.43)$$

where η – kinematic viscosity
k – capillary coefficient
t – time which sample needs to flow from level 1 to level 2

The kinematic viscosity can easily be converted into dynamic viscosity by equation (2.44).

$$\nu = \eta \cdot d \qquad (2.44)$$

where ν – dynamic viscosity
d – sample density

Dynamic viscosity can be determined directly by using the rotation viscometer. The scheme of such a viscometer is shown in Figure 2.67.

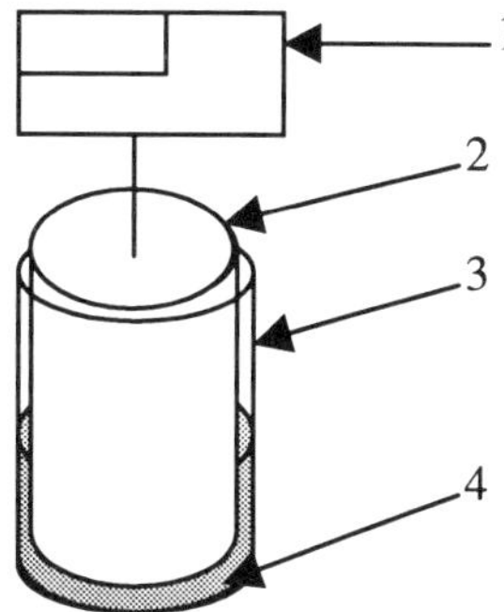

Fig. 2.67: Rotation viscometer:
1 – monitoring device
2 – static cylinder
3 – rotating cylinder
4 – sample

The rotation viscometer consists of two coaxial cylinders, between which the sample is introduced. The rotating cylinder is rotated with a constant velocity. The static cylinder is unmovable. The static cylinder is hung on a torsion wire, which is turned, depending upon the flow resistance, at a more or less large angle. This rotation angle can be read from a monitoring device. The rotation angle is a measure of the viscosity of the liquid. The viscosity should be extracted or read from calibration tables with regard to the measured rotation angle.

During the measurement of the viscosity by both methods, it is important to note that viscosity is very sensitive to temperature, and thus it is important to thermostat the viscometer before starting analysis.

2.5.3 Refractive Index Determination

Refractive index is an important physical property of crude oil and its products which is usually used for further evaluation of the characteristics of crude oil and its products. Such characteristics include the n-d-M method described in an earlier section of this chapter. Refractive index represents the ratio between two angles, the angle of incidence and the angle of refraction. The angle of incidence (α_1) is the angle of a light beam before hitting the sample, the angle of refraction (α_2) is the angle of the light beam after hitting the sample. An example of refraction is shown in Figure 2.68.

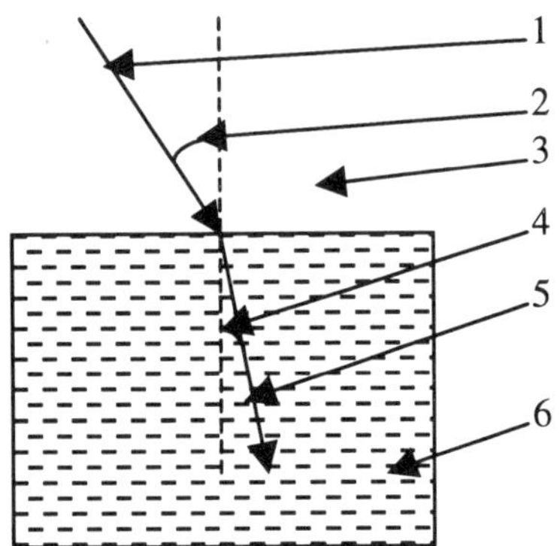

Fig. 2.68: Refraction
1 – beam of light
2 – angle of incidence
3 – atmosphere
4 – angle of the light beam after hitting the sample
5 – refracted beam
6 – sample

The refractive index is calculated by equation (2.45) with regard to the refraction scheme shown on Figure 2.68.

$$n_D = \sin(\alpha_1) / \sin(\alpha_2) \tag{2.45}$$

where n_d – refractive index
α_1 – angle of incidence
α_2 – angle of the light beam after hitting the sample

The refractive index is measured directly by refractometers. The most popular refractometers used in the laboratories use the functioning principle shown in Figure 2.69. This figure shows a combination of two different prisms. Bottom prism has a rough surface to create scattered light striking the liquid-glass interface. The incident beam relating to the critical angle. The refractive index of the sample is determined by measurement of the refraction angle of the refracted beam of light 1 (see Figure 2.69)

A few drops of the liquid sample are placed on the bottom prism. The refractive index can be read directly from the built-in scale, looking into the refractometer.

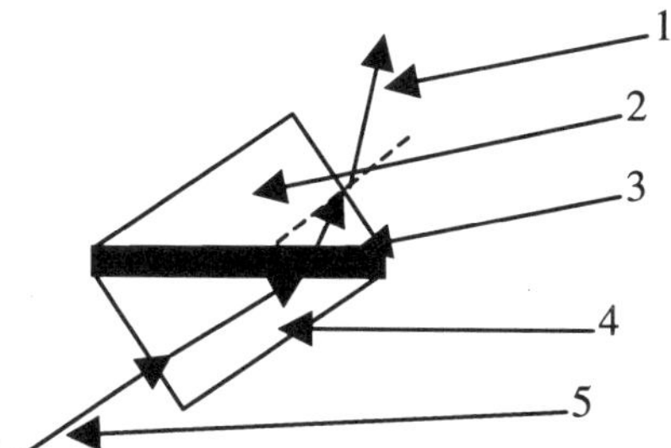

Fig. 2.69: Functioning principle of the refractometer:
1 – refracted beam of light
2 – upper prism
3 – sample
4 – bottom prism
5 – incidence beam of light

Refractometry is the last analysis method discussed in this part of the book. However, some more analysis methods can be found in chapter 8 of this book. The reason to describe these analysis methods in another chapter is the high importance attached to these methods for asphaltene chemistry, which is an important part of crude oil chemistry.

Bibliography

1 A. I. Bogomolov, A. A. Geile, V. V. Gromova, A. E. Drabkin, C. G. Nerucheev, V. A. Proskuryakov, D. A. Rozental, M. G. Rudin, A. M. Syroezhko. Chimiya nefti i gaza. Chimiya, 1995.

2 W. N. Erih, M. G. Rasina, M. G. Rudin. Chimiya i Technologija Nefti i Gaza. Chimiya, Leningrad 1977.

3 J. G. Speight. The Chemistry and Technology of Petroleum. Marcel Dekker, New York, 1980.

4 G. W. Mashrush, J. G. Speight. Petroleum products: Instability and incompatibility. Ed Taylor & Francis, 1995.

5 V. P. Gavrilov. Chernoe zoloto planety. Nedra, Moscow, 1990.

6 J. D. Saxby, K. W. Riley. Petroleum generation by laboratory-scaled pyroly-

sis over six years simulating conditions in a subsiding basin, J. Nature, 308, 177-179, 1984.

7 J. D. Saxby et al. Petroleum generation: simulation over six years of hydrocarbon formation from torbanite and brown coal in a subsiding basin, J. Organic Geochemistry, 9(2), 69-81, 1986.

8 B. R. T. Simoneit, P. F. Lonsdale. Hydrothermal petroleum in mineralized mounds at the seabed of Guayamas Basin, J. Nature, 295, 198-202, 1982.

9 B. M. Didyk, B. R. T. Simoneit. Hydrothermal oil of Guayamas Basin and implications for petroleum formation mechanisms, J. Nature, 342, 65-69. 1989.

10 T. A. Abrajano, N. C. Sturchio, J. K. Bohlke, G. L. Lyon, R. J. Poreda, C. M. Stevens. Methane-hydrogen gas seeps, Zambales ophiolite, Philippines: deep or shallow origin? J. Chemical Geology, 71, 211-222, 1988.

11 E. Bonatti, P. J. Michael. Mantle peridotites from continental rifts to ocean basins to subduction zones, J. Earth and Planetary Science Letters, 91, 297-311, 1989.

12 E. Bonatti. Long-lived oceanic transform boundaries formed above mantle thermal minima, J. Geology, 24, 803-806, 1996.

13 E. Bonatti, M. Seyer, N. M. Sushevskaya. A cold suboceanic mantle belt at the Earth's equator, J. Science, 261, 315-320, 1993.

14 M. Cannat, C. Mevel, M. Maia, C. Deplus, C. Durand, P. Gente, P. Agrinier, A. Belarouchi, G. Dubuisson, E. Humler, J. Reynolds. Thin crust, ultramafic exposures, and rugged faulting patterns at the Mid-Atlantic Ridge (22°-24°N), J. Geology, 23, 49-52, 1995.

15 J. L. Charlou, H. Bougault, P. Appriou, T. Nelsen, P. A. Rona. Different TDM/CH_4 hydrothermal plume signatures: TAG site at 26° N and serpentinized ultrabasic diapir at 15°05' N on the Mid-Atlantic Ridge, J. Geochim. Cosmochim. Acta, 55, 3209-3222, 1991.

16 J. L. Charlou, Y. Fouquet, H. Bougault, J. P. Donval, J. Etoubleau, P. Jean-Baptiste, A. Dapoigny, P. Appriou, P. Rona. Intense CH_4 plumes generated by serpentinization of ultramafic rocks at the intersection of the 15°20' N fracture zone and the Mid-Atlantic Ridge, Geochim. et Cosmochim. Acta, 62(13), 2323-2333, 1998.

17 R. G. Coleman, T. E. Keith. A chemical study of serpentinization - Burro Mountain, California, J. Petrol., 12, 311-328, 1971.

18 J. P. Donval et al. High H_2 and CH_4 content in hydrothermal fluids from Rainbow site newly sampled at 36°14' N on the Amar segment, Mid-Atlantic Ridge (diving FLORES cruise, July 1997), Comparison with other MAR sites, Eos Trans. (abstr), 78, F832, 1997.

19 J. Escartin, M. Cannat. Ultramafic exposures and the gravity signature of the lithosphere near the Fifteen-Twenty Fracture Zone (Mid-Atlantic Ridge, 14°-16.5° N), J. Earth Planet. Sci. Lett., 171, 411-424, 1999.

20 H. P. Eugster, G. P. Skippen. Igneous and metamorphic reactions involving gas equilibria, Researches in Geochemistry, 2. P. H. Abelson (Ed.), p. 492, Wiley, New York, 1967.

21 J. A. Haggerty. Evidence from fluid seeps a top serpentine seamounts in the Mariana Forearc: Clues for emplacement of seamounts and their relationship to forearc tectonics, J. Mar. Geol., 102, 293-309, 1991.

22 A. Hood, C. C. M. Gutjahr, R. L. Heacock. Organic metamorphism and the generation of petroleum, J. Am. Assoc. Pet. Geol. Bull., 59, 986-996, 1975.

23 H. D. Klemme, G. F. Ulmishek. Effective petroleum source rocks of the world: stratigraphic distribution and controlling depositional factors, J. AAPG Bulletin, 75(12), 1809-1851, 1991.

24 M. D. Lilley, D. A. Butterfeld, E. J. Olsomn, J. E. Lupton, S. E. Macko, R. E. McDuff. Anomalous CH_4 and NH_4 concentrations at an unsedimented mid-ocean ridge hydrothermal system, J. Nature, 364, 45-47, 1993.

25 B. Martin, W. S. Fyfe. Some experimental and theoretical observations on the kinetics of hydration reactions with particular reference to serpentinization, Chem. Geol., 6, p. 185, 1970.

26 M. J. Mottl. Metabasalts, axial hot springs, and the structure of hydrothermal systems at mid-ocean ridges, J. Geol. Soc. Amer. Bull., 94, 161-180, 1983.

27 C. Neal, G. Stanger. Hydrogen generation from mantle source rocks in Oman, J. Earth Planet Sci. Lett., 66, 315-320, 1983.

28 R. J. Poreda, K. Marti, H. Craig. Rare gases and hydrogen in native metals, in: From Mantle to Meteorites, K. Gopalan et al. (Eds.), pp. 153-172, Indian Ac. Sci., 1990.

29 P. A. Rona, S. D. Scott. Preface, a special issue on sea-floor hydrothermal mineralization: New perspectives, Economic Geology, 88(5), 1934-1976, 1993.

30 M. W. Schmidt, S. Poli. Experimentally based water budgets for dehydrating slabs and consequences for arc magma generations, J. Earth Planet. Sci. Lett., 163, 361-379, 1998.

31 J. E. Snow. Of Hess crust and layer cake, J. Nature, 374, 413-414, 1995.

32 A. P. Vinogradov, L. V. Dmitriev, G. B. Udintsev. Distribution of trace elements in crystalline rocks of rift zones, J. Phil. Trans. Roy. Soc. London, A 268, 487-491, 1971.

33 J. A. Welhan, H. Craig. Methane, hydrogen and helium in hydrothermal fluids at 21° N on the East Pacific Rise, Hydrothermal Processes at Sea Floor Spreading Centers, P. Rona et al. (Ed.), 391-409, 1983.

34 B. P. Tissot, D. H. Welte. Petroleum Formation and Occurrence, Springer-Verlag, Berlin, p. 538, 1984.

35 R. Littke, D. Leythaeuser, M. Radke, R. G. Schaefer. Petroleum generation and migration in coal seams of the Carboniferous Ruhr Basin, Northwest Germany. J. Adv. Org. Geochem. (B. Durand and F. Behar, Eds), Org. Geochem., 16, Pergamon Press, Oxford, 247-258, 1989.

36 G. H. Taylor, M. Teichmueller, A. Davis, C. F. K. Diessel, R. Littke, P. Robert. Organic Petrology. Gebr. Borntraeger, Berlin-Stuttgart, p. 704, 1998.

37 A. Lueckge, M. Boussafir, E. Lallier-Vergès, R. Littke. Comparative study of organic matter preservation in immature sediments along the continental margins of Peru and Oman. Part I: results of petrographical and bulk geochemical data. J. Organic Geochemistry, 24, 437-451, 1996.

38 C. Müller, F. Theilen, B. Milkereit. Large gas-prospective areas indicated by bright spots. J. World Oil, Jan., 60-67, 2001.

39 C. Yin, P. Weidelt. Geoelectrical fields in a layered earth with arbitrary anisotropy, J. Geophysics, 64, 426-434, 1999.

40 C. Yin. Electromagnetic induction in a layered conductor with arbitrary anisotropy, Ph.D thesis, Technical University of Braunschweig, Germany, 1999.

41 M. Wagner. Ergebnisse der mikrobiologischen Prospektion im Küstenbereich der Ostsee, In Vorträge des Internationalen Symposium "Erdölmikrobiologie", Bruno (Malek & Schwartz, Ed.), Akademie Verlag, Berlin, 1964.

42 M. G. Baum, M. Wagner et al. Application of surface prospecting methods in the Dutch North Sea, J. Petroleum Geoscience 3, 171-181, 1997.

43 M. Wagner, H. J. Rasch, J. Piske, M. Baum. MPOG - microbial prospection for oil and gas. Field examples and their geological background: Conference Cracov, Poland, AO-05, 118-121, 1998.

44 M. Wagner, H. J. Rasch, J. Piske, B Ziran. Mikrobielle Prospektion auf Erdöl und Erdgas in Ostdeutschland, J. Geologisches Jahrbuch, 149, 287-301, 1998.

45 T. T. Klubova. Clayey reservoirs of oil and gas, A.A. Balkema Publishers, Rotterdam, p.178, 1991.

46 G. I. Amurskii, G. A. Abramenok, M. S. Bondareva, N. N. Solov'ev. Remote sensing methods in studying tectonic fractures in oil- and gas-bearing formations, A.A. Balkema Publishers, Rotterdam, p. 146, 1991.

47 Aadnoy, S. Bernt. Modern well design, A.A. Balkema Publishers, Rotterdam, p. 256, 1996.

48 A. T. Hruschev. Geografiay promyshlennosti SSSR. Mysl, Moscow, 1986.

49 V. A. Kryukov. Polnye kanistry i pustye kormany. ECO, Nr. 1, 1994.

50 Neft i gaz v zerkale planety. Delovoj mir, 1-7 august 1994.

51 D. Zudkevitch, A. K. S. Murthy, J Gmehling. Thermodynamic aspects of reformulation of automotive fuels, Part 1. The effects of oxygenates on the vapor pressures and volatilities of gasolines. J. Hydrocarbon Processing, 93-100, June 1995.

52 W. J. Piel, R. X. Thomas. Oxygenates for Reformulated Gasolines, J. Hydrocarbon Processing, pp. 68-73, July 1990.

53 C. K. Westbrook. The Chemistry Behind Engine Knock, Chemistry & Industry (UK), pp. 562-566, 3 August 1992.

54 G. Schomburg, Gaschromatographie, VCH Verlagsgesellschaft Weinheim,

1987.

55 T. Wiedmann, K. Ballschmiter. Quantification of chlorinated naphthalenes with GC/MS using the molar response of electron impact ionization, Fresenius, J. Anal. Chem. 346: 800-804, 1993.

56 U. Weidlich, J. Gmehling. Extension of UNIFAC by headspace gas chromatography. J. Chem. Eng. Data, 30, 95, 1985.

57 K. Ballschmiter, A. Mennel, J. Buyten, Long chain alkyl-polysiloxanes as non-polar stationary phases in capillary gas chromatography, Fresenius, J. Anal. Chem. 346: 396-402, 1993.

58 H. Willsch, H. Clegg, B. Horsfield, M. Radke, H Wilkes. Liquid chromatographic separation of sediment, rock, and coal extracts and crude oils into compound classes. J. Analytical Chemistry, 69: 4203-4209, 1997.

59 D. A. Skoog, D. M. West, F. J. Holler. Fundamentals of Analytical Chemistry. Sixth Edition, Saunders College Publishing, 1992, Chapters 26, 27, 29, 30.

60 R. P. Haugland. Handbook of Fluorescent Probes and Research Chemicals; Molecular Probes Inc., Eugene, OR, 1985.

61 J. H. Knox, B. Kauer. High Performance Liquid Chromatography, P. R. Brown and R. A. Hartwick, Eds. Wiley Interscience: New York, 1989, Chapter 4.

62 Beckman Model 330 HPLC Manuel, Beckman Instruments, Fullerton, CA.

63 G. Hesse. Chromatographisches Praktikum. 2. Aufl. Akadem. Verlagsges., Frankfurt a. M., 1968.

64 G. Pataki. Dünnschichtchromatographie in der Aminosäure- und Peptid-Chemie. De Gruyter, Berlin, 1966.

65 K. Randerath: Dünnschicht-Chromatographie. 2. Aufl., 2. Nachdr. Verlag Chemie, Weinheim, 1975.

66 E. Stahl. Dünnschicht-Chromatographie. 2. Aufl. Springer, Berlin-Heidelberg-NewYork, 1967.

67 P. R. Griffith, J. A. Haseth. Fourier Transfom Infrared Spectroscopy, John Wiley & Sons, New York, 1986.

68 R. W. Hannah, J. S. Swinehart. Experiments in Techniques of Infrared Spectroscopy, Perkin-Elmer Corporation, 1974.

69 M. Hesse, H. Meier, B. Zeeh. Spektroskopische Methoden in der organischen Chemie, G. Thieme Verlag, 1987.

70 R. S. Drago. Physical Methods in Chemistry, Saunders, Philadelphia, 1977, Chapter 6.

71 L. A. Woodward. Introduction to the Theory of Molecular Vibrations and Vibrational Spectroscopy, Oxford Univ. Press, 1972.

72 D. A. Long, Raman Spectroscopy, McGraw-Hill, New York, 1977.

73 S. B. Dierker, C. A. Murray, J. D. Legrange, N. E. Schlotter. Characterization of order in Langmuir-Blodgett monolayers by unenhanced Raman spectroscopy, Chem. Phys. Lett., 137, 453, 1987.

74 S. M. Angel, T. F. Cooney, H. T. Skinner. Applications of fiber optics in NIR Raman spectroscopy, in Modern Techniques in Raman Spectroscopy, J. J. Laserna (Ed.), Wiley, Chichester, 1996.

75 J. J. Laserna. Combining fingerprinting capability with trace analytical detection: surface-enhanced Raman spectrometry, Anal. Chim. Acta, 283, 607, 1993.

76 E. B. Sandell. Colorimetric determination of traces of metals, Interscience, New York, 1950.

77 G. Charlot. Colorimetric Determination of Elements (Principles and Methods), Elsevier Publ. Co, Amsterdam-London-New York, 1964.

78 P. Pringsheim. Fluorescence and phosphorescence, Interscience Publishers, New York, 1949.

79 E. J. Bowen, F. Wokes. Fluorescenece of Solutions, Longmous, Green and Co., London, 1953.

80 L. Dunemann, J. Begerow. Analytik von Platinmetallen in Körperflüssigkeiten mit ET-AAS und hochauflösender ICP-MS. Platin Analytik Anwendertreffen, 1994 Nov 22-Nov 23; Stuttgart, 1994.

81 F. Peters. "Dual-Atomiser" - Ein neues Konzept in der AAS. Elementspektroskopie Usermeeting ´99 TJA-Unicam-VG Elemental. Oct-

1999, Offenbach, 1999.

82 R. K. Malhotra, K. Satyanarayana, G. V. Ramanaiah. Determination of Au, Pd, Pt, and Rh in rocks, ores, concentrates, and sulfide float samples by ICP-OES/FAAS after reductive coprecipitation using Se as collector. Atomic Spectroscopy, 20(3), 92-102, 1999.

83 R. Jenkins, R. W. Gould, D. Gedeke. Quantitative X-ray spectrometry, Marcel Dekker Inc., New York, 1981.

84 M. A. Blochin. Methods of X-ray spectroscopic research, Pergamon Press, Oxford-New York, 1963.

85 K. Van-Ness, H. Van-Westen. Sostav maslynyh fracziy nefti I ih analiz, IL, Moscow, 1954.

Part II

REGIONAL PETROLEUM INDUSTRY

OVERVIEW

It would be appropriate to say at this point that this chapter deals more with crude oil economy than with crude oil chemistry. However, the aim of this chapter is to show the importance of crude oil chemistry not only for crude oil chemists, but also for all of mankind. Almost everything around us is derived from crude oil. This includes plastic parts, car fuel, jet fuel, oils, and even asphalt on the road; these are all made from petroleum. It is hard to imagine what modern life would look like without these items, which began their existence from the oil well.

Petroleum is responsible not only for making our life more comfortable, but also it has a great influence on international politics. It is well known that crude oil is often called "black gold". This name emphasizes the importance of petroleum vis-à-vis the world economy.

The foregoing discussion explains why this chapter deserves a place in this book. It helps us to understand the importance of crude oil chemistry for the people making decisions about the future of petroleum education, for example. This chapter shows new impressive sides of this branch of study.

3

Petroleum Producing Countries: OPEC and Non-OPEC

3.1 INTRODUCTION

The many applications of petroleum have been known from ancient times. Initially, primitive ways were employed in petroleum operations. Examples include collecting petroleum from the ground surface and the processing of oil sands. The development of the petroleum industry started when mechanical drilling for oil wells for petroleum production was employed in 1859 in the USA. Practically all petroleum extracted in the world now makes use of mechanical drilling.

At the moment, there is a reserve of up to 140,000 million tons of economically exploitable petroleum referred to as proven reserves. About 3,400 million tons of crude oil is processed annually in refineries to high-quality finished products. Petroleum has played a very important role in the economy of many countries.

In the two chapters (chapters 3 and 4) in part II of this book, the reader will be taken on a trip from the west to the east involving some crude oil producing countries. The influence of the crude oil industry and petroleum organizations such as the Organization of Petroleum Exporting Countries (OPEC) is shown in these chapters.

3.1.1 Short Background on OPEC

Before the reader starts on the promised trip, a short history of the most important organization for every crude oil producer is given. This organization is OPEC.

All the countries that control the world petroleum market can be broadly classified as the west cartel and the east cartel. The east organization is called OPEC, acronym for Organization of Petroleum Exporting Countries. This organization controls up to 61% of the world petroleum export. OPEC is a multinational union of crude oil extracting countries. The aim of this organization is to coordinate the oil policy on the world market as well as control petroleum deliveries. This organization was founded in 1960 and now twelve states belong to OPEC: Algeria, Gabon, Indonesia, Iran, Iraq, Qatar, Kuwait, Libya, Nigeria, Saudi Arabia, the United Arab Emirates and Venezuela. Ecuador joined OPEC in 1973 and withdrew its membership in 1992. The main headquarters of OPEC is now located in Europe in Vienna, Austria.

The work of OPEC is controlled by half-yearly meetings attended by crude oil ministers or finance ministers of OPEC member countries. Since 1988, there has been the establishment of a ministerial supervisory committee. The aim of this committee is to control and develop new guidelines for crude oil extraction strategy. An economic commission checks the price strategy. Since 1994, a new organ of the OPEC general secretariat leads various research projects, and is also responsible for the legal and administrative questions inside the organization.

However, it can be said that the reason for the founding of OPEC was to control the huge crude oil supply in comparison to the demand towards the end of the 1950s. Oil prices increased as a result of this control of the difference between supply and demand. Consequently, the international currency (money) paid by international oil companies (for purchasing crude oil) to oil producing countries increased. Thus, the Organization of Petroleum Exporting Countries appears to have been founded to correct this payment imbalance. From 1973 to 1974, OPEC obtained almost a quadruple increase for the international selling price for crude oil at almost twelve US Dollars per barrel. A consequence of this price increase was the first world oil crisis. In 1979 and 1980, OPEC members started a second round of price increases, which raised oil prices to over 30 US Dollars per barrel, which led to a rapid rise in inflation in industrial nations. And thus, the second world oil crisis came into existence.

The following Table 3.1 compares crude oil extraction levels in OPEC countries in 1999. From the Table, it can be seen that Saudi Arabia is the largest crude oil producer among the OPEC countries. It must be emphasized that Saudi Arabia also extracts more crude oil than any other country in the world.

Table 3.1: Crude oil extraction in OPEC countries (1999).

Country	Oil extraction [Million tons per annum]
Nigeria	98.1
Kuwait	97.9
Indonesia	70.5
Libya	66.4
Qatar	35.5
Algeria	56.0
United Arab Emirates	109.6
Iraq	132.5
Venezuela	154.8
Iran	180.4
Saudi Arabia	413.4

3.2 NORTH AMERICA

3.2.1 United States [1-2]

The history of the petroleum industry in North America, in general, and the United States, in particular, can be considered to date back to the year 1846 when a Canadian archeologist, A. Hesner, developed a petroleum distillation process for petroleum rich porous minerals. Kerosene was the main product in this process. Fortunately, kerosene was the main fuel for lighting and was considered to be better than light oil. As a result, this simple type of distillation process was widely used. About thirty-four companies were already using the process by the end of the 1850s. By this time also, the USA was already producing kerosene at the rate of 8 million barrels per year.

In 1859, the former railway conductor E. Drake drilled the first oil well for petroleum extraction. This was the first known oil well in the world that used mechanical drilling. The method was so simple and effective that, shortly after it was introduced, it was employed by many companies. The major reason was that it could extract a large amount of petroleum in a less expensive manner.

By the 1860s, there was a rapid growth in the number of petroleum companies and oil refineries. The main product of these pioneer oil refineries was kerosene. In the little town of Cleveland, fifteen such refineries were in business. In 1865, the young John D. Rockefeller bought one such refinery. This was the beginning of the Rockefeller family and the oil company known as Standard Oil.

In 1873, Standard Oil became the biggest oil enterprise in the USA. In the 1880s, Standard Oil began to expand outside of the USA. In 1885, a major part of the company business (70%) was outside of the USA. In 1895, this company tried to share the world petroleum market with the Russian government, but Russia did not accept this offer. After the strong economic crisis in Russia, and the First World War as well as the Russian Revolution, the national economy of Russia crumbled. At this time, Standard Oil was the world's biggest petroleum enterprise. Practically, Standard Oil was a world monopoly. In 1911, the U.S. Supreme Court ordered the break-up of the Standard Oil Trust, resulting in the spin-off of 34 companies. It was from this break-up of the world's biggest petroleum monopoly that Exxon, Mobil, Texaco, SOCo (Amoco) and Soccal (Chevron) came into existence.

The next stage in the development of the petroleum industry in the USA began after the First World War. At this time, high quality gasoline became the main product of the processing of petroleum. This required the development of new technologies for oil processing and complete reorganization of the structure of refineries. The policy of the state was directed against monopolies and the government began to take steps to put petroleum business under state control. Federal taxes for fuels and other petroleum products increased. In 1928, the biggest petroleum enterprises in the USA such as Exxon, Mobil, Chevron, Texaco, Gulf, the British company British Petroleum, and British-Hollandaise Shell formed an international organization called Seven Sisters. This organization played a very important role in the development of petroleum extraction (used interchangeably with production) in the Middle East regions and in transportation of this petroleum to the United States.

The extensive development of the automobile industry in the 1930s helped to sustain the level of petroleum processing and market in the years of the great crisis. In 1933, the government, in conjunction with petroleum enterprises, tried to stabilize the existing prices of fuels and petroleum products. It is important to note that in the 1930s, the American petroleum companies, helped by the development of many petroleum fields in the Arabian East, aided further development of the Seven Sisters.

During the Second World War, the state control of the petroleum industry was much stronger. The following operations were centralized and controlled by the state: petroleum transfer from the Arabian East, petroleum products transfer in the land, and fixing the price for fuels and petroleum products.

After the Second World War, the world saw five petroleum crises. However, these had little influence on the American petroleum industry because of the good relationship that existed between the government and the petroleum companies. Because of this good relationship, the petroleum industry had the opportunity to extract cheap oil from the Arabian East. This explains why prices for petroleum products were constant in the time between 1947 and 1967. There was an increase in petroleum import during this period because the petroleum imported from the Arabian East was 15-20 times cheaper than domestic petroleum.

The period of cheap petroleum ended in the 1970s, at which time an organization known as Organization of the Petroleum Exporting Countries (OPEC) was formed. Member countries were the countries of the Arabian East, Nigeria, Algeria, Ecuador and Venezuela. All these countries stopped the selling of cheap petroleum to the United States and the Seven Sisters. This was the start of the increase in price for petroleum products in the USA and the beginning of the petroleum crisis. The export from the Arabian East was still at a very high level because, despite the relatively high prices of petroleum, the oil exported from OPEC countries was cheaper than domestic petroleum.

The introduction of new ecological laws became the basis for the second increase of petroleum prices. This required the development of new technologies for the production of fuels with a higher quality as well as the reorganization of the structure of petroleum refineries. The state price control played a positive role during the period of price increase such that during the 1980s, the prices were more or less stabilized. By 1986, petroleum prices were as low as 10 US$ per barrel. In the 1990s, there was a greater stability in the price of petroleum in the market. The price changed by only US$ 3-4 per barrel.

Nowadays, petroleum and natural gas are the biggest source of energy. Together, they supply 65 percent of the energy used in the USA. Figure 3.1 shows the breakdown of power sources as given by a U.S. Energy Agency.

About half of the oil consumed in the USA is produced in the United States. The rest is imported.

The United States has 3,013 million tons of proven petroleum reserves.

At the end of the twentieth century, petroleum reserves had been declining in the USA at an average of 2 percent per year. However, the fact that older reserves were added to these estimates and more negative revisions made the decline much more severe in 1998. If drilling resumes, it is expected that the 2 percent declining trend will be reestablished in the future. As the reserve base depletes through production, the price of exploration rises through the increased cost of deeper drilling.

At the present time, the following four areas account for 79 percent of U.S. crude oil proven reserves:

- Alaska 24%
- Texas 23%
- California 18%
- Gulf of Mexico Federal Offshore 13%

Of these four areas, California increased its reserves in 1998, while Alaska, the Gulf of Mexico, and Texas all had decreases in crude oil proven reserves.

The year 1999 saw a rapid recovery in oil prices, offsetting the weakness of the previous year. The annual average price of petroleum rose by up to 39% in 1998.

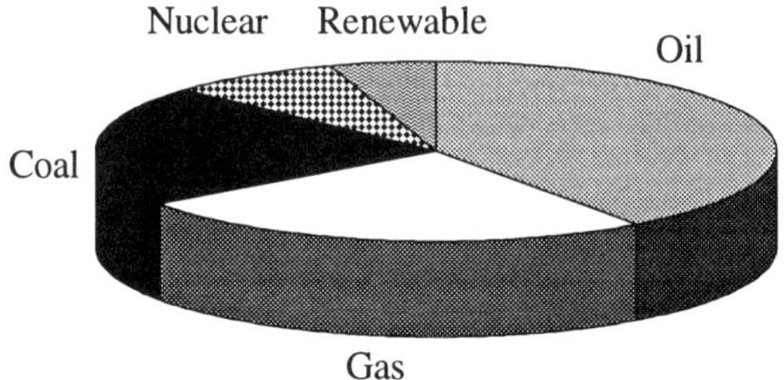

Fig. 3.1: The breakdown of power sources.

3.2.2 Canada [3]

Canada is another petroleum producer in North America. In comparison with other petroleum producers, Canada is considered to be the biggest non-conventional resource base (consisting of oil sands deposits in northern Alberta) in the world.

These non-conventional resources can only be treated by secondary processes, which include thermal, catalytic cracking, and/or visbreaking. Thus, different technologies and secondary production processes are used for the treatment of non-conventional resources. Consequently, Canada has a reputation as a pioneer country in non-conventional oil production. The Canadian industry has been very successful at reducing the cost of developing these resources. The production from oil sands is almost a major component in crude oil production in Canada.

Canada is the world's third largest producer of natural gas and eleventh largest producer of crude oil. Canada produces much more petroleum and natural gas than is consumed in the domestic market. Consequently, petroleum and natural gas export plays a very important role in the Canadian economy. Another factor that influenced the activity in the Canadian petroleum industry was the pipeline expansion in the year 2000. This opened up new markets. A major fraction of petroleum products are exported to markets in the United States.

3.3 RUSSIA [4-6]

In Russia, the first set of oil wells was drilled in Kuban in 1864. In 1866, one of the wells produced a petroleum fountain with a volume of more than 1,300 barrels per day. Then, the extraction of petroleum was conducted by the main monopolies that were dependent on foreign capital. There was not much mechanization of the oil extraction process. Therefore, a lot more wells had to be produced to maximize the income. In the beginning of the 20th century, Russia occupied first place in the production of petroleum.

The main areas of oil extraction in the first years of the Soviet regime in Russia were Baku and Northern Caucasus. However, the oil wells of these old areas failed to satisfy the needs of the developing industry. Construction started in Bashkiria and resulted in the creation of the large Volgo-Ural petroleum area. The new oil wells in Central Asia in Kazakhstan were discovered for which the production of petroleum reached approximately 210 million barrels per day. The Second World War (1939 – 1945) brought serious damage to the areas of Northern Caucasus. This essentially reduced the volume of petroleum extraction. However, the period after the war brought a parallel restoration of the oil-extraction complexes of Groznyj and Maikop, marking the development of the largest reserves in the Volgo-Ural petroleum area. And by 1960, the reserves had already produced approximately 71% of the petroleum extracted in the country. Fifty years after the Second World War, about 270 million barrels has been extracted. In the 60th year, this value is expected to increase up to 990 million barrels. In 1974 on the other hand, the unique West-Siberian oil fields attained a leading position, and even overtook the level of petroleum extraction in Tataria.

3.3.1 The Role of the Petroleum Industry for Russia

Before the political reorganization, the so-called "Perestroyka", petroleum and gas were the backbone of the Soviet Union economy. The cheap power provided by oil and gas prevented the structural reorganization of the energy dependent industry in USSR. Petroleum and gas also united the countries of the eastern block. Foreign currency from the export of gas and oil was used to provide the consumer market with imported goods.

Much has changed since then. The internal state structure has been radically changed. The process of reorganization in the Russian administration has reached an advanced stage. Meanwhile, oil and gas still remain the major source of foreign currency for the country.

The oil and gas industry significantly strengthened the national economy even during the years of reforms. After its initial disorder, the oil and gas industry attained independence. As regards the Russian economy, the oil and gas industry sector was damaged to a much smaller degree by the recession than the manufacturing and other industry sectors.

It is important to note that a majority of the industries in the processing sector are unprofitable even though they consume power in excess of that prescribed world wide during the fuel and energy crisis in 1970s. In the scenario of manufacture decline, social problems and unemployment, the stable and export-oriented oil and gas industry became the much-needed backbone for the economy of Russia. Even now, the processing sector is still in deep crisis.

Thus, the oil and gas sector is the forte of Russia. The oil industry of the Russian Federation is closely connected to all sectors of the national economy and has a huge importance for the Russian economy. The demand for petroleum and gas is rather stable, though it is subject to crisis and decrease in price. Thus, in the face of tax increase such as prescribed in Russia, this export-oriented operation can be brought to liquidation. For these reasons, practically all advanced states of the world and Russia, in particular, are interested in the successful development of the Russian oil and gas industry.

3.3.2 Reforms in the Russian Oil Industry

As is well known, the more economically developed the state is, the more advanced will its scientific and technological base be. For this state, most of the imported goods will essentially be raw material. On the other hand, export goods will comprise of expensive finished products and technological know how.

The construction of a similar system in Russia was the underlying basis for Perestroyka. The structural reorganization of the Russian economy at the beginning of the reforms were as follows: (1) to provide modern and high technological economy, (2) to develop a political economy where ownership of property is possible, and (3) to make demonopolization and other socio-economic transformations. The economic reforms were also to provide liberalization of foreign trade activities, as well as introduce modern energy-saving technologies. In other words, the task was to restructure the national economy, in the shortest time, to a strong industrial base that is founded on modern technologies. As such, the economy would be competitive.

Unfortunately, the model of Russian economy got oriented towards the export of raw materials initially during the reforms, contrary to original intentions. This was so partly because the processing sector was basically noncompetitive and was in a very difficult economic situation. Consequently, the former central uniform economic complex in Russia broke up into separate industrial and territorial-industrial corporations (including a fuel and energy complex) during the reforms. This exclusive old monopoly was very non-uniform by its territorial-sector structure. There was constant struggle for privileges and sources of state financing between the enterprises and associations. Simultaneously, the new monopolies and commercial groupings closely connected to external market were promptly formed during the years of reforms in the country. These were the commercial structures that had interest in the reformed Russian economy.

As was noted previously, the petroleum industry was stable even under crisis conditions in Russia. However, problems existed in the oil industry as well. These problems perhaps provided the challenges and opportunities for future developments. The petroleum and gas industry is represented in the gas sector by Gazprom, and in the petroleum sector by such petroleum corporations as LUKOIL, YUKOS, SIDANKO, ROSNEFT, etc. These companies were formed during the structural reorganization of the oil and gas industries. This reorganization had the following basic stages:

1. 1987-1990: The development of self-financing enterprises in the oil and gas sectors, the expansion of economic independence for the enterprises, and being direct exporters of power sources to the world markets.
2. 1991-1994: The associations of the enterprises as vertical integrated structures, the beginning of privatization of holding petroleum companies, the creation of an infrastructure for the shares market, coming into existence of financial and industrial groups and consortia for participation in international projects.
3. Since 1995: The concentration of economic authority in vertically integrated companies, the transformation of companies from state controlled enterprises into private companies by means of the mechanism of money auctions, reorganization of the capital of the companies.

In the first stage (i.e., by the end of the 1980s), Russia introduced some elements of market driven economy. A first step became the transition of the enterprises to being self-financing. State ownership of the manufacturing sector was kept. However, in the long term, it was expected to transform into private or collective enterprises over the transitive period. The first stage of the structural reorganization of the national economy could not eliminate crisis. Instead, it was aggravated. There was a strategic miscalculation in the first stage of reform. The fact that privatization could lead to a high degree of natural monopolization of the oil industry was not considered. The splitting of the petroleum industry into separate enterprises, which actually occurred after easing the state control, aggravated the crisis in the oil industry.

In 1992, during the second stage of the accepted Russian program for privatization, a mechanism for easing State control for the petroleum sector was worked out. The essence was that the control package (the shares) of the enterprises remained the property of the state for a fixed term. Privatization of the sector proceeded in two stages. At first, all the enterprises entering into the system known as "Rosneftgas" and "Glavnefteprodukt" were transformed to open joint-stock companies. After that, the creation of the basis for the joint-stock vertically integrated holding companies began. The state enterprise, Rosneft, was created for commercial management of the state shares of the enterprises that were not included in one company on the basis of Rosneftgas. For the transport enterprises holding companies, "Transneft" and "Transnefteprodukt" were created. This distinctive feature of the second stage of the oil and gas industry reforms led to the escalation of the structural crisis. This was also promoted by the macro-economical preconditions. With liberalization of the prices in the beginning of 1995, inflation was no longer hidden, and the rates of inflation exceeded 30-40% per month. It affected the strong growth performance of petroleum and gas export, and simultaneously there was recession of internal consumption.

The main characteristic of the third stage of the reforms in the oil and gas industry in Russia was the new redistribution of property rights to the active enterprises and companies based mostly on political considerations. The second stage of privatization of enterprises began after 1995. This was the key moment for the realization of auctions pledged. The state gave the opportunity for commercial banks to participate in crediting the federal programs by giving share packages belonging to the state as pledges. The pledging auction represented the transition of control share packages of the companies to banks with the prospect of their repayment. By the end of 1996, the banks had received the right of repayment for the shares. This created the prospect for transformation of all petroleum companies into private enterprises.

Two opposing opinions have emerged as to further organizational development of the gas industry. One of them proposed monopoly by the Russian power since this would provide the stability needed by the gas industry even under today's adverse conditions. The other prefers splitting into enterprises such as in Gazprom and similar to the existing enterprises in the petroleum industry. The former opinion is based on the thought that splitting the gas industry will weaken its competitiveness in the world market.

3.3.3 Russian Petroleum and Gas in the World Market

Russia is not actively involved in formulating the world energy policy. However, the slightest socio-economic and political instability in Moscow or Tyumen is reflected in the petroleum price in the stock exchanges in New York or in London.

Up till now, petroleum policy was defined by two cartels - west and east. The first unites the six largest petroleum companies that extracts 40 percent of oil (by volume) from countries that are not OPEC members. The cumulative volume of sales for these companies in 1991 was almost 400 billion dollars. The eastern cartel (OPEC) includes thirteen countries, which makes up 38 percent of world oil production and 61 percent of world petroleum export. The oil production in Russia makes up 10% of world petroleum production. Therefore, it is possible to say with confidence that Russia occupies a strong position in the international petroleum market. For example, OPEC declared before the crisis that the member states of this organization could not fill any shortage of petroleum, should the world market lose Russia.

It may not be possible for petroleum to be completely replaced by alternative power sources in the foreseeable future. According to forecasts, the world demand will grow at the rate of 1.5 percent per year, and the supply will essentially not increase, unless Iraq gets a new quota to increase the volume of petroleum for export. Before the energy crisis in 1973, the world's oil extraction was practically doubled every ten years during the past 70 years. But now, only four member countries of OPEC (which hold 66% of the world oil reserves) can increase their volume of oil extraction. These are Saudi Arabia, Kuwait, Nigeria and Iraq.

There was recently a very difficult situation in the world petroleum market. The prices for petroleum had fallen to less than 10 dollars per barrel for the first time in a long time. OPEC countries wanted to see an increase of even up to 30 dollars per barrel. With this in mind, it became necessary to reduce the export of petroleum to the world market, even though each country did not want to do so at the expense of its overall export.

The military actions of the USA and Great Britain in December 16-18 of 1998 on Iraq also had positive consequences on Russia. As a result of the aerial bombardment, the Iraq factory for petroleum processing as well as ports for petroleum export were destroyed. It is necessary to note that, although Iraq produces only about 3% of the world's petroleum due to sanctions, this petroleum influenced world petroleum prices. Iraqi petroleum is of very high quality but attracts a very low price. As a result, prices increased sharply on the stock exchanges in London and New York following the incident. Experts forecasted that the price would reach 30 dollars per barrel. All this could have a positive effect for the economy of Russia.

Irrespective of the situation regarding energy sources in the world market, petroleum and gas always will play an important role in the home market because they are the most valuable raw material for the petrochemical industry. More than 2000 kinds of products are made from this raw material. Russia is interested that domestic petrochemical synthesis becomes an independent and powerful sector of industrial manufacture, and also becomes competitive in the world market. It is also interested that the country exports not only crude oil, but also much more value-added products of petroleum processing. This will bring appreciable income into the country.

3.3.4 Structure of the Petroleum Sector in Russia

The main problems of the petroleum sector in post Perestrojka is that of the reduced proven oil fields and rising cost of development for the available oil fields. This problem exists because of insufficient financing of geological prospecting, which implies that exploration of new oil fields has been partially suspended. According to expert forecasts, geological prospecting can potentially increase production for the Russian Federation from 5 billion up to 7.5 billion barrels per year (about 2 billion barrels was extracted in 1997). In 1992, Russia occupied second place in the world for reconnoitered oil wells (after Saudi Arabia on which territory a third of the world oil reserves is concentrated). The oil reserve in Russia in 1995 was 140 billion barrels of petroleum.

It should be noted that the degree of conformability of foreseen oil fields is very low and there are still a large share of oil wells with high costs of development (of all the oil fields, only 55% have high efficiency).

Even in Western Siberia where the most gain of reserves is expected, about 40% of this gain will not be efficient oil fields with extraction volume from new oil wells less than 70 barrels per day. This is the limit of profitability for this region.

Therefore, the present condition of the petroleum industry in Russia is characterized by volume reduction of industrial petroleum reserves, a decrease of both quality and rates of processing, reduction in prospecting and operational drilling, and an increase of quantity of idle oil wells. It is also characterized by the absence of significantly large reserves; need to be involved in industrial operation of oil wells which are located in difficult areas; increasing technical and technological difficulties by the sector; insufficient attention to the requirements of social development and ecology.

Experts have given reasons for the difficult economic situation in the Russian petroleum sector:

- Production in the biggest oil fields that are state funded and are the main components of the resource base in Russia are substantially completed.
- New developed oil fields are worsened by oil conditions. No large or productive oil fields has recently been opened.
- The financing of exploration has been reduced. So, in Western Siberia, where the degree of development for foreseen resources was about 35 percent, the financing of geological works, which began in 1989, was reduced to 30 percent. The volumes of prospecting drilling have decreased as well.
- There is not enough highly-efficient technology and equipment for production and drilling. The main part of this production has deteriorated by more than 50 percent. Only 14 percent of machines and equipment reach world standards. With the disintegration of the USSR, the situation has worsened with deliveries of drilling equipment from the countries of CIS.
- The low internal prices for petroleum do not provide self-financing for the oil-extracting enterprises. This situation has persisted because after a series of increases, there are declines in the prices for petroleum in the world and domestic markets. As a result, there has been a serious deterioration of technical and financial maintenance of the sector.
- The shortage of effective and ecological equipment creates pollution problems in the oil sector. Significant material and financial resources have been spent on solving these problems. Unfortunately, these do not contribute directly to the increase in the production of petroleum.
- As the oil and gas market is controlled by the State private companies, there is presently no central oil and gas agency to broker deals between domestic and international organizations as well as private individuals.
- Debts of republics for the supplied petroleum.

The decline of the oil-extracting industry is caused by complex and interconnected reasons. The exit from the present situation is complicated by the global nature of other problems. Therefore, if the economic crisis in the country persists and the process of political reform in the former Soviet Union does not proceed smoothly, the production of petroleum will be reduced.

In the petroleum sector as well as the gas industry, the problem of lack of foreign investments is very serious. It is necessary to note that the flow of investments in Russia was never large during the years of reforms, in comparison with China and other former socialist countries. Investment is necessary for the energy sector of the economy, though this money and this amount of the joint projects is obviously not sufficient to extricate the sector from crisis. The reasons for the small cash flow are related to both economic and political situations in Russia as well as the absence of a leader among the oil companies in the petroleum sector.

On the other hand, Russian companies have begun to carry out projects in other countries (Prikaspiy, Kazakhstan, Iraq, etc.). In fact, these projects have continued despite the last global crisis in Russia. This can only illustrate the power of the oil sector.

There are other methods to increase petroleum processing and stabilize the petroleum industry. Two methods can be facilitated by the state: these are the increase of prices for petroleum in the domestic market and an improved taxation system for oil-extracting enterprises. Enterprises have obligations as well. These are in technical equipment, extraction of petroleum from oil wells with small petroleum reserves (it is possible only with an increase in prices for petroleum in the domestic and/or world markets), increase in the level of oil processing and sale of produced petroleum and petroleum products.

The Tyumen area is the main oil-extracting region of Russia. Today almost 80 percent of extraction in this area is performed by six enterprises (Yuganskneftegas, Surgutneftegas, Nizhnevartovsneftegas, Noyaborskneftegas, Kogalymneftegas and Langepasneftegas). But in the near future, the absolute volumes of production will be reduced. According to expert forecasts, reduction in Nizhnevartovsk will be to 60%, and in Yugansk to 45%. Presently, both the Russian policy and economy are determined practically by complicated interactions between various independent oil enterprises. There is no recognized leader among them, and there is also presently no competition. Such a disintegration creates many problems. Integration has been postponed until the future because of (i) the large dynamics of the sector, (ii) the decrease of production for one enterprise leads to the increase of production for another enterprise, and (iii) strong competition by LUKOIL, YUKOS and SIBNEFT for influence in the region between.

3.4 ARABIAN EAST [7-9]

Modern Arabian East includes thirteen Arabian countries and the state of Israel. Despite its modest population, the Arabian East occupies a rather prominent position in world culture, politics and economics. The Arabian East owes this prominence to two factors: first, a unique geographical location and second, extreme riches via petroleum. The Arabian East holds the largest petroleum reserves in the world. At least four countries of this sub-region (Saudi Arabia, Iraq, Kuwait and United Arab Emirates) have, in the last few decades, constantly been in the top ten of oil extractors in the world. The eastern regions of the Arabian Persian Gulf constitute the basic crude oil "rich" territory. Opposite this area is the western Mediterranean region that is essentially lacking in petroleum reserves. Extraordinarily high levels of petroleum production, coupled with a small native population in the Persian Gulf is responsible for this level of affluence in these countries.

In the 1980s, the developed countries of the West developed complex measures to prevent a relapse of the increase in the prices of petroleum similar to the level that occurred in 1973, as was stated earlier. The price of petroleum had begun to reduce after the crisis. There was now a strong interest in alternative energy sources. This was a clear indication to exporters of petroleum that the prominence or value attached to petroleum can decline. Thus, even with the exploration of new reserves that added to the very large reserves of petroleum in the region, the possibility of alternative energy sources meant that the value of those reserves could decline. First, Bahrain, that had the least reserves of petroleum among the countries of this region, had once seen the possibility of exhausting their petroleum reserves. However, new reserves are being discovered all the time since the exploration of new oil fields is a continuous process.

Petroleum exporters are usually interested in raw material, spare parts, equipment, furnishings and especially of technologies delivered by transnational corporations. The form of business cooperation between transnational corporations on the one hand and the locals on the other hand is typically tuned towards a mixed enterprise with more than 50% participation of local capital. As well, there are also "non-joint"-stock forms of activity involving license agreements, training of personnel, management and advertising, contracts on a "turn-key basis", and engineering.

The developed countries that benefit from export of petroleum from the Arabian East have definitely succeeded in their investments. The transnational corporations and large international financial organizations (for example, the International Monetary Fund) start from a simple investment of < 1 percent, to more purposeful investments by proposing personal participation in management and forcing acceptance of their decisions. The volume of foreign investments in Saudi Arabia is estimated at hundreds of billions of dollars, and most of this ends up in the USA. Kuwait owns a part of many British, Canadian, American, Western European and Japanese oil companies, including British Petroleum, General Motors, IBM, Kodak, Total, Sony, etc. Besides Kuwait's three refineries, it also owns three European refineries located in Denmark, Netherlands and Italy. It also owns a thousand tank stations in Denmark, Sweden, Great Britain, Italy and Norway as well as ten air-refueling stations in large Western European airports. Kuwait has 25 of its own oil tankers. Kuwait today owns a marketing network (involving the full cycle from production until sale to the ultimate user) for a significant part of produced petroleum. Foreign investments in Kuwait by the developed western countries achieve profits of not less than 100 billion dollars. These profits from foreign capital investments have become an essential source of foreign currency in Kuwait.

3.4.1 Oman

Oman began oil extraction and export much later than the other countries of this region. Therefore, although Oman is well endowed with petroleum reserves (400 tons of petroleum per citizen), it remains a relatively not highly developed country. Modern infrastructure is yet to fully take shape in Oman.

3.4.2 Iraq

The process of nationalization of the petroleum industry was completed in 1975. Today, Iraq provides approximately 3% of the world's petroleum supplies. Analysts attribute the high petroleum activity and capacity to the availability of an oil-extracting complex in Iraq. More than half the number of oil fields in the world are concentrated in the Persian Gulf. A considerable fraction of this is located in Iraq.

Iraq occupies second place (after Saudi Arabia) in the world in terms of petroleum and gas reserves. It has a proven reserve of about 112.5 billion barrels. However, in view of the UN sanctions, the rate of production of petroleum has been low. For example, in the period from January - August 1998, the production rate was approximately 2 million barrels per day. In the absence of UN sanctions, Iraq projects that in the first year the level of production would be 3 million barrels per day. After 3-5 years, the production rate would increase to 3.5 million barrels per day, and to 6 million barrels per day in 10 years. If it uses the full pipeline loading capacity, Iraq is capable of exporting 1.4-2.4 million barrels per day. This will be made up of 0.8-1.6 million barrels per day over the pipeline at Circuc-Jeheyn and 0.6-0.8 million barrels per day over port of Mine Al-Bacr).

Iraq nationalized the mineral industry in 1972 by putting the sector under the control of the ministry responsible for the petroleum industry and the Iraq State Petroleum Company. The Iraq nationalization was completed by confiscation of the Iraqi petroleum participating company, co-owners of which were British Petroleum, Total, Shell, Exxon, Mobil, and Partex. Before the Iran-Iraq war, Petrobras and Elf Aquitaine also operated in Iraq under contracts alongside with the Iraqi Petroleum Company. While these companies were operating in Iraq, many of the largest oil fields, such as the ones in Kircooc (1927), Rumeyla (1953), Buzurgan (1969), Abu Jirab (1971), Megun (1976) and Nahr Umar (1977) were opened.

The main company now in Iraq is the Iraq State Petroleum Company. The following companies are working independently but are still subordinates to the State Petroleum Company:

- The State Company for Oil Projects (SCOP) is responsible for work connected to the development of upstream and downstream projects.
- The Oil Exploration Company (OEC) is responsible for prospecting and geophysical works.
- The State Organization for Oil Marketing (SOMO) is engaged in the trade of petroleum and is responsible for connections with organization of the countries to which they export Iraqi oil.
- The Iraqi Oil Tankers Company (IOTC) - transport tanker company.
- Northern Oil Company (NOC).
- Southern Oil Company (SOC).

The last two companies are engaged in extracting petroleum in the northern and southern parts of Iraq, respectively.

Finally, it has to be noted that as a result of the use of modern technologies (horizontal and multilateral drilling) for petroleum extraction from Iraq oil wells, the estimated proven petroleum reserves in Iraq are sure to increase in the near future. It is also necessary to emphasize that a major part of the research and development work in Iraq is still in progress. There is the intention to research on the deep oil wells on Jurassic and Triassic levels (mainly in the Western Desert). A successful outcome can open up additional oil reserves. However, research in this area has not been carried out up till now.

3.4.3 Iran

The proven petroleum reserve estimates in Iran are 90 billion barrels (not less than 9% of all world reserves). In 1998, 3.6 million barrels of petroleum were extracted daily from Iranian oil fields. Daily consumption of petroleum is 1.13 million barrels. Capacity of petroleum processing is 1.45 barrels per day. The importers of Iranian oil are essentially Japan, South Korea, Great Britain, China, Turkey, Thailand, India, Brazil and Taiwan. The petroleum and gas industry of Iran is under the complete control of the state. The State Petroleum Company is called National Iranian Oil Company NIOC. It conducts research on development of both petroleum and gas wells. It is also engaged in providing transportation for both the raw petroleum and its products. The National Iranian Gas Company (NIGC) is engaged in extracting, processing, transporting and the export of gas, whereas the National Petrochemical Company (NPC) is responsible for petrochemical manufacture in Iran. The most active foreign companies in Iran today are Gazprom, Petronas, Shell and Total. The main oil fields in Iran are Gagaran, Marun, Avaz, Bangistan, Aga Gari, Raga-i-Safid, Pars and Hakim. The main oil refining factories are Abadan (capacity 477,000 barrels per day), Isfagan (251,000 barrels per day), Bandar Abas (220,400 barrels per day), Teheran (213,750 barrels per day), Arak (142,500 barrels per day), Tebriz (106,400 barrels per day), Shiraz (38,000 barrels per day), Kermanshah (28,500 barrels per day) and Lavan (20,000 barrels per day).

In August of 1996, the American Congress approved a law that was aimed at imposing sanctions against Iran and Libya (so-called ILSA). The provisions of the law sought to ban non-American companies investing not less than 40 million dollars annually from performing the development of petroleum and gas wells in Iran. In 1997, the investment limit was lowered to 20 million dollars. ILSA is not the first sanction imposed by the USA against Iran. For example, in the beginning of 1995, President Clinton signed two governmental orders, according to which American companies and their subsidiaries are not permitted to finance the development of any petroleum sector in Iran. One of the large-scale contracts cancelled as a result of these orders was the contract between Iran and the American company Conoco for the development of blocks A and E of the Sirri oil field. The project was estimated to be 550 millions dollars. French Total and Petronas were quickly engaged to replace Conoco. In the middle of August of 1997, President Clinton signed one more governmental order (Nr. 13059), which stipulated interdiction for any investment actions of American citizens in Iran. Despite ILSA, a consortium comprising Total, Petronas and Gazprom has continued with the project of developing gas wells called the Southern Pars (volume of investments was 2 billion dollars).

From a purely objective reasoning in May of 1998, the White House made it clear that the ILSA sanctions were not applicable to oil and gas pipelines that were only connecting through the territory of Iran. This move probably resulted from the construction of major pipelines and gas pipelines from Kazakhstan, Azerbaijan and Turkmenestan. It has to be noted that Iran occupies a strategic geopolitical location for connection of petroleum transportation routes. This unique location allows a considerable decrease in transportation price for transfer of raw material to the world markets through connecting pipelines in Iran in comparison with other routes (for example, through Turkey).

In December of 1998, the President of Iran declared that the main thrust of government in the petroleum sector was the re-structuring and modernization of the petroleum industry, and the opening up of new oil fields. In January of 1999, the parliament of Iran ordered the Ministry of Petroleum of Iran to report monthly concerning work that was carried out. In turn, the National Iranian Petroleum Company (NIOC) concentrated efforts on prospecting work. Accordingly, NIOC's plan was to drill 61 prospecting oil wells in the sea and on land by 2000.

By the year 2000, Iran was able to offer about 20 projects in gas and oil sector to investors. These included measures for development of sea deposits and for finishing the modernization of a number of refineries (for example, on an island Lavan). In March of 1998, Bow Valley Energy (Canada) and British Premier Oil signed a contract for 270 millions dollars for providing development of a sea oil field called Balal valued at an estimated 80 millions barrels of petroleum. But at the end of 1998, Bow Valley Energy left the project, with the explanation of financial problems resulting from the Asian crisis. In February of 1999, the government of Iran gave the rights for the development of the sea oil field at Dorud and gas field near Harg Islands to a French company Elf Aquitaine and an Italian enterprise ENI. In the opinion of the Iranian party, the prospecting of Dorunda (the project is estimated to be half a billion dollars) will increase their proven petroleum reserves to 700 million barrels and an increase in the daily production of petroleum from 90 to 220 thousand barrels.

In January of 1999, British Petroleum and Amoco began negotiation with the Iranian government about the development of Avaz oil field. The same oil field also attracted the French enterprise, Total. At the same time the Norwegian company Saga Petroleum initiated negotiations with the Iranian government concerning development of Dehl Uran and Cheshmen-Kosh oil fields. Saga Petroleum has signed an agreement with NIOC for 2.7 millions dollars for seismic data in a number of prospective places, including the block of Dara and the sea oil well Handidgan. In December of 1998, Gazprom and NIOC - Naftgaran Engineering Services Co. created a joint venture aimed at carrying out prospecting and drilling work in Iran.

3.4.4 Qatar

The main foreign trade partners for Qatar are Japan, the USA, Great Britain, Germany, France and Italy. Oil export accounts for 80% of total export.

The proven petroleum reserve estimate for Qatar is 3.7 billions barrels. More than 650,000 barrels of petroleum is extracted in Qatar daily. The country exports more than 600,000 barrels of petroleum per day. Approximately 70% of petroleum in Qatar is exported to Japan and less than 10% to other countries of Southeast Asia. The capacity for oil processing in Qatar is 57.5 thousand barrels of petroleum per day. The petroleum sector of Qatar is under the total control of the state. The state petroleum company of Qatar - Qatar General Petroleum Corporation (QGPC) - is engaged in prospecting work and the production of petroleum products. The National Oil Distribution Company (NODCO) carries out the processing of petroleum whereas Qatar Petrochemical Company (QAPCO) is engaged in petrochemical manufacture. Qatar Fertilizer Company (QAFCO) produces fertilizers. On the other hand, Qatar Liquefied Gas Company (Qatargas) and Ras Laffan LNG Company supervise manufacture and marketing of liquefied natural gas (LNG).

The "overland" oil field of Qatar is in Duhan. About 2.2 billion barrels of petroleum is concentrated there. It should be remembered that the proven petroleum reserve of Qatar is about 3.7 billion barrels. More than 1.5 billion barrels of petroleum are concentrated in six shelf oil fields (Bul Hanan, Meydan Mahzam, Id Al-Shargi, Al-Shahin, Al-Rayan and Al-Halig). Practically all extracted petroleum from Qatar is exported to the countries of Southeast Asia.

One more significant oil field - Al-Holig - is located near the sea border between Qatar and Iran. The work on this field began in March of 1997 (it has to be noted that the French company Elf Aquitaine planned to begin the work in 1991). In March of 1998, this French Company extracted petroleum at the rate of up to 30 thousand barrels per day from this field. Contract "production sharing" signed between Elf Aquinaine and Qatar was made for 25 years. Elf Aquitaine has 55% of individual share while 45% share belongs to Italian Agip. The industrial reserves of petroleum in Al-Haliga are estimated at 70-80 millions barrels. Nowadays, Al-Shahin is considered to be one of the most productive oil fields in Qatar. Approximately 100 thousand barrels of petroleum per day are extracted from this field. In 2000, the Danish company was going to finish a project to increase the level of production to 150 thousand barrels per day.

3.4.5 Kuwait

The history of the petroleum industry in Kuwait can be considered to have started in 1934. At this time, sheyh Ahmad Al-Gdaber Al-Salah allocated a place for business collaboration between Anglo-Persian Oil (now British Petroleum) and Gulf Oil Corporation. Drilling began in 1936 and in 1938, the Burgan oil field was opened. But its development only began after the Second World War, and export started from 1946. In 1960, the Kuwait national company was established. The share of the state was 60% at first, and became 100% since 1975. In 1980, Kuwait Petroleum Corporation (KPC) was established. Included as sectors were the Kuwait national company and many others. Kuwait has been a member of OPEC since its creation in 1960. At the time of the Iraq-Kuwait war in August of 1990, the production of petroleum in the country was approximately 100 million tons, from which 10 million tons were extracted from oil wells in shared zone taking place under joint administration of Saudi Arabia and Kuwait. Production in the territory of the country has been conducted on 12 basic deposits.

A major portion of the oil production is exported to more than 30 countries. The main importer of Kuwaiti petroleum is Japan. In Kuwait, there are four oil refineries with a total capacity of 40 million tons per year. In 1989, 38.6 million tons of petroleum were produced, from which six million tons were used inside the country. The rest was exported.

During the Iraq-Kuwait war, 800 or more than half of the oil wells in Kuwait and the shared Zone were destroyed. One-third of the petroleum extracted in the period preceding the war was burned during the war. It is estimated that by the time the last fires on oil wells were extinguished petroleum worth approximately 40 billion dollars was lost. After restoration of the oil wells, there was significant reduction of pressure in the oil wells. As a result, secondary and tertiary methods of oil extraction were required for production. The consequence has been a four-fold increase in the cost of extraction (four dollars per ton before the war). As a result, the capacity for oil processing has decreased to 6.5 million tons per year (more than six times lower).

The proven petroleum reserve estimate in Kuwait is 96.85 billion barrels. This accounts for 10% of world reserves. Natural gas wells are 1500 billion cubic meters. The annual volume of natural gas extraction is 7.8 billion cubic meters. The oil-extracting capacity is 2.35 million barrels per day. In 1998, Kuwait extracted 101 million barrels of petroleum from its wells. It was expected that there would be the potential to bring the extraction to 2.5 million barrels per day at the end of 2000 and up to three million barrels per day by 2005. Two thirds of the extracted petroleum is exported, 20% is exported to the USA, 50% to Southeast Asian and Japan.

The oil and gas industry of Kuwait is subordinated to Kuwait Petroleum Corporation (KPC). Its structure includes five companies. (1) Kuwait Oil Corporation (KOC) looks at both the production and export of petroleum and gas. (2) Kuwait National Petroleum Company looks after oil processing and manufacture of liquified natural gas, as well as their marketing and export. (3) Petrochemical Industries Company (PIC) is engaged in the manufacture and export of petrochemical products. (4) Kuwait Oil Tankers (KOTC) engages in the transportation of petroleum, petroleum products and liquefied gas. (5) Kuwait Foreign Petroleum Exploration Company looks at both the extraction of petroleum and gas abroad. In 1998, the Supreme Council of Kuwait accepted the decisions to reorganize the KPC, as well as privatize the PIC and KOTC.

The activity of foreign oil companies is limited by the agreement concerning technical support. KPC has signed agreements with Chevron, British Petroleum, Shell, Exxon and Total. Generally, there are 150 foreign firms, mostly from the USA and England, involved in the petroleum business. About 15 billion dollars of investments is required in order to realize the "ten years program".

Western oil companies insist on production sharing for capital investments and "know-how" in the development of oil wells. The Supreme Council of petroleum accepted cooperation with foreign firms on the conditions of production sharing, but progress in this direction was suspended because of strong opposition from the National Assembly. As a compromise, the oil ministry developed a model of cooperation with foreign companies with agreements only for technical service. According to the arrangement, the foreign company will be completely responsible for an oil well, including the investment, its development, as well as applying the required technology and equipment. All petroleum remains the property of Kuwait, but the foreign company, besides indemnification for operational expenses and deductions for the capital investments, receives a certain percentage from the sale of additional volume of raw material extracted with the help of their know-how. This model of attracting investors to the local market has been referred to as the Kuwaiti project.

In the first stage of the project, the model will be used for the development of five northern and two western oil wells. As a result of the application of advanced western technologies, it is expected that by 2005, the extraction of petroleum will increase as given below. From 400,000 to 900,000 barrels per day in the northern oil wells, 225,000 to 515,000 barrels per day in the Raudatein oil wells, 95,000 to 250,000 barrels per day in Sabrie, 3,000 to 30,000 barrels per day in Bahrahe, and 78,000 to 110,000 barrels per day in Ratge and Abdali. Western oil wells should give an extraction volume of 270,000 barrels per day. The Kuwaiti project also includes the construction of a new export terminal on the island of Bubiyan.

Readiness to cooperate in the Kuwaiti project was expressed by Mobil, British Petroleum, Amoco, Shell, Chevron, Texaco, Conoco, Philips, Arco, Elf, Total and Lasmo. In view of the large number of potential partners, Kuwait will not give the whole field to only one foreign company, but to a foreign enterprise to form a consortium of two or more firms.

One of the strategic directions of Kuwait petroleum policy is the creation of infrastructure for oil processing and marketing of petroleum products abroad. This is achieved by purchasing shares by foreign oil companies that will allow an effective control of the complete cycle of extraction, processing, and marketing of petroleum and petroleum products. This is expected to increase the capacity of oil refineries abroad to 700 thousand barrels per day, for those in Europe up to 300 thousand barrels per day, and in Asia up to 400 thousand barrels per day. The overall capacity of oil refineries owned by Kuwait in Denmark, Netherlands and Italy is 230 thousand barrels per day. Kuwait supervises 6.5 thousand gas stations in Western Europe. By the signing of an agreement with Swedish companies, Kuwait supervises 26% of the oil market in this country. The agreements for the creation of joint refineries with China, Pakistan and Thailand with capacity of 300 thousand barrels per day have already been signed.

Nationalization of the petroleum sectors by the largest producers in OPEC such as in Iraq, Venezuela, Saudi Arabia and Kuwait resulted in the reduction of investments by Kuwait in these countries. The extra funds freed up were spent for the development of new sources of raw material in other regions of the world. According to a Middle East Economic Survey, 350 billion dollars was spent to increase the volume of petroleum extraction in nontraditional regions in 1980-1995.

The application of modern technologies for prospecting, developing and operating petroleum wells has essentially reduced financial expenses for these needs (approximately 40%). This scenario has also resulted in the increase of petroleum export and a decrease in the price of crude oil. The planned increase in the volume of petroleum extraction will be twelve million barrels per day in 2005, from which the part for OPEC is nine million barrels per day, and the part for other countries is about three million barrels per day. After 2005 it is predicted that there will be recession of oil manufacture in the so-called independent exporters States.

3.4.6 United Arab Emirates

The proven petroleum reserve estimate of the United Arabian Emirates is approximately 98 billion barrels (slightly less than 10% of world oil reserves). The biggest part of oil wells is concentrated in Emirate Abu Dabi. The extraction of petroleum in United Arab Emirates exceeds 2.3 million barrels per day. Nowadays, approximately 2.2 million barrels of petroleum is exported. The main importers of petroleum from the United Arab Emirates are Japan (more than 60% of general export) and other countries of Southeast Asia (not less than 20%). The capacity for oil processing in the United Arab Emirates is approximately 287 thousand barrels per day.

The petroleum sector in each Emirate is controlled by the government. The State Petroleum Company of United Arab Emirates is known as the Abu Dhabi National Oil Company (ADNOC). This includes three oil and gas operational companies, five service companies, two transport companies (sea transportation), etc. The main wells are in Abu Dabi-Asab, Beb, Bu Hasa and Al-Zakum.

3.4.7 Saudi Arabia

The major portion of exports from Saudi Arabia is petroleum and petroleum products. The income from petroleum export is about 24 billion dollars (90% of the general income). It should be noted especially that, despite competition from Mexico, Venezuela and Canada, Saudi Arabia remains the main exporter of petroleum to the USA and Japan.

The proven oil reserve estimate for Saudi Arabia is 261.5 billion barrels (almost a quarter of the proven oil reserves in the world). The daily extraction of petroleum in Saudi Arabia exceeds 8 million barrels. Oil refining capacity is 1.6 million barrels per day. The petroleum industry in the country was nationalized in the 1970s. The petroleum sector is operated by the Supreme Petroleum Council and the state companies of Saudi Arabian Oil Co. (Saudi Aramco) and Petrochemical Saudi Basic Industries Corporation (SABIC). The main oil fields are in Gavar, Safaniya, Nazhd, Berry, Manifa, Zuluf, Shaybah, Abu Saafa, Hursaniya and Abgeyg. The main foreign companies operating in Saudi Arabia are the Arabian Oil Company from Japan (AOC), Mobil, Shell and Texaco. However, the Arabian Oil Company has not been operating in Saudi Arabia since the beginning of 2000.

In Saudi Arabia there is approximately a total of 77 oil and gas fields. However, the main oil fields of the country are concentrated in eight areas. One of them is Gavar, the biggest overland Oil field with reserves estimated at 70 billion barrels of petroleum. Another is Safania, the largest shelf oil field in the world, that has reserves estimated at 19 billion barrels of petroleum.

The Japanese company AOC operates on two shelf oil wells, Hafgy and Hut, and extracts about 300 thousand barrels per day. American Texaco developed three oil wells on the ground, Vafra, Southern Favaris and Southern Um Gudayr, and extracts more than 200 thousand barrels per day. Texaco has signed a contract for its operation to last till 2010, and the American company plans to considerably increase the volume of production. In September of 1998, the Minister of Oil and Natural Resources of Saudi Arabia met with heads of a number of American oil companies (Chevron, Mobil, Texaco, Arco, Conoco, Phillips Petroleum). As a result of these negotiations, officials of Saudi Arabia have asked the companies to present their bids for possible joint projects in the gas and petrochemical sectors.

4

International Petroleum Companies

4.1 BRITISH PETROLEUM [9]

British Petroleum is one of the biggest companies in Britain and one of the world's largest oil and petrochemicals groups. At the beginning of its history it was known as the Anglo-Persian Oil Company.

The history of the British Petroleum or the Anglo-Persian Oil Company began with the work of William Knox D'Arcy shortly after the turn of the twentieth century (1901). William Knox D'Arcy obtained a concession from the Shah of Persia to explore the oil resources of the country, excluding the five northern provinces that bordered Russia. Having been granted the concession, D'Arcy employed an engineer, George Reynolds, to undertake the task of exploring for oil in Persia.

Meanwhile, the costs mounted, stretching D'Arcy's resources to the point where he sought outside financial assistance. This came in 1905 from the Burmah Oil Company, which provided new funds for his venture.

More exploration in Persia followed without success until eventually, in May of 1908, Reynolds and his associates struck oil in commercial quantities at Masjid-i-Suleiman in southwest Persia. It was the first commercial oil discovery in the Middle East. This signaled the emergence of that region as an oil producing area.

After the discovery had been made, the Anglo-Persian Oil Company was formed in 1909 to develop the oilfield and exploit the concession. At the time of Anglo-Persian's formation, 97% of its ordinary shares was owned by the Burmah Oil Company. The rest were owned by Lord Strathcona, the company's first chairman.

Although D'Arcy was appointed a director and remained on the board until his death in 1917, he was not to play a major part in the new company's business. His role as the initial risk-taking investor was past and the daunting task of developing the oil discovery into a commercial enterprise shifted to others, amongst whom one particular person stands out: Charles Greenway. Greenway was one of Anglo-Persian's founder-directors, becoming managing director in 1910 and chairman, after Strathcona, in 1914.

Greenway, anxious to avoid falling under the domination of Royal Dutch-Shell, also turned to another potential source of revenue and capital: the British government. The basis of an agreement to their mutual advantage lay in Greenway's desire to find new capital and an outlet for Anglo-Persian's fuel oil; and, on the government's part, in the desire by the Admiralty to obtain secure supplies of fuel oil, which had advantages over coal as a fuel, for the fleet of the Royal Navy.

After long negotiations, agreement was reached in 1914 shortly before the outbreak of the First World War. Anglo-Persian was contracted to supply the Admiralty with fuel oil and the government injected two million pounds of new capital into the company, receiving in return a majority shareholding and the right to appoint two directors to Anglo-Persian's board.

Although the government undertook not to interfere in Anglo-Persian's normal commercial operations, its shareholding introduced an unusual political dimension to the company's affairs. In later years, the government shareholding was reduced and, apart from a tiny residual holding, ended in 1987.

Further expansion followed in the decade following the First World War. New marketing methods were introduced, with curbside pumps replacing two-gallon tins for the distribution of motor spirit (gasoline). Anglo-Persian also marketed its products in Iran and Iraq; it established an international chain of marine bunkering stations, and in 1926 began to market aviation spirit. New refineries, much smaller than the plant at Abadan, also came on stream at Landarcy in South Wales in 1921 and at Grangemouth in Scotland in 1924. Moreover, the company's majority-owned French associate had a refinery at Courchelettes, near Douai. On the other side of the world, in Australia, a new refinery at Laverton, near Melbourne, was commissioned in 1924.

Exploration was carried out not only in the Middle East, but also in other areas, such as Canada, South America, Africa, Papua New Guinea and Europe.

By the time Greenway retired as chairman in March 1927, he had realized his main strategic goal of establishing Anglo-Persian as one of the world's largest oil companies, with a substantial presence in all phases of the industry. In 1935, the company was renamed the Anglo-Iranian Oil Company.

After the Second World War, Europe had to be reconstructed. This demanded a high amount of oil that enabled Anglo-Iranian to expand its business greatly. The company's sales, profits, capital expenditure and employment all rose to record levels in the late 1940s. By this time, the refinery at Abadan was the largest in the world.

While the company was expanding its operations in the late 1940s, it was also engaged in negotiations with the Iranian government concerning the terms of its oil concession. Long and complex negotiations failed to produce an agreement, and in 1951 the Iranian government passed legislation nationalizing the company's assets in Iran. It was then Britain's largest single overseas investment. The nationalization precipitated a major international crisis in which the British government became deeply involved. The company's operations in Iran were brought to a halt.

Only after three years of intensive negotiations was the crisis resolved by the formation of a consortium of oil companies, which, by agreement with the Iranian government, re-started the Iranian oil industry in 1954. Anglo-Iranian was renamed The British Petroleum Company in 1954 and held a 40% share in the consortium.

At first, it was noted that the company focused its business in the Middle East, first of all in Persia (Iran). From the late 1960s the center of company interests shifted westwards, towards the USA and Britain itself.

Although all of these events were important for the company, it was hydrocarbons under the North Sea and under the permafrost of Alaska that were to play the key role in transforming British Petroleum into the company it is today. Earlier, in 1959, the Dutch had discovered a giant gas field on the edge of the North Sea at Groningen. This discovery encouraged others to begin searching for hydrocarbons offshore. British Petroleum had their first success in British waters when, in 1965, it found the West Sole gas field, which it brought on stream two years later. The search for oil spread further north, and in 1970 British Petroleum discovered the Forties field, the first major oil and gas commercial discovery in the UK sector.

At this time in Alaska, USA, British Petroleum was rewarded for ten years' exploration effort when, in 1969, it announced a major oil discovery at Prudhoe Bay on the North Slope. When it became clear that through its large share in Prudhoe Bay, British Petroleum acquired the rights to part of the biggest oilfield in the USA, the company decided that its Alaskan oil could best be handled by a well-established US refining and marketing company. It signed an agreement with the Standard Oil Company of Ohio in August of 1969. This company, the original John D. Rockefeller Standard Oil, was the market leader in Ohio and was strongly represented in neighboring states.

Under the agreement, which became effective from January 1, 1970, Standard took over British Petroleum leases at Prudhoe Bay and some East Coast downstream assets that British Petroleum had acquired in 1968. In return, British Petroleum acquired 25% of Standard Oil equity, a stake that would rise to a majority holding in 1978 when the Standard Oil share of Alaskan production passed 600,000 barrels a day.

The 1970s saw the great petroleum crisis that was to have serious effects on the world's economy. British Petroleum lost direct access to most of its supplies of OPEC oil as the OPEC countries took control of production and prices.

In 1973, the price explosion had a dramatic effect on the demand and the sales of British Petroleum. By 1978, sales were a little higher, but then came the Iranian revolution and another major rise in the price of oil. In 1979, BP suffered further blows when its assets in Nigeria were nationalized and its supplies from Kuwait cut back. By 1980, its sales dropped again.

The company was rescued, thanks to high investments outside of the Middle East. Since the early 1980s, British Petroleum has developed many more oil and gas fields in the North Sea. Among these have been, in the UK sector, Magnus, the Village gas fields at Miller and Bruce and, in Norwegian waters, Ula and Gyda. In Alaska, the construction of the Trans-Alaska Pipeline System enabled the Prudhoe Bay field to come on stream in 1977. In 1981, the Kuparuk field also started production, and towards the end of 1987, the world's first continuous commercial production was recorded from an offshore area in the Arctic when the Endicott field was commissioned.

In 1989, the company launched a campaign to introduce a stronger corporate identity, featuring a restyled British Petroleum shield and an emphasis on the color green. In a complementary program that was to prove highly successful, British Petroleum started to re-image its global network of service stations in a new design and livery.

At the same time, British Petroleum explorers, in the quest to find new sources of oil and gas, began to focus their skills more and more on regions of the world that for political or technical reasons remained relatively unexplored. The regions included Colombia, the Republics of the former Soviet Union, and the deep water areas of the Gulf of Mexico.

The process of integration following this major transactional phase was very well advanced in the year 2000 and British Petroleum acquired a new base from which to take the next step forward. The combination of British Petroleum, Amoco, ARCO and Burmah Castrol provides the area with the skills and the people necessary to deliver a distinctive rate of performance growth on a sustainable basis.

British Petroleum at the beginning of the twentieth century is an international company, having operations in over 70 countries. Its key businesses are oil and gas exploration and production; the refining, marketing and supply of petroleum products; and the manufacturing and marketing of chemicals.

4.2 CASTROL [10]

The history of Castrol can be considered to have begun in the twentieth century, when Charles Cheers Wakefield founded the specialty lubricant company. This company played a key role in the development of the transport industry.

Many of developments in the area of lubricants can be credited to Castrol. In 1909, the motor oil based on castor oil appeared on the market. Twenty-four years later, the company became the first company in the world to use additives in motor oil (organic compounds of chrome).

In 1949, the Deutsche Castrol GmbH company introduced motor oils with anti-corrosive and anti-oxidative additives to the market. Three years later, the company developed an absolutely new type of motor oil with low viscosity for sport cars. This was the reason why Castrol was the chosen oil for many of the sport events where world speed and endurance records, on land, sea and air, were broken. The land speed record alone has been broken 21 times by cars using Castrol lubricants.

In 1975, the company became the first in the world to introduce a new oil, SAE 15W-40, to the market. One year later, Castrol developed an oil using a hundred percent synthetic oil as base.

In 1986, Castrol developed the first oils with low phosphor content to protect catalysts. And within nine years, Castrol created a synthetic oil with the lowest viscosity.

At the beginning of the twentieth century, the products of Castrol helped machines and instruments to achieve greater reliability, endurance and cost effectiveness. The new technologies developed by Castrol have resulted in technological achievements that make Castrol lubricants the cost-effective choice across a wide range of applications.

4.3 EXXONMOBIL [11]

The history of the ExxonMobil corporation began when Exxon and Mobil companies were, at the beginning the 20th century, components of one company (Standard Oil, see the first section of chapter 3). At the end of the century, they came together as a single organization. For most of the years in between, they blazed separate trails as independent, competing enterprises. Each company placed a singular imprint on the energy industry and on a dynamic era of world history.

Both Exxon and Mobil have their roots in the late 19th century with the Standard Oil Trust. Standard Oil Company of New Jersey and Standard Oil Company of New York were the chief predecessor companies of Exxon and Mobil.

For both companies, the remainder of the 19th century was a time of expansion beyond America's shores. The large kerosene market enabled overseas shipments of products in large quantities. Affiliates and sales offices of both companies spread across Europe and Asia. Standard Oil's MEI FOO kerosene lamps introduced illumination across China and opened a vast new market.

After the dissolution of Standard Oil Trust, the American kerosene output was eclipsed for the first time by a formerly discarded byproduct - gasoline. The growing automotive market ultimately inspired the product trademark Mobiloil, registered by Socony in 1920. Jersey Standard and Socony separately faced rising competition. Both companies were not fully integrated. Over the next twenty years, each expanded across the U.S. and abroad.

Large acquisitions and mergers helped Jersey Standard acquire a 50 percent interest in Humble Oil and Refining Company, a Texas oil producer. Socony purchased a 45 percent interest in Magnolia Petroleum Company, a major refiner, marketer and pipeline transporter. In 1931, Socony merged with Vacuum Oil Co., an industry pioneer dating back to 1866 and a growing Standard Oil spin-off in its own right.

In the Asia-Pacific, Jersey Standard had oil production and refineries in Indonesia, but no marketing network. Socony-Vacuum had Asian marketing outlets supplied remotely from California. In 1933, Jersey Standard and Socony-Vacuum merged their interests in the Asian region. Standard-Vacuum Oil operated in 50 countries, from East Africa to New Zealand, before it dissolved in 1962.

The intensive expansion of the companies was slowed down by the Second World War. Each company improved the refining output to supply the war effort. This was the main motivation for the new technologies developed, such as Jersey Standard's groundbreaking process for increasing the fuel octane number and Socony-Vacuum's synthetic lubricants. Both companies suffered wartime casualties. Many refineries and other facilities in Europe and Asia were destroyed.

Over the next few years after the war, ExxonMobil's predecessor companies started to process refinery by-products into many basic petrochemical and numerous derivatives. Since the end of the Second World War, the two companies each already had advanced technologies and expanded business in more than 100 countries.

Mobil Chemical Company was formed in 1960. In 2000, the principal products included all the basic aromatics and olefins for the petrochemical industry, ethylene glycol and polyethylene. The company produced synthetic lubricating oils, additives, propylene packaging films and catalysts. Manufacturing facilities were located in 10 countries.

Exxon Chemical Company became a worldwide organization in 1965 and in 2000 was a major producer and marketer of olefins, aromatics, polyethylene and polypropylene along with specialty lines such as elastomers, plasticizers, solvents, process fluids, oxo alcohols and adhesive resins. The two chemical companies combined their operations within ExxonMobil Chemical.

In 1955, Socony-Vacuum became Socony Mobil Oil Company and, in 1966, Mobil Oil Corporation. Ten years later, a newly incorporated Mobil Corporation embraced Mobil Oil as a wholly owned subsidiary. Jersey Standard changed its name to Exxon Corporation in 1972 and established Exxon as an uncontested trademark throughout the United States. In other parts of the world, Exxon and its affiliated companies continued to use its long-time Esso trademark and affiliate name.

During the oil crisis in the 1970s, Exxon and Mobil escalated exploration and development of oil wells outside of the Middle East: in the North Sea, the Gulf of Mexico, Africa and Asia. By the early 1980s, oil was in surplus, and prices fell. In the mean time, Exxon and Mobil continued to operate at a relatively low price. Each company continued to advance new technologies, introduce marketing innovations and extend its reach into emerging, high-growth markets. The two companies became more efficient, reduced costs and increased shareholder value.

In 1998, Exxon and Mobil signed the historic document (for both companies): a definitive agreement to merge and form a new company called ExxonMobil Corporation. After shareholder and regulatory approvals, the merger was completed in November 30th, 1999.

The modern world marketing of the ExxonMobil products proceeds under the symbol of the tiger. This tiger was spotted in England in the mid-1930s, when it was the Esso symbol. This tiger was stopped in its tracks by the Second World War. The Exxon brand tiger returned in 1953 when competitive gasoline marketing resumed. In England and elsewhere in Europe, the powerful cat helped dispel memories of the low-quality fuels available during the war years.

In 1959, as tiger advertisements waned in Europe, the tiger came to life in Chicago, USA, where an advertising copywriter sat at his typewriter thinking up symbols of power for a local Esso campaign. At the same time, a famous spot slogan (which still exists now) appeared: "Put a Tiger in Your Tank". At first, this tiger was the friendly cartoon character. After the Arab oil embargo in 1973, the world began to focus on the importance of energy conservation. Esso marketers realized this serious issue deserved a more serious symbol than a cartoon. In 1975, Esso marketers in Britain introduced the first television advertisement featuring a real tiger to depict strength and reassurance.

The modern ExxonMobil Corporation conducts business in more than 200 countries around the world, whether it is exploration and production of oil and gas, manufacturing and marketing of fuels, lubes and chemicals, electric power generation or coal and minerals operations.

The upstream business of the corporation is organized into five global companies:

- Exploration
- Development
- Production
- Gas Marketing
- Research

The ExxonMobil corporation is the world's largest non-government producer and reserves holder for petroleum products.

The proven reserves of the corporation stand at 21 billion oil-equivalent barrels, about 13 years of production at current levels. The non-proven portion of the resource base is approximately 48 billion oil-equivalent barrels.

ExxonMobil Exploration company has activities in 48 countries around the world and undeveloped acreage holdings in excess of 120 million acres. In 1998, the companies added resources of more than 1.7 billion oil-equivalent barrels.

ExxonMobil Development Company's works are geographically and technically diverse, ranging from heavy oil in Venezuela to deep-water development in the Gulf of Mexico, West Africa (e.g. Nigeria) and Indonesia to liquefied natural gas in Qatar. The portfolio includes both company-operated projects and major projects operated by others.

The ExxonMobil Production Company is responsible for one of the largest industrial portfolios of worldwide producing oil and gas operations. ExxonMobil produces 4-4.5 million oil-equivalent barrels of oil and gas per day from 24 countries.

The ExxonMobil Gas Marketing Company is the world largest non-governmental marketer of gas. In 1998, ExxonMobil sold gas and liquefied natural gas in 25 countries. The organization maintains gas-marketing offices in 19 countries.

The ExxonMobil Research Company provides the successful technical expertise and large work experience of both companies that has historically contributed to Exxon and Mobil's successes. The Upstream Research Company facilitates efficient and effective technology development and transfer to ExxonMobil upstream companies. Exxon and Mobil have historically shared a strong commitment to upstream technology research and have pioneered many key technologies in use today. Continued development of breakthrough and proprietary technologies will allow ExxonMobil to access and develop new resources at a lower cost.

ExxonMobil downstream business includes refining, retail marketing, lubricant basestock production and sales, finished lubricants, petroleum specialty products and downstream technology. The global downstream business is divided into four companies:

- Refining and Supply
- Fuels Marketing
- Lubricants and Petroleum Specialties
- Research and Engineering

The ExxonMobil Refining and Supply Company operations include supply, marine and pipeline transportation and refining and fuels terminaling. The company has operations in North, Central and South America; the Caribbean; Europe; the Middle East and the Asia Pacific region.

The ExxonMobil Fuels Marketing Company provides the marketing and sales of fuel products to retail customers, industrial and wholesale customers and aviation and marine customers. The Exxon Mobil Corporation brings together three world known brands Exxon, Esso and Mobil.

ExxonMobil Lubricants and Petroleum Specialties Company is organized along eight discrete business lines: Passenger Vehicle Lubricants, Commercial Vehicle Lubricants, Industrial Lubricants, Marine Lubricants, Aviation Lubricants, Basestocks, Petroleum Specialties (wax, process oils and so on), and Asphalt for roads and roofing.

ExxonMobil Research and Engineering Company provides the research, development and use of the process, product, and engineering technology to support downstream and selected upstream and chemical segments of ExxonMobil worldwide businesses.

4.4 NESTE/FORTUM [12]

Neste company was created in 1948 to secure the oil supply in Finland. In that year, the company owned the first oil tanker and started oil import.

Even before 1966, Neste started crude oil refining operations in Naantali and Porvoo. The annual capacity of both refineries was about fourteen million tons. At that time, these two refineries were the most advanced oil refineries in Europe. Four years later, the company began petrochemicals and plastics production. This led to that company becoming the largest company in Finland to play an important role in balancing the former Finland-Soviet trade.

By 1980, Neste was already in the international oil and chemical market. Ten years later, the business of the company expanded to the North Sea and the Middle East. Neste service stations appeared in the Baltic Sea region states. The petrochemicals joint venture with the Russian company – Gazprom – was created.

In 1995, Neste's shares appeared on the Helsinki Stock Exchange. Three years later, a new company, Fortum, was founded and Neste became its subsidiary. In the same year, the share of Fortum appeared on the Helsinki Stock Exchange.

Fortum manufactures petroleum products for use in traffic, heating, industry, agriculture and energy generation. Additionally, Fortum manufactures methyl tertiary-butyl ether (MTBE) in Finland, Portugal, Canada and Saudi Arabia and tertiary-amyl methyl ether (TAME) at its Porvoo refinery. MTBE and TAME are essential components in reformulated gasoline. Fortum is the leading producer of reformulated fuels in northern Europe.

The company supplies only reformulated gasoline and diesel fuels to the market. Reformulated gasoline is improved with regard to octane number by using oxygenates. City Diesel made by Fortum is sulfur-free. Neste and later Fortum was the first company in Europe to start the manufacture and marketing of reformulated gasoline. The range of gasolines comprises two unleaded grades: Futura 95 ER and Futura 98 ER for vehicles fitted with catalytic converters.

Futura CityDiesel is a very high quality, sulfur-free diesel fuel (or fuel with sulfur content less than 0.005 wt%). Futura CityDiesel has a minimum cetane number of 53. Another kind of diesel fuel manufactured by Fortum is conventional diesel with a low sulfur content.

Fortum is well known on the market with its high quality oils that use polyalphaolefins (PAO) and extra high viscosity index (EHVI) oils as base oils. PAO are synthetic oils of very high quality that are used for oils manufacture for the transport, food and cosmetic industries. Neste was one of four companies in the world that manufactured PAO oils. EHVI oils are manufactured using the hydro-isomerization process. The quality of this oil is very close to the quality of PAO and much higher than the quality of the normal mineral oils.

The well known high quality motor oils of Neste, and later Fortum, are the Neste City Pro oils. All these oils are based on synthetic base oils and they exhibit a very high quality in a wide temperature range. Neste Torbo oils are the high quality oils for diesel motors working with low sulfur diesel fuel. Neste Hydraulic oils are high quality hydraulic oils working in a wide temperature range. The Neste Biohydraul is the synthetic bio-degradable oil for working in the natural environment.

4.5 SHELL [13]

The history of this oil company (one of the biggest) began in 1833, in especially unusual circumstances for an oil concern, when Englishman Marcus Samuel opened a little shop in London, selling antiques, curios and sea shells to Victorian enthusiasts. Shells were wildly popular then for decoration, and it did not take long before the trading of shells turned into a thriving general import/export business.

The history of Shell as a real oil enterprise began in early 1890, when Marcus Samuel Junior made a visit to Batum on the Black Sea from where Russian oil from Baku was imported. The businessman was very impressed by the scale of operations. He saw a large market for kerosene in the Far East where it was used for lamps and cooking. Unfortunately, American Standard Oil Trust had a monopoly on the business. So, Marcus had to find a way to undercut prices. Quite separately, the Dutch company, Royal Dutch Petroleum Company, was formed to develop an oilfield in Pangkalan Brandan in Sumatra. Established in the Hague, it enjoyed the support of King William III of the Netherlands.

The solution was found in the Suez Canal. In 1892, Marcus Samuel commissioned the first special oil-tanker, SS Murex, launched at Hartlepool, which delivered 4,000 tonnes of Russian kerosene to Singapore and Bangkok.

Marcus Samuel and Company and the Royal Dutch Company competed with each other and with the US giant Standard Oil, and by 1897 Samuel's oil business had become so extensive that he formed a separate company to operate it. Taking the name from the original business, the new company was registered as the Shell Transport and Trading Company.

In 1901, Shell Transport had become the first oil company to draw its supplies from all around the world, with owned or contracted production from Borneo, Russia, Romania and Texas. Its markets were also very widespread in the world, with its products kerosene for lighting and heating, lubricants and fuel oil.

In 1903, the Shell Transport and Trading Company and Royal Dutch merged together into the enterprise called Asiatic Petroleum Company. The partnership between these two enterprises worked so well that four years later, in 1907, it was extended to operations world-wide, with the creation of the Royal Dutch/Shell Group of Companies. The two parent companies retained their separate businesses and own the Group, with more than 1,700 active companies, in the proportion of 60% to Royal Dutch Petroleum and 40% to Shell Transport and Trading Company.

Shell UK was created when Shell-Mex and British Petroleum, which had combined forces during the Great Depression of the 1930s, decided to break up.

Shell-Mex Limited had been the distribution organization of the Shell and Eagle Groups, the latter being a Mexican company in which Shell took an interest in 1919.

During the Second World War, all oil companies in Great Britain set aside their usual competition, working together under the guidance of one of Shell Transport's senior directors, Andrew Agnew, and providing unbranded "Pool" products. This arrangement continued into peacetime. When more normal conditions had been restored, Shell was determined to re-establish its pre-war market.

The world marketing for Shell proceeds under the logo of the pecten, or scallop shell. This sign is derived from the time when Marcus Samuel and his brother Sam Samuel founded The Shell Transport and Trading Company in London on the 18th of October 1897. They named the new venture after their late father's most profitable trade - decorative sea shells. The new company's first trade mark, registered on the 10th of October 1900, was a picture of the mussel. In 1961 the pecten was evolving into a pure emblem. It became so in 1971 and the word SHELL vanished from its body - a silent, but strong sign of Shell.

The many developments of Shell scientists and researchers are remarkable. In the 1950s, Shell scientists studied the introduction of the first generation of detergent additives.

In the 1960s, a second generation of Shell detergents was introduced leading to a better performance by engines, greater fuel economy and lower emissions. In 1984, a new state of the art detergent and an entirely new concept in fuel technology - the spark aider - was developed. This helped cars become more driveable, especially when cold.

Shell launched its advanced fuels range, the first range of fuels all containing detergent additives, in 1988. The launch followed more than 25 million miles of road tests - one of the most thorough.

A Shell engine test showed Shell Advanced fuels were 375 times cleaner than standard market fuel and the competitors had to respond by adding detergent to their gasoline.

Additive technology continued improving. In 1994, Shell launched an even better detergent package, enabling drivers to gain an average of nine extra miles to a tankfull of gasoline, faster acceleration and still lower emissions.

Shell continued its development of new fuels in 1995 with the introduction of Shell Advanced Low Lead 4 Star and the first low sulfur diesel to be manufactured in the UK. These two new fuels continued to give drivers more ways of reducing the impact of their vehicles on atmospheric quality.

Since 1999, the Shell businesses in the UK became a part of the Royal Dutch/Shell Group of companies.

The main types of product offered to the market under the sign of Shell are:

- Shell Premium Unleaded and Shell Super Unleaded gasoline - improved gasoline.
- Shell Lead Replacement gasoline, the new improved mark of gasoline has been on the market since the 1st of January, 2000.

- Shell Pura Diesel is a market leading ultra low sulfur diesel which, in the year 2000, has already met the known year 2005 emissions standards.
- The range of motor oils providing superior performance in passenger car engines. The range includes fully synthetic, semi-synthetic and mineral oils for different types of engines, age and operation.
- Oil formulated for heavy diesel use and tailored to exceed both industrial specifications and equipment manufacturers' requirements.
- The range of high temperature and high performance greases for lubrication of automotive wheel bearings.
- Automatic transmission fluids suitable for automotive hydraulic, power steering and some manual transmission applications.
- Fuel efficient gear oils which deliver ultimate performance.
- Synthetic air compressor oils which give outstanding performance in rotary and reciprocating compressors.
- Oils that are specially formulated to give excellent performance in rotary and reciprocating compressors.
- The range of conventional and high performance, milky soluble cutting fluids that ensure high cooling rates, good lubrication and excellent anti corrosion properties.
- Low smoking, low misting range of neat oils suitable for all neat and metal-working applications giving excellent tool life and component finish.
- Corrosion protectors for machine components from one week to two years.
- High performance industrial gear oils for use in all helical, bevel and spur gear boxes.
- High quality lubricant for bearing and circulation systems.
- The high performance multipurpose greases suitable for use in high temperatures and in the presence of moisture.
- Lithium base greases designed to give high performance in industrial bearings.
- The high technology grease that enables optimum performance for grease lubrication in industrial bearings, giving extended regreasing intervals and improved equipment life.
- High performance anti-wear hydraulic oils that are universally acknowledged as the market leader in the field of industrial hydraulic and fluid power lubrication.

4.6 TOTAL / FINA / ELF [14,15]

The first part of this corporate company, PetroFina, was founded in the year 1920. Within four years, PetroFina was already a well known company that had its own distributor, PurFina, for sales in Holland and Belgium. In 1924, the Compagnie Francaise des Petroles, one of the ancestors of Total, was formed. In 1927, this company discovered the first oil field near Kirkuk in Iraq. Two years later, the shares of this company appeared on the Paris Bourse.

In the meantime, the new company - Compagnie Francaise de Raffinange- together with Compagnie Francaise des Petroles and in conjunction with the French state and several private French companies got involved in the French petroleum business.

In 1931, Compagnie Navale des Peroles was formed from Compagnie Francaise des Petroles. Two years later, a company, Compagnie Francaise de Raffinage, was formed. This company was responsible for petroleum processing. This company had already owned its own petroleum refinery only within two years of formation.

It was only in 1945 that the successors of the third part, Total Fina Elf companies, was born. These were the French companies Régie Autonome des Pétroles (RAP), the Société Nationale des Pétroles d'Aquitaine (SNPA) and the Bureau de Recherches de Pétrole (BRP). The actual start of the formation for these companies can be considered to be the year 1939, when the exploration of the oil field at Saint-Marcet in Aquitaine yielded some success.

In 1947, the first marketing subsidiary of the petroleum enterprises, the ancestors of Total, was created. This company was named Compagnie Francaise De Distribution des Petroles en Afrioque (CFDPA).

The Société Industrielle Belge des Pétroles (Belgian Industrial Petroleum Company) - SIBP - was founded in Antwerp in 1949 as a result of the industrial integration through the control of the refining process after the Second World War and began production in 1951. In the same time, PetroFina undertook exploration and production in Mexico, Canada, Angola, and Egypt. It was during these years that the company started its world-wide growth. In 1954, PetroFina started its petrochemical production with its first plastic manufacturing.

In 1954, the parent company of Total Petroleum participated in the Iranian Oil Consortium. In the same year, the first trade mark – TOTAL – was created and the first marketing company for Total products was thus founded.

Two years later, the Hassi-Messaoud oil field and Hassi R'Mel gas field in the Algerian Sahara were discovered, and within three years Compagnie Francaise des Petroles began the commercial production and operation on oil wells in these fields.

In 1964, the Compagnie Francaise des Petroles started exploration operations in the North Sea and four years later started its involvement in Indonesia.

In 1965, the historic decision for the future of Elf Company was accepted. RAP and BRP merged to form Enterprise de Recherches et d'Activités Pétrolières (ERAP). ERAP was the real foundation stone for the future Elf Aquitaine group, which comprised a chain of specialties from the oil well to the pump. It was not until this year that the ancestors of this company discovered oil fields all over the world. The company was assured oil production and treatment. It was therefore decided to add refining and marketing as essential supplements. The Union Générale des Pétroles (UGP) was formed, and then the Union Industrielle des Pétroles (UIP).

Since 1967, all the products of the Elf ancestor companies have come to the market with only one short name: ELF.

In 1970, the petroleum company which is well known nowadays as Total Petroleum was founded. This foundation was actually the renaming of an already existing company: the French Petroleum Company of Canada. Within three years, the share of this company had already appeared on the London Stock Exchange, and within one year Total Indonesia started commercial production in East Kalimantan.

After three years, Total Petroleum began the commercial production from the Frigg gas field in the North Sea. In the same year, the historical document, the agreement of the merger of the Compagnie Auxiliare de Navigation and the Compagnie Navale des Petroles to form the Total Compagnie Francaise de Navigation, was signed.

In 1972, PetroFina bought back all their shares in Société Industrielle Belge des Pétroles. The company focused on its refining activities. Fina Raffinaderij Antwerpen became the center of an integrated petroleum and petrochemical network.

In 1980, agreement was signed for the start of exploration and production in China. In 1981, work started in the refinery at Victoria in Cameroon that was built and operated with technical assistance from Total.

In 1983, there existed the new merger of Minatome and Total Energie development to form Total Compagnie Miniere. Two years later, the Compagnie Francoise des Petroleum was renamed Total CFP.

In 1987, Total owned the hydrocarbon assets held by TIPCO in the United States, as well as those of FRANCAREP ITALIA. In the same year, Total owned the petroleum refinery at Denver in the United States.

In 1988, the products of Fina chemical production were sold around Europe and in the USA.

Three years later, Total started the discovery of the oil field at Cusiana in Colombia and the gas field at Peciko in Indonesia. In the same year, the company name – Total – was adopted as the official trade mark and the international company name. In the same year Total shares appeared on the New York Stock Exchange and the direct share holding of the French government in Total was reduced from 31.7 to 5.4%.

Five years later, the shares of Elf Aquitaine appeared on the New York Stock Exchange. As of the end of the twentieth century, Elf Aquitaine had three strong company sectors: hydrocarbons, chemicals and health.

In 1995, Total was selected to lead the Yemen gas liquefaction project. One year later, a further 4% of Total capital was divested by the French government; that reduced government stake to 0.97%.

In the twenty-first century, TotalFinaElf has combined the strengths of Total group, PetroFina and Elf Aquitaine, and has thus become the fourth largest oil and gas company in the world.

The Mitteldeutsche Erdoel Rafenerie Middle German Petroleum Refinery (MIDER) was built and operated by Elf in 1995. This refinery is especially interesting, because of the application of cutting age technologies and processes in the refinery. Some details concerning this refinery are given below.

The refinery has been designed for an operating life of approximately 30 years. At the end of the twentieth century, MIDER was the most modern petroleum refinery in Europe. This was built based on the old Leuna Refinery and projected to use Russian oil first of all. New construction included the product pipeline Leuna-Harimannsdorf/Bohlen and the rail- and road-loading facilities as well as parts of the effluent treating plant. On the supply side, a refinery owned power station was built by STEAG (power plant operator) whereas Linde built the hydrogen plant.

The first and therefore unique distillation process developed by Elf together with Technip in the 1980s operates at the MIDER refinery.

Energy consumption of the rectification process is reduced via integrated atmospheric and vacuum rectification as well as optimal utilization and operation of heat flows. MIDER claims to save some 50,000 tons of fuel oil per annum compared with a traditional distillation process. The process is characterized by the use of five instead of the usual two distillation columns. The process development was based on the objective of avoiding unnecessary overheating of the light components. Additionally, it avoids degrading the thermal levels associated with the drawing off of heavy fractions.

The processing of Russian crude in the above mentioned unit results in a vacuum residue with boiling point of over 585°C. Progressive distillation yields the following fractions:

- Light gasoline (IBP-80°C)
- Naphtha as feedstock for the petrochemical industry (80-95°C)
- Heavy naphtha (96-160°C)
- Kerosene (160-225°C)
- Light and heavy gas oil, vacuum gas oil (225-400°C)
- Medium, and heavy VGO as catalytic cracking feed (400-585°C)
- Vacuum residue for bitumen production and feed for visbreaker

Maurice Promager, manager of R&D of Elf Antar France, quotes the fuel consumption of the progressive distillation as being 1.25% fuel oil equivalent (FOE) for the Soviet Export Blend and 1.15% FOE for the Arabian heavy oil. This compares with 1.7% to 2.0% for the Arabian light oil by conventional processing.

The selection of process units and individual processes was determined by projected product specifications, required range of products and the overriding necessity to minimize residue yield.

In the partial oxidation for the production of methanol (POX/methanol) complex, a major portion of the visbreaker residue (approximately 670,000 tons per annum) is converted to gas. Part of the syngas produced in the POX is used to cover the hydrogen demand of the refinery. Therefore, a plant producing high purity hydrogen has been installed and is operated by Linde. Another part of the syngas is added to the refinery fuel gas system for consumption in the process furnaces, which are exclusively gas fired.

Downstream of progressive distillation, light distillates are processed in an alkylation unit (sulfuric acid catalysts), a hydro-treater for naphtha, and a continuous catalytic reformer (CCR). The alkylation unit is fed essentially with olefins from catalytic cracking. The middle distillates are fed to two identical gas oil hydrofiners with integrated hydrogen sulfide removal for vacuum flash gas and a dual train (each 60% capacity) sulfur unit utilizing Sulfreen to achieve a guaranteed 99.5% conversion.

Vacuum distillate undergoes hydrogenation in the VGO hydrogenation unit before being fed to the catalytic cracking (FCC), whilst vacuum residue passes to the visbreaker for further conversion. FCC residue is used as a fuel component in the refinery's own power station, and it supplies the refinery, in return, with steam and electrical power. It includes three oil- and one gas-fired boilers and is designed for 110 MW output.

There are several levels of automation in the refineries. These embrace not only the vital process control system, but also the laboratory information management system (LIMS) and the general company data processing. These are distinguishing features of the refinery systems concept. In this way, the central databank incorporates relevant updated production data, information from LIMS and other refinery internal data systems to serve the overall management needs.

The process control system includes almost all refinery plants, covering not only the process plants but also tankage, pipeline supervision and effluent treatment. The POX/methanol complex and the loading facilities are, however not included in the system. The dispatch computer has been tailored to the details and requirements of the loading operation. The POX/methanol complex is equipped with its own system that is linked with the refinery control room for the purpose of data exchange. Similarly, the refinery power station exchanges data with the refinery control room in order to match the dynamics of supply and demand.

There are a total of five panel operators who manage the central refinery control room. Outside personnel carry out solely supervisory tasks.

Panel operators handle a total of 40,000 individual data points visualized in four control groups containing four display units in each case, and an extra group containing five display units. Foxboro supplied the control system. The number of units attended to by each individual panel operator is obviously related to the total number of control loops involved. At the same time, consideration also had to be given to the physical boundaries of the individual units to facilitate effective communication with outside operators.

The safety system of the refinery is called "AES" and has been applied to units with high danger potential in addition to the usual shut down system. The AES system was developed originally for use in steam crackers and ethylene plants, and is initiated using both simultaneously pressing and turning a switch.

AES offers the process operator three stages of action. In the first stage, the unit is put into a safe stationary condition, essentially by cutting off or, if need be, rerouting of feed/product streams together with heat and steam inputs to ensure that no additional risks can be brought into the unit. Under this condition, it would be possible to re-stream the unit with little effort. Initiation of stage two would effect isolation of the unit and relief to flare at a predefined pressure level. With this, the relevant section of the unit would be completely sealed off from neighboring sections. Finally, by initiating stage three the whole unit would undergo sealing off and relief.

Furthermore, the AES system can be programmed to any relevant time functions on a sequential shut-down. Such a system appears at a first glance not to be too complicated, but when one realizes that all the units are connected with each other, then shutting down one unit in an emergency situation could be seen as a significant disturbance to the integrated production. Therefore, in all cases where immediate action is needed, neighboring units have to become involved and are put into an appropriate safe condition.

The refinery is primarily supplied with Russian crude oil via pipelines. However, this crude oil can also be supplied via the Baltic ports of Rostock and Danzig. For this purpose the MVL (Schwedt pipeline) is available together with the raw material pipeline Rostock-Bohlen (RRB), which belongs to the Dow Chemical Company, and is used for supplying its olefin group (BSL) in Bohlen with ethylene cracker feedstock in a roundabout way via Leuna. The refinery tank farm, consisting of a total of 63 tanks, has a crude oil storage capacity of roughly 300,000 m^3. In addition, 490,000 m^3 capacity is available for intermediate products and 265,000 m^3 for finished products. LPG tankage totals 11 tanks with a capacity of 24,000 m^3.

Finished products are distributed by rail, road and pipelines. Rail and pipeline distribution each account for roughly 4 million tons per year. Road transport accounts for a total of 3 million tons per year of products for customers within a 150 km radius of the refinery. The Central German Product Pipeline (MIPRO) has a design capacity of 3.2 million tons per year and is used for supplying the Hartmannsdorf depot near Chemnitz.

With an annual production of over four million tons of diesel fuel and domestic heating oil in addition to two million tons of motor gasoline, the refinery is in a position to almost fulfill local demand.

MIDER will process Russian export blend crude oil almost exclusively. Arabian heavy crude oil will form the exception in the summer months, whilst producing bitumen. Segregated vacuum residues from Arabian heavy oils and Russian export blend crude oil will be blended to meet local bitumen specifications and demands.

There are abundant connections between the new refinery and other petrochemical sites. Up to 700,000 tons per year of naphtha (LDF) will be charged to the feedstock pipeline Rostock-Bohlen for processing in Bohlen. Refinery produced methanol, amongst other things, goes to two customers in Leuna, namely Atochem for their formaldehyde based glue production and UCB for the manufacture of amines and formamides.

The refinery yields include 50% middle distillates, 23% gasoline, 7% LDF, and 2% LPG. Most of the C_3 LPG (70%) is propylene, which is mainly used for cumol synthesis via the caprolactum process, operated by Caproleuna in Leuna.

Regarding product quality, the gasoline will be produced with a benzene content < 1%, sulfur content < 0.01% and aromatics content of 35%. Diesel fuel will have a sulfur content of < 0.05% with a lower capability of < 0.025%, and a cetane index of 52.

4.7 LUKOil [16]

OAO LUKOil is the leader of Russia's fuel and energy complex. LUKOil is the first Russian integrated oil company operating in all the petroleum processing spheres from petroleum extracting to marketing of petroleum products. LUKOil was formed in 1991 in the form of a business concern, based on the three biggest oil and gas producing enterprises in Western Siberia - Langepasneftegaz, Uraineftegaz and Kogalymneftegaz. The first letters of these names were combined to form LUKOil. Subsequently, LUKOil absorbed other oil-producing, oil-refining, sales, petrochemical, transportation and other oil business enterprises.

In the year 2000, the LUKOil concern was expected to expand its operations in 40 regions of Russia and 25 countries outside of Russia. This company owns one of the biggest proven petroleum reserves in the world that is owned by a private oil company. LUKOil proven reserves in 2000 exceed two billion tonnes. This company takes first place among Russian companies in terms of the volume of oil produced.

In 1999, the company produced 75.6 million tonnes of oil and 4.7 billion cubic metres of gas. A large proportion of the oil is extracted by its basic oil-producing subsidiary, OOO LUKOIL-Western Siberia, which includes:

- LUKOil-Langepasneftegaz
- LUKOil-Uraineftegaz
- LUKOil-Kogalymneftegaz
- LUKOil-Pokachevneft

In the western part of Russia, oil production is carried out by the main oil- and gas-producing subsidiaries of the company:

- OOO LUKOil –Astrakhanmorneft
- OOO LUKOil- Kaliningradmorneft

- OOO LUKOil- Nizhevolzhskneft
- OOO LUKOil-Permneft
- ZAO LUKOil-Perm
- OAO KomiTEK

In 1999, the oil production in European Russia that was carried out by LUKOil amounted to 24 million tonnes. In the same year, OAO LUKOil acquired the oil company KomiTEK, the result being a substantial increase in the company's raw material base: proven oil reserves increased by 400 million tonnes and those of gas by 47 billion cubic meters.

The ultimate assessment of the reserves based on the results of tests of deep prospecting wells is expected at the end of the year 2000.

LUKOil is the first Russian petroleum enterprise to start the manufacture of high quality half-synthetic and fully synthetic oils. The products of this company are used not only on the domestic market, but also on the European and American markets.

The quality of LUKOil oils has been certified and recommended by the German companies Mercedes, BMW and VW.

4.8 YUKOS [17]

YUKOS was formed on the 15th of April 1993 according to the government decree No. 354. The acronym YUKOS was chosen as the name for the company, with its letters taken from the names of its original two primary operating units:

- Yuganskneftegas - one of Russia's largest oil production units, located in the Tyumen region
- Kuibishevnefteorgsintes - a major petrochemical holding located in the Samara region

In that year, YUKOS comprised one production entity, Yuganskneftegas; three refineries, Kuibishev, Novokuibishevsk and Syzran; and eight petroleum product suppliers located in the Russian regions of Samara, Penza, Voronezh, Oryol, Bryansk, Tambov, Lipetsk and Ulyanovsk.

Two years later, the company owned a second production entity, Samaraneftegas, along with a number of product marketing and research and development organizations.

As a result of the financially difficult transition period, the Russian government decided to sell YUKOS'S state-owned shares to private investors. Through a series of tenders and auctions held in 1995-1996, YUKOS became the first fully privatized Russian oil company. The major purchaser was a group of private investors led by YUKOS Chairman Mr. Mikhail Khodorkovsky.

Under Mr. Khodorkovsky's leadership, YUKOS completely repaid its debts (accumulated while it was a government-owned company) to the Russian federal and regional governments. In the first year following its privatization, YUKOS received 1 billion US$ in loans and investments primarily due to the efforts of the company's new modern management team.

In 1997, YUKOS acquired a controlling stake in the Eastern Oil Company (Russian acronym VNK), thus adding to its existing assets a third production entity. The configuration and profiles of the VNK enterprises complemented those already under YUKOS control and allowed the company to expand both in terms of operational capacity and eastern geographical reach. At the same time, YUKOS initiated systematic efforts to integrate VNK and its subsidiaries into its corporate, financial and operational structures.

In 1998, YUKOS signed an agreement establishing a strategic alliance with Schlumberger, one of the world's largest oilfield service companies. Under the agreement, YUKOS outsources a significant part of its oilfield service operations to Schlumberger. This has enabled YUKOS to fundamentally restructure many of its own service enterprises. Moreover, the cost of the oil well being served has been reduced by 22 percent, yielding a significant reduction in overall production costs.

In 1999, a production enhancement network was established, a reservoir enhancement network was developed, and a technology network set up to optimize equipment utilization. About 800 wells were treated during the year, and a total of 1,500 well candidates were identified.

In the twenty-first century, YUKOS has become the second largest oil company in terms of oil production and reserves in Russia.

The company products include all the ranges of fuels and oils made in the petroleum industry. The products are sold on the domestic and international markets.

4.9 TNK [18]

Tyumenskya Neftyanaya Kompaniya (Tyumen Oil Company) or TNK was formed as an open joint-stock company in 1995 and is known as the most stable petroleum company in Russia.

Until 1997, the company and its subsidiaries operated as an autonomous production and financial division resulting in a decline of overall production and lack of financial control. In July 1997, Novy Holding acquired a forty percent stake in TNK through an investment tender. Novy Holding spent more than US$600 million to regain control of the shares and to acquire additional shares in the subsidiaries. By February 1998, Novy Holding together with Novy Petroleum increased their joint stake in TNK's charter capital to 50.1 percent. The new shareholders launched a large-scale restructuring program to create an efficient and responsive vertically integrated production and management structure.

In 1998, the restructuring of TNK began with the first step taken by Novy Holding to replace the management of TNK with individuals educated in the west, having extensive experience with international oil giants and possessing comprehensive knowledge of the Russian oil industry.

In October of 2000, Tyumen Oil Company received the American Society for Competitiveness's (ASC) annual award for leadership in developing globally competitive practices in an emerging economy.

The products of the company include:

- Gasoline with octane numbers of 92, 95, 98; all kinds of gasoline made by TNK are unleaded; the company has started the production of reformulated gasoline
- Diesel fuel with a sulfur content of 0.2%, and an ecological diesel with a sulfur content of 0.05%
- Domestic fuel
- Residual fuels
- Industrial oils
- Motor oils
- Additives for motor oils
- Catalysts for the Petroleum treatment processes, such as reforming and hydrotreating

Bibliography

1 U.S. Crude Oil, Natural Gas, and Natural Gas Liquids Reserves. 1998 Annual Report, December 1999, Energy Information Administration Office of Oil and Gas, U.S. Department of Energy.

2 Internet Publication: BP Amoco, Chicago, IL. BP Amoco in Alaska: Badami Factsheet. http://www.bpamoco.com.

3 P. Seidel. Schweres Erdöl - ein alternativer Rohstoff zur Erzeugung von Treibstoffen. Expert Verlag, Renningen-Malmsheim, 1994.

4 E. N. Kokotchikova. Macroeconomicheskoe znachenie neftegazovogo kompleksa v economike Rossii. GANG, 1996.

5 N. A. Kruglov. Vchera, segodnya, zavtra neftyanoj i gazovoj promyshlennosti. INGIRGI, 1995.

6 Internet: homepage of BP Company. http://www.bp.com/default.asp.

7 V. A. Dinkov. Neftyanaya promyshlennost vchera, segodnya, zavtra. Moscow, VNIIOENG, 1988.

8 Nezavisimoe neftyanoe obozrenie. Moscow, VNIIOENG, Nr. 1, 1993.

9 V. L. Berezin. Neft i gaz zapodnoy Sibiri, Moscow, 1990.

10 Internet: homepage of Castrol Company. http://www.castrol.com.

11 Internet: homepage of ExxonMobil Company. http://www.mobil.com/index_flat.html.

12 Internet: homepage of Neste/Fortum Company. http://www.neste.com/.

13 Internet: homepage of Shell Company. http://www.shell.com/.

14 Internet: homepage of Total / Fina / Elf Company. http://www.totalfinaelf.com/fr/html/index.htm.

15 Stand der Technik bei Raffinerien im Hinblick auf die IPPC-Richtlinie. Bericht IB-610, Umweltbundesamtes.

16 Internet: homepage of LUKOil Company. http://www.lukoil.ru/.

17 Internet: homepage of Yukos Company. http://www.yukos.ru/.

18 Internet: homepage of TNK Company. http://www.tnk.ru/.

Part III

MAIN PROCESSES IN THE PETROLEUM REFINING INDUSTRY

OVERVIEW

Every modern refinery is uniquely designed to process a given crude petroleum into selected products. It is therefore necessary to know the chemistry of the crude to be processed into selected products in order to make an appropriate design of the refinery processes to use. The process designer needs to produce a design to meet the business objectives of the refinery by selecting from an array of basic processing units. In general, these units together perform three functions:

- separating the many types of hydrocarbons present in crude oils into fractions of more closely related properties,
- chemically converting the separated hydrocarbons into more desirable products, and
- purifying the products by removing unwanted elements and compounds.

The following processes are employed in almost every modern refinery to perform these three functions:

- petroleum rectification,
- vacuum rectification,
- hydroprocessing,
- gas processing plant,
- sulfur recovery unit (SRU),
- reforming (platforming),
- isomerization and hydroisomerization,
- alkylation,
- polymerization,
- catalytic cracking,
- hydrocracking,

- thermal cracking,
- visbreaking,
- residue conversion processes,
- blending.

These processes can be classified as primary or secondary processes. In this book, primary processes are used to describe processes in which crude petroleum/products undergo only physical treatment. On the other hand, secondary processes describe processes in which petroleum/products undergo chemical transformations.

Furthermore, additional processes are required in order to provide the petroleum processing units with steam, electric power and cooling water that are needed for these units to function. In modern refinery complexes, the energy consumption is low (lower than 5% of the crude oil throughput). Water consumption is low as well (below 1 m^3 per ton of crude oil).

5

Crude Oil Distillation

5.1 PETROLEUM AND GAS PREPARATION

As a rule, the initial stage of petroleum extraction from oil wells occurs under the natural layer pressure practically without contamination with water. With time, however, the water layer gets produced together with the petroleum layer. About two-thirds of all petroleum is extracted in high humidity conditions. Water layers can differ considerably by their chemical and bacteriological contents. During the extraction of this mixture of petroleum with water layer, an emulsion is formed. Emulsions can be considered to be a physical mixture of two immiscible liquids, one of which is distributed in another phase as droplets of various sizes. The water present in petroleum results in an increase in the transportation price because of an increase in both volume and viscosity of the resulting liquid to be transported.

The presence of corrosive mineral salts in solution in the water leads to fast deterioration of petroleum pumps and oil refining equipment. The presence of up to 0.1% of water in petroleum leads to intensive foam formation in the rectification tower of oil refineries. This results in an alteration of the technological scheme for oil processing. Besides, it also affects the condensation equipment.

The light fractions of petroleum (gases from methane to butane) are valuable raw materials for the chemical industry from which products such as solvents, liquid motor fuel, alcohol, synthetic rubber, fertilizers, artificial fiber and other products of organic synthesis are made. Therefore, it is necessary to reduce the loss of light fractions during petroleum processing. Thus, all hydrocarbons derivable from petroleum need to be preserved for subsequent processing.

Modern petrochemical complexes manufacture various high-quality oils and fuels as well as new types of chemical products. The quality of these products depends on the quality of the initial raw material, i.e. crude petroleum. In the past, the technological processing scheme of older oil refineries could be used to process crude petroleum with mineral salts contents of 100—500 mg/L. Petroleum with a lower salt content is required in modern refineries. Thus, frequently before oil processing, it is necessary to completely remove salt from the petroleum by desalting.

The presence of different types of salt impurities in petroleum affects pipelines and oil pumps. This complicates the petroleum-processing scheme because the solids form scales in coolers/refrigerators and furnaces. This leads to a reduction in the heat transfer coefficient of the scaled surface. Solid impurities also promote emulsion formation.

The presence of mineral salts in the form of crystals in petroleum and in solution in water leads to extensive corrosion of metallic parts of equipment and pipelines. It also leads to an increase in emulsion stability, thereby adding to the complication in the petroleum-processing scheme. The quantity of mineral salts dissolved in water divided by its volume is called General Mineralization.

Under favorable conditions, a part of magnesium chloride ($MgCl_2$) and calcium chloride ($CaCl_2$) present in the water layer can be hydrolyzed with the formation of hydrochloric acid. Also, hydrogen sulfide is formed during petroleum processing as a result of decomposition of sulfur compounds. This increases the intensity of corrosion of metals in the presence of water. Hydrogen chloride in water solution also corrodes metal parts. Corrosion is especially intensified in the presence of hydrogen sulfide and hydrochloric acid in water. The requirement for petroleum quality is, in some cases, very stringent: the contents of salts should not be more than 40 mg/L in the presence of water, whose content should only be up to 0.1%.

These requirements and other reasons make the preparation of petroleum before processing very necessary. Petroleum preparation includes drying (removal of water or dewatering) and desalting of petroleum, and complete or partial removal of dissolved gas.

5.1.1 Formation of Petroleum Emulsions and Their Basic Properties

In order to select the proper drying process, it is necessary to know both the mechanism of formation and properties of petroleum emulsions.

Usually, no emulsions are formed within the petroleum layer. Emulsion formation begins during the movement of petroleum to the mouth of the oil well and intensifies during further transport of petroleum in pipes (i.e. emulsions are predominant where there is the potential for continuous mixing of petroleum and water). The intensity of emulsion formation in an oil well depends on the method of petroleum extraction. This, in turn, is defined by the character of the oil wells, time of its operation and physical-chemical properties of the petroleum.

When petroleum is extracted from oil wells using the natural layer pressure (which is typical in the initial period of oil well operation), there is usually a very high rate of extraction of the oil from the oil well. The intensity of petroleum mixing with water in elevating pipes of the oil well increases due to dispersion of the solved gases at decreasing pressure. This leads to emulsion formation at the early stages of movement of the petroleum-water mixture.

During deep pumping extraction of petroleum, emulsion formation occurs in the valves, in the pump cylinders and in the elevating pipes during the reciprocating movement of pump bars.

In the compressor oil well, the base of the emulsion formation is the same as at extraction using natural layer pressure. Air sometimes mixes with the gas in an oil well and oxidizes a part of the heavy hydrocarbons to form asphaltenes-resinous materials. These adversely influence emulsion formation negatively. Thus, the presence of salts of organic acids as well as asphaltenes-resinous materials leads to emulsion formation. This type of emulsion has a very high stability.

It is important to distinguish two phases in emulsions – the discontinuous and continuous phases. The continuous phase is the liquid in which very small droplets of another liquid are dispersed. The discontinuous phase is the liquid that is dispersed in the form of fine droplets in the continuous phase.

Two types of emulsions are distinguishable – "oil in water" and "water in oil". The type of emulsion formed basically depends on the volume ratio of the two phases. The continuous phase is the liquid with the greater volume. In practice the most frequently known emulsion is the water in oil emulsion.

The presence of emulsifiers (materials that promote emulsion formation) influences the ability to form an emulsion between petroleum and water. Emulsifiers act by lowering the interfacial tension between the phases and creating a strong adsorbed layer around the surface of the internal phase. Emulsifiers that are soluble in water (hydrophilic) promote the creation of oil in water emulsion. Alkaline soaps, starch and so on are such hydrophilic emulsifiers. Hydrophobic emulsifiers (i.e. soluble in petroleum) promote the formation of water in oil emulsions. Hydrophobic emulsifiers include resins dispersed in particle form within soot, clay and other substances. Petroleum emulsions can be characterized using properties such as viscosity, dispersion, density, electrical properties and stability. The viscosity of petroleum emulsion changes within wide ranges and depends on the viscosity of petroleum, temperature, and amounts of petroleum and water.

Petroleum emulsions that are disperse systems have unusual properties under certain conditions, i.e. they are Newtonian liquids. Thus, petroleum emulsions like all Newtonian liquids can be characterized by an effective viscosity.

Dispersion of the emulsion is the degree of distribution of the droplets of the discontinuous phase in the continuous phase. Dispersion is characterized by the diameter of the droplets d, and by D = 1/d (= specific surface area) calculated by dividing the total surface area of the particles by their total volume.

The sizes of droplets can change depending on the physical-chemical properties of petroleum and water as well as the conditions of emulsion formation. Sizes range from 0.1 micron up to several tens of mm. Disperse systems which consist of droplets of the same size (single diameter) are referred to as mono-disperse. On the other hand, systems that consist of droplets of different sizes are called poly-disperse systems. Petroleum emulsions are poly-dispersed systems.

The critical droplet size that can exist in the flow at a given thermodynamic mode is determined by the velocity of movement of the mixture of water and petroleum, interfacial tension between the phases and pulsation of flow.

In the turbulent flow regime, caused by non-uniformity of pulsation and variable cross section of pipeline, the existence of droplets of various diameters is possible. Fine droplets that move in the section of pipeline of lower velocity gradients and smaller pulsation tend towards agglomeration, whereas those that move in the zones of high velocity gradients and large pulsation tend towards splitting. The presence of additional factors (such as heating, addition of emulsifier, etc.) under certain hydrodynamic conditions can lead to the separation of the emulsion phases in the pipelines.

A higher degree of emulsion stability depends on the structure of the compounds that constitute the protective layer formed on the surface of the droplet. The emulsifier adsorbs on the surface of the droplet and covers it with a preserving layer thereby stabilizing the droplet. This layer prevents droplets from merging with each other (i.e. promotes emulsion formation and stability).

The chemical nature of water in the petroleum layer has an influence on the stability of petroleum emulsions. All the water layers contain chemically different materials. However, they can all be divided into two basic groups: the first group is hard water containing calcium chloride, magnesium chloride, etc.; the second group is alkaline water. The presence of increased amounts of acid in the water layer leads to the formation of more stable emulsions. Adding alkali to the emulsion can reduce the amount of acid in the water layer.

5.1.2 Separation of Water–Oil Emulsions

The process sequence of breaking petroleum emulsions follows the consecutive steps: agglomeration and flocculation of droplets, destruction of the protecting layers, coagulation of the droplets of the disperse water up to a size sufficient to further merge by the influence of gravity, and then falling down to the bottom of the separator.

If the droplets have sufficient energy for the destruction of the protecting layers to occur, then the droplets will merge.

A number of technological methods are applied for petroleum drying. The choice of method for petroleum drying and performance scheme employed substantially depends on the amount and condition of water present.

Water contained in crude petroleum appears in free form, i.e. non-dispersed form, in some cases. Such water can be removed from petroleum directly by coagulation.

More often however, water in crude petroleum is present in its dispersed form, emulsion of water in oil. There are two versions of this emulsion: mechanical non-stabilized and stabilized by interfacial-active substances. This distinction for emulsions is essential for petroleum drying. Water from non-stabilized emulsions can be easily separated by the usual settling method or by settling with moderate heating. More complex processes, such as intensive heating, chemical processing, electrical processing, and combinations of these methods are required for separation of water from stabilized emulsions.

For designing the schemes for drying of petroleum under industrial conditions, it is necessary to determine for the petroleum to be dried the water content in the petroleum, type and quantity of impurities in water, and also the conditions in which water is present.

The processes of drying and desalting are very similar. In desalting, however, the water is removed from the petroleum together with the dissolved mineral salts. If it is desired, more complete desalting is achieved by introducing additional fresh water to the petroleum to dissolve the mineral salts and subsequent drying.

Settling, centrifugation and filtration are mechanical methods for petroleum drying.

Settling is applied if unstable emulsions are processed. Here, the droplets are stratified due to the difference in the densities of the materials in the phases.

The droplets' settling speed for designing separators is calculated using the formulas:

$$\omega^2 = \frac{4}{55.5} gd \frac{\rho_d - \rho_k}{\rho_k} Re^{0.6} \tag{5.1}$$

for $2 < Re < 500$:

$$\omega = 3gd\sqrt{\frac{\rho_d - \rho_k}{\rho_k}} \tag{5.2}$$

for $Re > 500$:

where: Re – Reynolds number $\{(d\omega\rho_d)/\mu\}$
μ – viscosity of medium
ω – settling speed
ρ_d – density of the discontinuous phase
ρ_k – density of the continuous phase
d – diameter of the droplet or particle
g – acceleration due to gravity (9.8 m/s^2)

These equations imply that the major factors that influence the efficiency of emulsion separation are:

- Densities of the phases in the emulsion (essentially, the difference in density of the phases is responsible for gravitational separation);
- The viscosity of the emulsion phases, especially the viscosity of the continuous phase (this is a very significant factor and affects efficiency of petroleum drying by affecting transport of droplets or particles through the medium);
- Diameter of droplets or particles of the discontinuous phase (this factor is very important because the settling speeds of the droplets or particles grow proportionately to the square of their diameters);
- The acceleration of the droplets or particles is due to their presence in the natural gravitational field. The value of the acceleration is approximately 9.8 m/s^2;
- The area available for settling.

These factors can therefore be used in design to improve the efficiency of emulsion separation. The favorable ways to apply these factors are as follows:

- Increase of temperature of the emulsion. This reduces the viscosity of the emulsion phases, and reduces the interfacial tension between the phases (the thermal methods of petroleum drying are based on this principle);
- Increase of droplet sizes of the discontinuous phases via various methods such as using chemicals and applying an electrical field (the chemical and electrical methods of petroleum drying are based on this principle);
- Increase of the settling speed of the droplets by replacing the natural gravitational force by the more powerful centrifugal force. Centrifugal force influences water and mechanical impurities in that their densities are higher than the density of petroleum. The method of centrifugation is not very productive but is complicated, and therefore has not found wide application in the oil industry;
- Increase of the useful settling area without increase of the general area of separator. The use of parallel plates in horizontal separators and separated disks in separators is based on this principle.

The efficiency of emulsion separation is reduced in the case where the density between the dispersed droplets or particles and the continuous phase is very small. Also, stabilized emulsions cannot be separated by mechanical methods. A significant proportion of "water in oil" emulsions is in this category.

Adverse hydraulic conditions of settling such as turbulence, convective flows, mixing, etc. represent negative influences on emulsion separation. Significant improvement of separation efficiency for petroleum emulsions is achieved by the use of gravitational settling in combination with thermal, chemical and electrical methods.

5.1.3 Mechanical Petroleum Drying

The most basic method in mechanical petroleum drying is gravitational settling. There are two types of settling modes: periodic (i.e. batch) and continuous. These are carried out in batch and continuous separators, respectively.

Cylindrical separating tanks (settling tanks), similar to petroleum storage tanks, are usually employed as batch separators. Crude petroleum intended for drying is pumped into the tank through the distribution pipeline. After filling the tank, the content is allowed to stay for a predetermined period known as settling time. The water then settles to the bottom section while petroleum floats on the top section of the tank. The settling is carried out under a mild petroleum treatment condition. Petroleum and water are taken out separately from the tank at the end of the petroleum drying process. A positive result is achieved from the tank only if the petroleum obtained is free of water.

In the case continuous separation, two types of separators can be distinguished: horizontal and vertical continuous separators. Horizontal separators are subdivided into longitudinal and radial. Longitudinal horizontal separators can be rectangular or round depending on the geometry of the cross section.

In gravitational continuous separators, the settling is carried out by the continuous flow of liquid through the separator. The separator is designed to achieve a desired degree of settling in a predetermined holding or settling time.

5.1.4 Thermal Petroleum Drying

One of the modern ways of drying petroleum is thermal drying or thermal processing. In this case, the petroleum for drying is heated up before the settling step. Heating destabilizes the "water in petroleum" emulsion and promotes the merger of fine droplets of water into larger droplets. Usually, protection layers consisting of asphaltene-resinous substances and paraffins are formed on the emulsion surface of water droplets. At normal temperatures, these layers create a very stable structure that prevents the merging of droplets. By increasing the temperature, the viscosity of the materials that form the protective layer is considerably decreased. This leads to a reduction of the stability of the disperse structure, thereby facilitating the merging of the water droplets. Besides, the viscosity of petroleum is lowered as a result of heating. This promotes the acceleration of the water separating from petroleum by settling. Thermal methods alone are applied rarely in industry but they are used in combination with settling. In modern methods, thermal processing is usually used as a component of a more complex set of methods of petroleum drying, for example, in combination with treatment with chemical additives and settling (thermo-chemical drying), as well as in combination with electrical processing, etc.

The heating of petroleum intended for drying is carried out in special heating sets (heaters). A large variety of such heaters have been developed. The heaters are installed in a technological scheme of petroleum drying after the section that deals with the separation of gases from petroleum, but before introducing the petroleum into the separator.

5.1.5 Chemical Methods of Petroleum Drying

The chemical drying methods are widely employed in the modern petroleum industry for petroleum drying. The fundamental principle of such methods is the destruction of the "water in petroleum" emulsions by using chemical additives. Many types of such additives have already been developed in the industry. The efficiency of chemical petroleum drying depends substantially on the type of additive used. The choice of an effective additive, in turn, depends on the type of emulsion that is to be destabilized. In each case, the choice of additive is made after the petroleum is analyzed in special laboratories.

Like in other combined petroleum drying methods, chemical drying is followed with settling of the emulsion under the influence of gravity. Heating of the petroleum is employed in some drying systems in combination with the use of additives. The additive is added and blended with the emulsion. This creates the conditions for removing the water from petroleum by settling. It is possible to apply both periodic and continuous emulsion separation, but nowadays preference is given to the continuous process. There are three locations where the chemical drying of petroleum could be implemented:

1. Drying and separation carried out inside the petroleum oil well;
2. Drying and separation carried out in the collector's pipeline;
3. Petroleum drying carried out directly in separation tanks where additives are added to the tank that is filled with petroleum.

The first two methods have many advantages and are more effective than the third method.

Filtration that is based on the selective adsorption of different substances is employed for the separation of unstable emulsions. The material of the filter layer can be dry sand; glass; aspen, maple, poplar and other types of non-resinous wood; and metal particles. Glass fiber, which is usually well moistened with water and not with petroleum is used quite often. Petroleum drying by filtration is applied very seldom because of low productivity as well as the need to change the filter material frequently.

5.1.6 Thermal Chemical Petroleum Drying

The stability of the protective layers is reduced or completely destroyed by thermal chemical methods. This method accelerates the process of petroleum emulsion separation. For more than 80% of petroleum processed, the thermal chemical method is used. This method has become widely used due to the opportunity it presents to treat petroleum with varying water contents without changing the technological processing scheme, as well as the ability to change the additives depending on the emulsion properties without replacing any of the devices used in a standard technological scheme. However, the thermal chemical method has a number of drawbacks. These include high additive prices and high heat consumption.

Desalting and drying processes are usually conducted in the industry at temperatures ranging from 50 to100°C. If a higher temperature is to be used, then the process must be carried out at a high pressure because of the need to keep the emulsion in the liquid phase. For this purpose, it is necessary to use separators with thicker walls. This leads to an increase in the price for the hardware.

The reduction of the protective influence of the interfacial layers on the water droplets is essentially affected by the presence of additives. All additives can be divided into:

- electrolyte
- non-electrolyte
- colloidal additives

Some organic and mineral acids (sulfuric, hydrochloric and acetic acids), alkali and salts (table salt, chloride ion, etc.) can be used as electrolytes. The electrolytes can form the insoluble products which reduce stability of the protection layer or promote their destruction. Electrolytes as additives are applied extremely rarely because of their high price and, particularly, their corrosive properties as it affects the separators. The non-electrolytes are organic substances which are capable of dissolving the protective film on the water droplets and reducing the viscosity of petroleum. This leads to the acceleration of the water droplets. This type of additive can be gasoline, acetone, alcohol, benzene, phenol, etc. Non-electrolytes are not used in industry because of their high price.

Colloidal additives are interfacial active substances which can destroy emulsions or weaken the protective film and can transform the native water in oil emulsion into the opposite type (oil in water); i.e. it can promote emulsion inversion.

Most effective additives are formed by reaction of ethylene oxide with organic substances; they are widely applied in the industry. Using the ethylene oxide/organic substance ratio involved in the reaction to produce the additive can control the efficiency of this additive group. The additive solubility in water increases with the lengthening of the ethylene oxide chain. If it is necessary, it is possible to make these additives with hydrophobic properties by reactions with propylene oxide, i.e. there is the opportunity to create various additives with different properties.

The additives should dissolve well in one of the emulsion phases (in water or petroleum), i.e. they should be hydrophilic or hydrophobic, in order to be able to destroy the protective film of the water droplets. Also, they should be inert to metals, should not worsen the quality of petroleum, should be cheap, and should be of universal applicability with respect to use for various emulsions.

The sooner the additive is added to the emulsion of water in petroleum, the easier the separation. For separation, however, just adding the additive to petroleum is not sufficient. It is necessary to ensure the best possible contact between the additive and the water droplets. Intensive mixing and emulsion heating can achieve this.

Electrical desalting and drying of petroleum is especially widely applied in the industry, but less often in the oil wells. The opportunity created by the application of the electrical method in combination with many other methods can be attributed to one main advantage of this method.

It is known that the separation of petroleum using an electrical field with variable frequency is more effective than the separation with a constant electrical field.

Viscosity and density of the emulsion, dispersion, water content, electrical properties as well as the stability of the protective films considerably influence the efficiency of electrical separation. However, the main factor is the intensity of the electrical field. The electrical separators are operated at the current industrial frequency (50 Hz for Europe and 60 Hz for North America), rarely by constant current. The voltage on the electrodes in the separators ranges from 10,000 to 45,000 V.

The electrical separators can be spherical and cylindrical in geometry, and can be installed in horizontal or vertical orientation.

5.1.7 Stabilization of Petroleum

The extracted petroleum can contain dissolved gases (nitrogen, oxygen, hydrogen sulfide, carbonic acid, argon and others) and light hydrocarbons in various quantities. During petroleum transportation from the oil wells to the oil refinery the gases are lost from the oil because of insufficient hermetic sealing of the pipelines and storage tanks. This leads to significant losses of the light petroleum fractions. During evaporation of light fractions such as methane, ethane and propane, the relatively heavier hydrocarbons (butane, pentane, etc.) can be evaporated as well. It is known that the more often petroleum contacts with the atmosphere and the longer these contacts are, the higher the loss of the light fractions.

To prevent the loss of petroleum, it is necessary to achieve complete hermetic sealing during petroleum transportation by all methods. However, the existence currently of non-perfect hermetic pipelines does not allow this.

Hence, it is necessary to remove the gases and light fractions from petroleum at the oil wells and to direct them for further processing. This consequently lowers the evaporation ability of petroleum.

It is also possible to prevent the loss of the light fractions of petroleum by application of rational systems of petroleum extraction, gas treatment, and petroleum stabilization before its subsequent transport and storage. It is necessary to understand that stabilization of petroleum in this case is the extraction of light hydrocarbons (which under normal conditions are gases) for further processing in the petrochemical industry.

Thus, separation is used in this case as the main method for petroleum stabilization at the oil wells. Separators of the widest applicability are of gravitational and centrifugal designs.

In gravitational separators, separation of droplets and firm suspensions from a gas flow occurs by the influence of gravity. A high degree of gas and liquid separation can only be achieved at a very slow flow speed. The optimum speed as established in practice is 0.1 m/s at a pressure of 6 MPa. The proportion of petroleum suspension separation from the gas at this flow speed is 75—85%.

In centrifugal separators, separation of the gas from petroleum occurs by the influence of a centrifugal force.

5.1.8 Technological Schemes for Petroleum Preparation

The extraction and preparation of petroleum and gas, which begin at the opening of the oil wells and end at the preparation units, follow a uniform technological system. There are many technological schemes of petroleum preparation. However, they are usually considered together with the petroleum extraction systems at the oil wells.

5.1.9 Pressure Extraction System

The pressure extraction system (see Fig. 5.1) is operated as follows. The quantity of petroleum from each oil well (produced with the natural layer pressure) is serially measured using an automatic group measurement unit. Then, the petroleum from various oil wells in the field is transported to the local separation station. The quantity of the petroleum from the oil wells is measured after the preliminary gas separation with the centrifugal separator. After that, the petroleum and gas are directed to a local separation unit, where gas is separated from the petroleum by the first stage separator at the pressure 4—5 bar and then directed to the gas processing station. Petroleum with layer water and solved gases will be pumped to the central processing station, where the second step of separation begins by the end of the separators. The gas from the second step of separation is directed to the gas processing station using a set of compressors.

This pressure extraction system is completely hermetically sealed. Therefore it excludes loss of gas and light petroleum fractions. The pressure extraction system allows for petroleum preparation at a central processing station for oils from several oil wells located in an area up to a 100-km radius. However, long distances for petroleum transport can lead to the creation of stable emulsions. With high humidity of the petroleum, this can lead to an increase in operations and transport costs. Nevertheless, it is one of the promising systems of petroleum extraction that is widely applied.

There are many technological schemes for petroleum preparation. However, technical and economic considerations determine which scheme is used and the location of installation. It is known that the lowest capital investment and operational costs for petroleum preparation are for installations in locations of the greatest petroleum concentration (collector stations, commodity parks, and head offices).

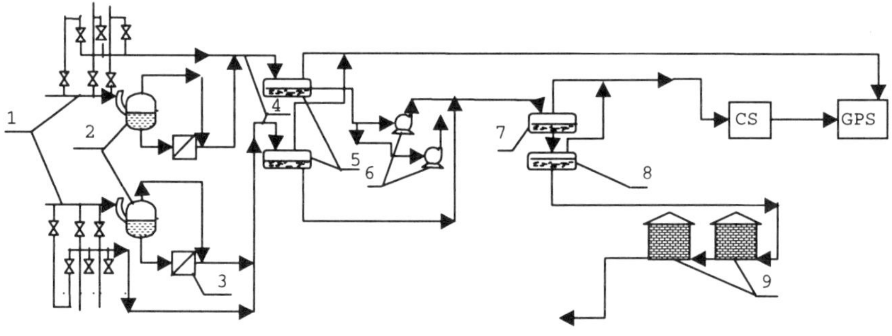

Fig. 5.1: Pressure extracting systems:
1 – bypasses, 2 – separators, 3 – measurement devices, 4 – collectors, 5 – first stage separators, 6 – pumps, 7 – second stage separator, 8 – third stage separator, 9 – reservoirs, CS – Compressor station, GPS – Gas Processing Station.

The optimal technological scheme for petroleum preparation is obtained with a scheme that allows petroleum to be processed to the allowable contents of water and salts, and with the necessary depth of stabilization at the lowest costs, in the shortest possible process time.

In the modern petroleum industry, complex petroleum preparation is carried out in areas close to the oil wells. Therefore, complex installations for petroleum preparation at the petroleum wells also integrate the processes of drying, desalting and stabilization.

Figure 5.2 shows the basic technological scheme of the installation for the thermal chemical petroleum preparation.

Petroleum from the oil well is directed to the combined separation installation 2, in which hot water from settling tank 6 containing additives is directed to mixer 1. This begins the partial separation of gas, petroleum and water. The separated water is directed to petroleum traps 20, whereas the separated gas moves to the gas processing station. The petroleum from separator 2 together with the remaining water is pumped by pump 3 through heat exchanger 4 and vapor heater 5.

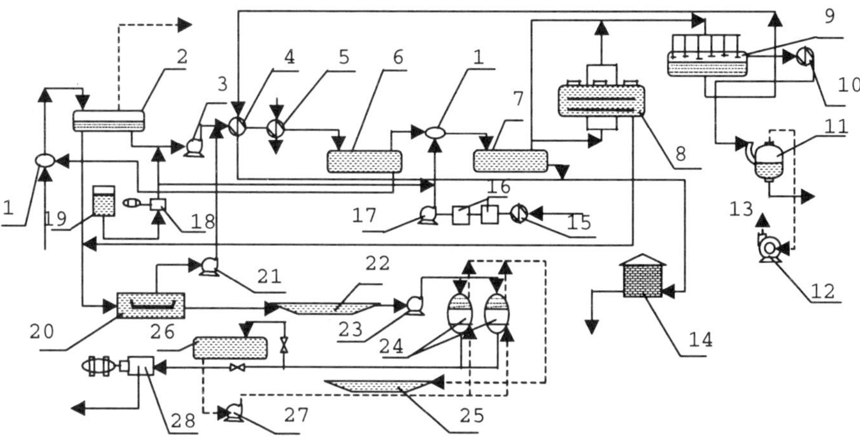

Fig. 5.2: Technological scheme of the complex for thermal chemical petroleum preparation

1. Mixer, 2. Separator, 3. Pump, 4. Heat exchanger, 5. Vapor heater, 6-7. Settling tanks, 8. Dryer, 9. Vacuum separator, 10. Refrigerator, 11. Hydro-centrifugal separator, 12. Pump, 13. Water line, 14. Oven, 15. Vapor heater, 16. Oxygen tank, 17. Pump, 18. Compressor, 19. Separator, 20. Oil traps, 21. Pump, 22. Vessels, 23. Pump, 24. Separators, 25. Vessel, 26. Tank, 27. Pump, 28. Compressor.

Then the heated petroleum is moved to settling tank 6 for the final separation of petroleum from water. The separated water is carried away from the petroleum with a major fraction of the mineral salts. For better desalting, the petroleum from settling tank 6 is directed for mixing with hot fresh water pumped by pump 17 after pre-heating in the vapor heater 15 and removing the oxygen in tank 16. After mixing fresh water with the petroleum, the emulsion is directed to settling tank 7, where the required salt concentration is achieved. After desalting and separation of water, the petroleum is directed, if required, from settling tank 7 to additional desalting and drying in electrical dryer 8. If the water and salts contents are within the allowable limits after electrical dryer 8, the petroleum is moved directly to vacuum separator 9. Vacuum compressors 12 transport the gases from separator 9, from which after refrigerator 10 and hydro-centrifugal separator 11, a major fraction of light hydrocarbons is separated. The condensed product from separator 11 is directed to gas and gasoline collectors. Additives are introduced into the petroleum before heat exchanger 4. This influences the interfacial property of the protective films between the two phases in the emulsion.

This system allows the cleaning of waste water and its subsequent direction by pumps for the flooding of the oil layer.

5.2 DESALTING

Even with the preparation of petroleum at the oil wells, the crude oil often contains water, inorganic salts, suspended solids, and water-soluble trace metal compounds. As a first step in the refining process (to reduce corrosion, plugging, and fouling of equipment and to prevent the poisoning catalysts in the processing units), these contaminants must be removed by processes of desalting and dehydration or drying.

The methods of desalting and dehydration are based on the same fundamental principles as those for petroleum preparation at the oil wells. The two most typical methods of crude oil desalting: chemical and electrostatic separation, use hot water as the extraction agent. In chemical desalting, water and chemical surfactant (demulsifiers) are added to the crude, heated so that salts and other impurities dissolve in the water or attach themselves to it, and then held in a tank where they settle out. Electrical desalting is the application of high-voltage electrostatic charges to concentrate suspended water globules at the bottom of the settling tank. Surfactants are added only when the petroleum has a large amount of suspended solids. Both methods of desalting are continuous. A third and less-common process involves filtering heated crude using diatomaceous earth.

The feedstock crude oil is heated to between 110 and 160°C to reduce viscosity and surface tension for easier mixing and separation of the water. The temperature is limited by the vapor pressure of the petroleum. In both methods, other chemicals may be added. Ammonia is often used to reduce corrosion. Caustic or acid may be added to adjust the pH of the water wash.

Wastewater and contaminants are discharged from the bottom of the settling tank to the wastewater treatment facility. The desalted crude is continuously drawn from the top of the settling tanks and sent to the petroleum rectification unit. All the apparatus for petroleum desalting and drying can be classified in two big groups:

- spherical desalter
- cylindrical desalter, which can be sub-classified into:
    - horizontal
    - vertical

During electrical desalting, electricity is used to increase the rate of movement the water droplets with the solved salts as well as to accelerate the merging of small droplets to form bigger ones. These cause the separation of the droplets from the petroleum emulsion.

In modern refineries, the sections for petroleum desalting and drying are combined with atmospheric and vacuum rectification.

5.3 ATMOSPHERIC RECTIFICATION

The petroleum prepared at the oil well comes to the petroleum refinery and the first process at modern refineries (excluding the refineries working only with non-conventional feed) is atmospheric rectification. The first refinery, which was opened in 1861, produced only kerosene and this was possible by using simple atmospheric distillation alone. The by-products of this refinery included tar and naphtha. For the next thirty years, kerosene still remained the main product that consumers wanted. Two significant events changed this situation:

- invention of electric light decreased the demand for kerosene,
- invention of the internal combustion engine created a demand for diesel fuel and gasoline.

Distillation is the process involving the conversion of a liquid into vapor that is subsequently condensed back to a liquid. It is exemplified when steam from a kettle becomes deposited as droplets of distilled water on a cold surface. A simple example of a distillation unit is shown in Figure 3.3.

Distillation is used to separate volatile products from non-volatile substances. The early experimentalists also employed distillation. Aristotle (384-322 BC) mentioned that pure water was made by evaporation of seawater.

Most methods of distillation used by industry are variations of simple distillation. The basic operation of the industrial process requires the use of the same principal installations as the above mentioned example of simple distillation: a still (flask and gas burner) in which a liquid is heated and partly vaporized, a condenser to cool the vapor, and a receiver to collect the distillate. In the still is charged a mixture of substances with different boiling points. The lowest boiling products distill first, followed by others in order of increasing boiling temperatures. In comparison to our example shown in Figure 5.3, larger equipment made of metal or ceramic is employed for industrial applications.

In the petroleum industry, the method of fractional distillation, differential distillation, or rectification is utilized for the primary separation of crude oil into fractions with regard to their boiling temperatures, because simple distillation is not efficient for separating liquids whose boiling points lie close to one another.

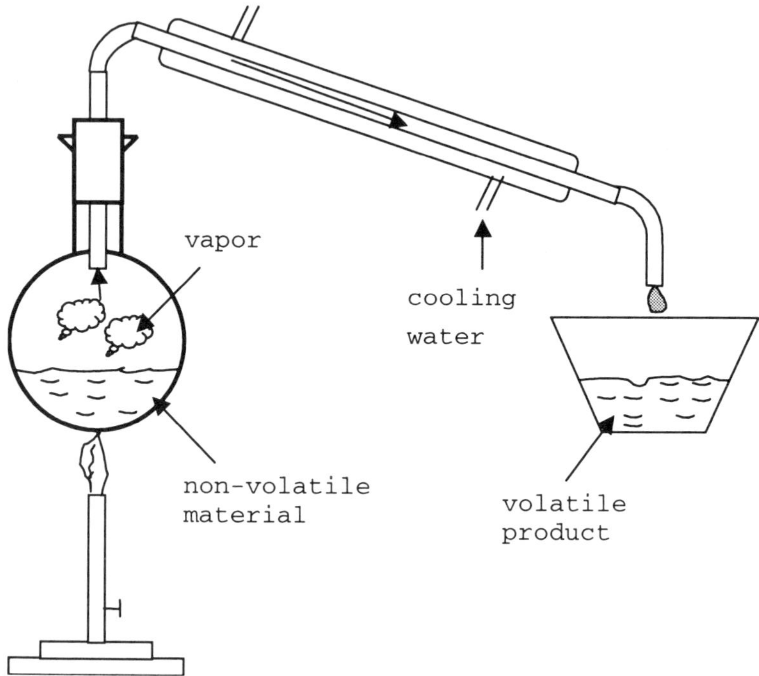

Fig. 5.3: A simple distillation unit.

In this operation, the vapors from distillation are repeatedly condensed and re-vaporized in the next evaporator. Figure 5.4 shows the principle of differential distillation or rectification. This installation works on the basis of the reverse or counter-current flow principle. For a steady state functioning of the rectification unit, it is important that there is equilibrium in every evaporator. This means that for the example shown, the following equations have to be obeyed:

- $y_0 = x_I$
- $y_I = x_{II}$
- $y_{II} = x_K$

Instead of the example shown, rectification towers are used for rectification in the petroleum industry. Rectification towers can be classified as follows:
On the basis of the area of application:

- atmospheric towers
- vacuum towers

On the basis of the type of tower internals:

- Tray towers
- Trickle or packed towers

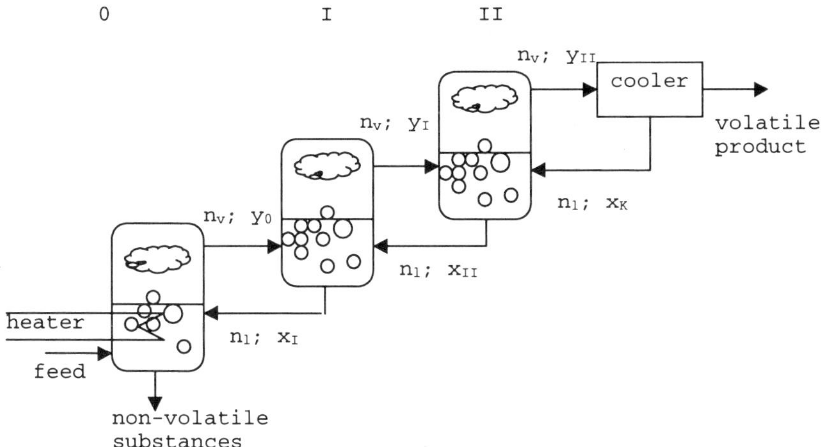

Fig. 5.4: Differential distillation, rectification.
n_v – stream of vapor product
n_l – stream of liquid product
y – concentration of the light component in vapor
x – concentration of the light component in liquid

On the basis of the type of function:

- simple towers
- complicated towers

One example of the simple tray tower is shown in Figure 5.5. From Figure 5.5, it is clearly seen that for the effective functioning of all types of rectification towers, it is especially important to return back to the tower part of the head condensed vapor and bottom product. The objective is to achieve the closest possible contact between the rising vapor and the descending liquid so as to allow only the most volatile material to rise in the form of vapor to the receiver, while returning the less volatile material as liquid towards the still.

A comparison of the towers shown in Figures 5.5 and 5.6 illustrates the difference between simple and complicated towers. The main difference between these two types of rectification towers is that in a complicated tower, many volatile products or side products are obtained. The addition of small towers called strippers are used to improve the purity of the side products. The function of the stripper is the same as for the big rectification towers.

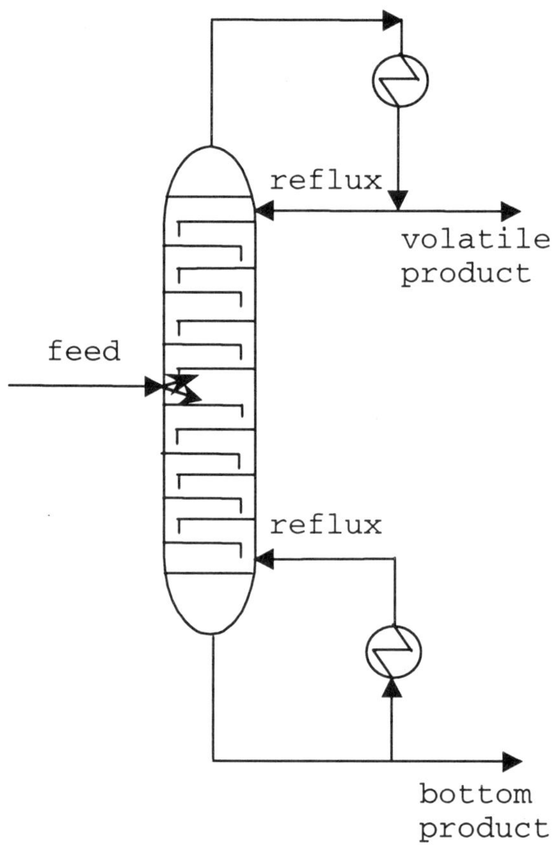

Fig. 5.5: The simple tray tower.

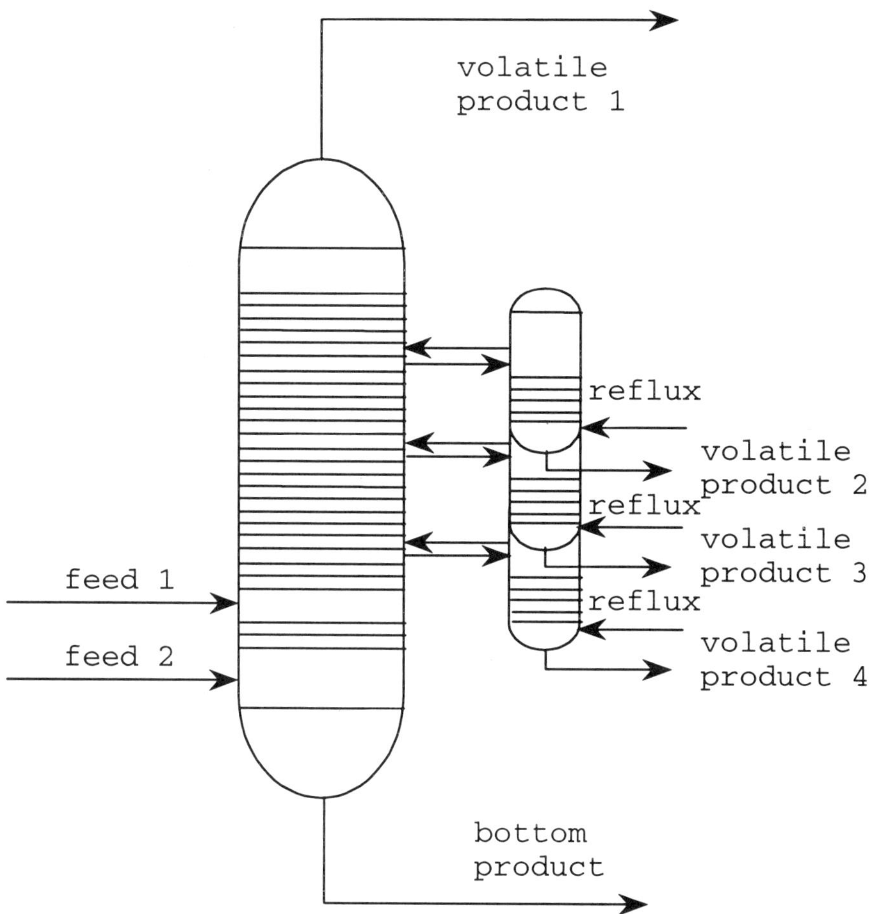

Fig. 5.6: Complicated rectification tower.

The schematic that illustrates the function of the tower tray is shown in Figure 5.7. It is shown that the tray tower has the same function as shown in Figure 5.4. The only difference is that in the actual rectification tower, the trays also play the role of the evaporators. The vapor goes through the liquid on the trays and there is material transfer in the bubble regime. The liquid coming from the top tray (phlegm) carries the material exchanged between the liquid and vapor phases in the tower.

Sieve trays consist of punched metal sheets with holes of diameters of 4-13 mm with a division of the holes of 2.5 to 4 times d_{hole} (see Fig. 5.8). The proportion of liquid on the tower trays ranges from 0.05 to 0.2 times the feed. Although there are many types of trays, only three are used in the petroleum industry. These are:

- sieve trays
- bubble cap trays
- valve trays

The liquid does not flow through the holes, but is held by the pressure loss during vapor flow through the tray holes. This determines that there is limited range of feed load.

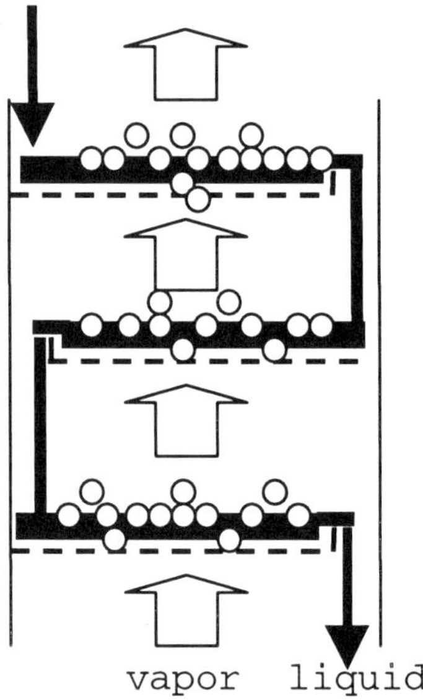

Fig. 5.7: The principle of the tray tower function.

In valve trays, the holes in the tray are bigger. Also, the valves are mobile and cannot be completely closed. These valves ensure that the flow of the liquid through the tray holes is impossible even with varying feed loads. The hole diameter of this type of tray ranges from 20 to 30 mm, valve diameter from 40 to 50 mm and division of the tray holes 1.5 to 3 times d_{valve}. Figure 5.9 shows an example of the valve in the valve tray.

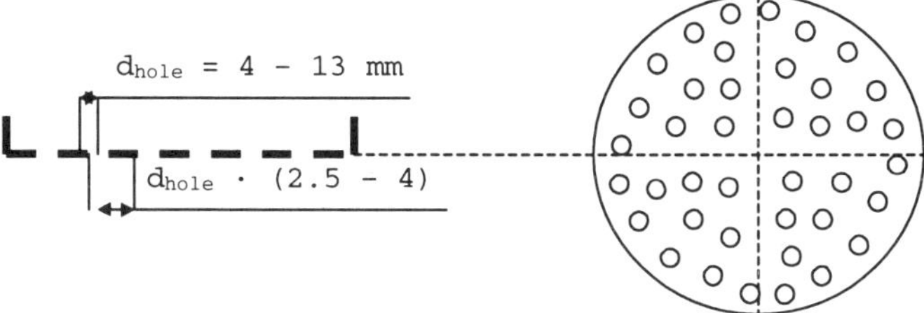

Fig. 5.8: Sieve tray.

The bubble cap trays have the largest range of feed load possible. However, it is also the most expensive. By immersing the bell edges into the liquid (this being necessary for bubble formation), the pressure loss can be characterized by the static proportion for the bell tray. During very low gas load, all the bells are no longer functioning, so that the liquid does not have sufficient contact with the vapor phase. This leads to a decrease in the tray efficiency. An example of the bell for the bubble cap tray is shown in Figure 5.10.

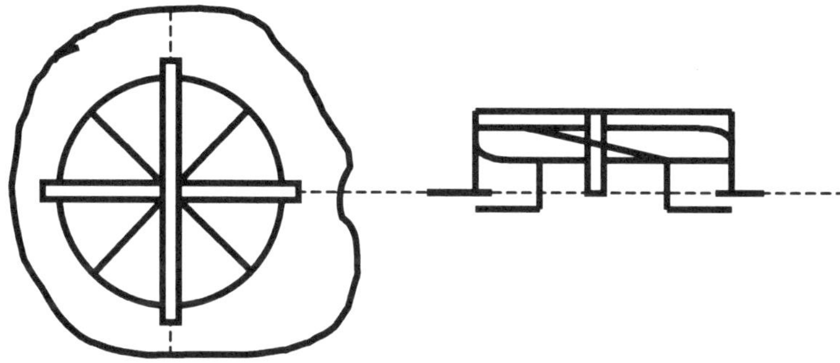

Fig. 5.9: Valve in the valve tray.

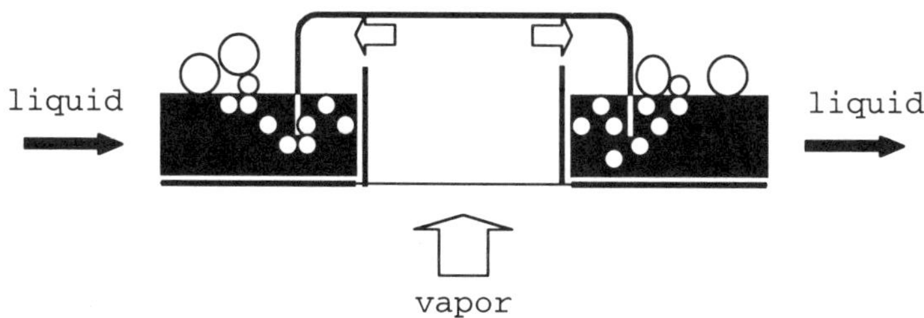

Fig. 5.10: The bell of the bubble cap tray.

Packed or trickle towers have similar disadvantages as tray towers. Packing materials are used in the packed tower instead of trays as in the tray tower. An example of the trickle or packed tower is shown in Figure 5.11.

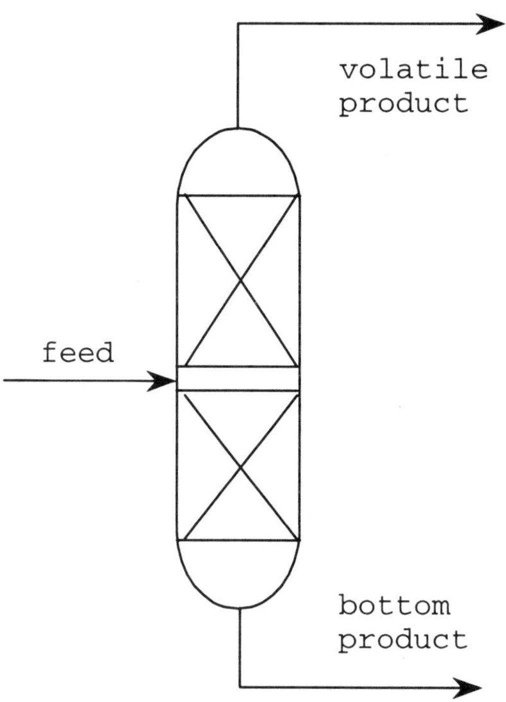

Fig. 5.11: The packed or trickle tower.

In comparison with tray towers, the packed or trickle towers have the following advantages:

- low pressure loss
- low installation cost
- more homogenate mixture of the vapor and liquid phases
- better material exchange between the vapor and the liquid phases

There are many types of packings for packed towers. However, the most frequently used are as shown below:

The main property of the packing for rectification towers is the specific surface area, which determines the efficiency of the packing. The larger the specific surface area, the more effectively the packing operates.

For both types of towers, it is important to determine the number of theoretical trays required to perform fractionation to specified compositions of fractions. For packed or trickle towers, the theoretical tray is replaced with the height equivalent of the packed layer equivalent to one theoretical tray.

There are a few methods to evaluate the number theoretical trays. Tray to tray calculations is one the most popular methods.

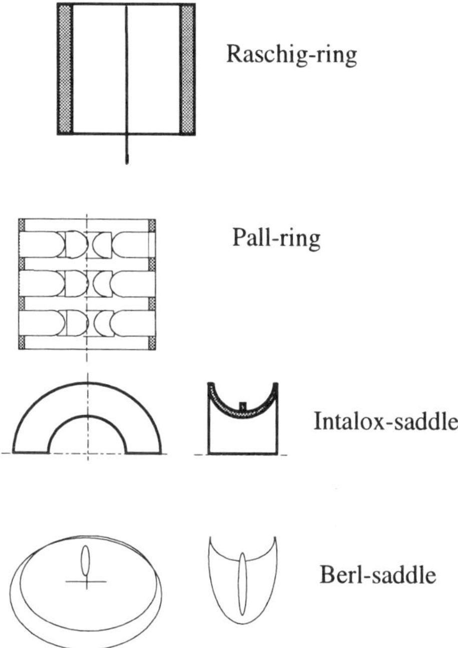

An illustration of this method is given below for a mixture containing four components. Components 1 and 2 have to be present in the volatile product, whereas components 3 and 4 have to be present in the bottom product. Therefore, the separation cut is between components 2 and 3. Moreover, components 2 and 3 have different boiling temperatures and are present in different amounts in the two products. They are called "key components (key - K)". Component 2 is the lighter or volatile component (LK - light key) whereas component 3 is the heavier or bottom component (HK - heavy key).

In the rectification tower, four equations are available from material balances (overall and component material balances) for the separation of the four components. If the feed or input amount and its composition are known, then one has two unknown streams H (head product) and B (bottom product) and three unknown concentrations in each stream. Thus, generally there are eight unknown values for a four component mixture. Only two of these unknown values are to be determined. Thus the material balance of the rectification tower can be evaluated to obtain the top and bottom product rates. The vapor stream in the stripping section can be determined as the vapor stream ascending from the tower bottom. This stream can be taken as constant throughout the stripping section of the tower. This vapor stream plus the fraction of feed that is vapor will yield both the head product stream as well as the reflux stream that will be returned to the tower. The reflux ratio can be determined based on the minimum reflux. There is a method for the evaluation of the minimum reflux. This method will be shown in a later section.

The next step for the calculation of the number of theoretical trays is to make a material balance around the tower bottom (it can also be started from the tower top). The concentrations of the components in the ascending vapor stream are calculated from equilibrium equation (Dalton Raoult laws) for the existing mixture composition in the tower bottom (5.3).

$$y_i^* = \frac{p_{Si}}{p} x_i \tag{5.3}$$

where y^*_i – the concentration of component *i* in the vapor
x_i – the concentration of component *i* in the liquid
p_{Si} – saturated vapor pressure of component *i*
p – pressure in the rectification tower

For hydrocarbons, these equilibrium relations are often represented by $y_i = K_i x_i$ where $K_i = P_{si}/P$ = K-factor for component i. K-factors are functions of both pressure and temperature but assumed to be independent of composition. Also, for a multi-component mixture, a relative volatility (α_i) can be defined for each component with one of the components, say 3, as a basis such that $\alpha_i = K_i/K_3$. Charts for K-factors for various components are available in Handbooks.

For a four component mixture, the above equilibrium relations will yield three independent equations thereby providing the three extra equations required to determine all the eight unknowns. The equilibrium equations together with the material balance equations are applied to each plate to calculate the vapor and liquid compositions for each plate as follows.

The liquid flow from the first tray at the tower bottom results from balances (5.4, Fig. 5.12):

$$L_1 = V_S + BP \tag{5.4}$$

The boiling temperature of liquid L_1 is calculated on a trial and error basis. This must be lower than the temperature of the tower bottom. This boiling temperature is calculated on the basis that at a specific pressure this temperature must satisfy the relation $\Sigma y_i = 1.0$. For a mixture of components 1, 2, 3 and 4 with component 3 as the basis, $\Sigma y_i = \Sigma K_i x_i = K_3 \Sigma \alpha_i x_i = \Sigma(P_{Si}/P)x_i = 1.0$. The trial starts with a guess of a temperature. Then the values of α_i are calculated from the values of K_i (read from a chart) at this temperature. The value of K_3 is calculated as $K_3 = 1.0/\Sigma\alpha_i x_i$. Then the temperature corresponding to the calculated value of K_3 is compared to the assumed temperature. If the values are different, the calculated temperature is used for the next iteration. After the final temperature is known (i.e. convergence is reached), the vapor composition is calculated from equation 5.1 or its equivalents.

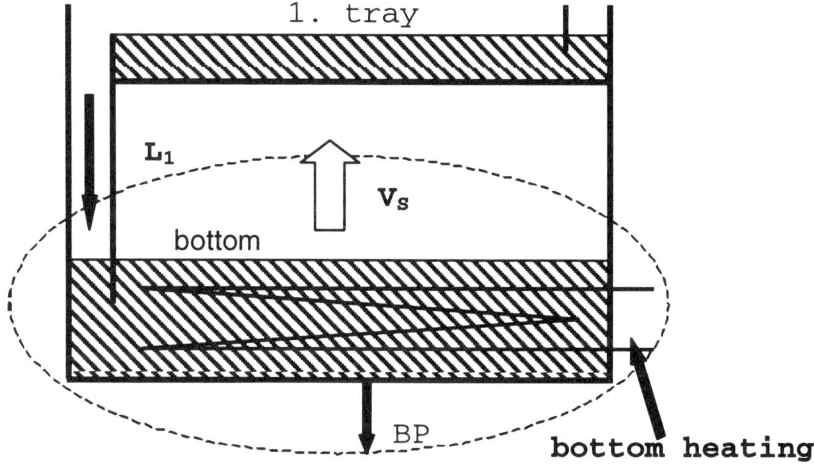

Fig. 5.12: Tower bottom.

The calculation of the temperature difference between the bottom and the first tray makes use of heat balance at the tower bottom.

The next step is the calculation of the liquid composition in tray 2 using material balance equations and the vapor composition for the first tray of the down part of the tower. During this calculation, it is important to remember that the composition of the liquid phase on the first tray is the same as the vapor composition in the tower bottom. This is because the liquid on the first tray is the product of the condensation of the vapor produced at the bottom of the tower. The material balance is shown in equation (5.5):

$$L_1 + V_1 = V_S + L_2 \tag{5.5}$$

The balance element for the equation (5.5) is shown in Figure 5.13.

These tray-by-tray calculations are continued until we reach the input or feed tray where the calculated composition of the input is similar to the actual feed composition.

The calculation of the upper section of the rectification tower proceeds in the same way as the bottom section. In this case, the calculation could be started from the top tray and proceeds towards the feed tray.

The disadvantage of this method is the need for iteration. Thus, the calculations are cumbersome and time-consuming if hand calculators are used. Nowadays all calculations of theoretical number of trays are carried out by use of the computer and many software packages for such calculations are available in the market. Two approaches can be used for distillation calculations:

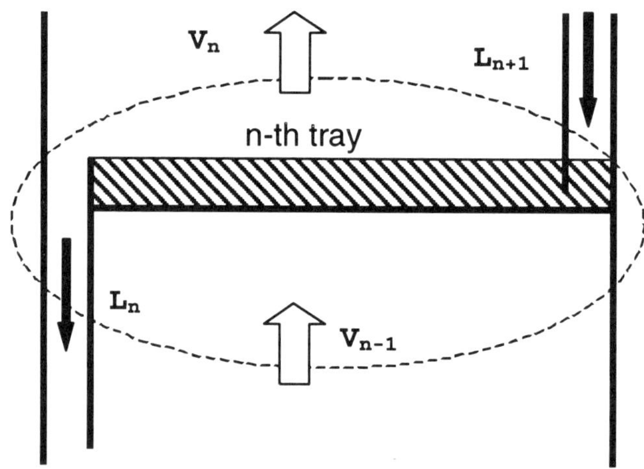

Fig. 5.13: Material Balance element of the rectification tower.

One is if the product compositions are known, then tray-by-tray calculations are used to determine the number of trays needed to obtain those product compositions. The other is if the number of trays is given (such as in an existing unit), then calculations are made to determine the compositions of products that could be obtained from such a unit.

If the distillation unit provides for side products and/or multiple feed locations, then the tray to tray calculations will use material balances for each tray that account for these additional streams. Furthermore, it is important to note that fraction cuts for crude oil fractionation are based more on boiling temperature ranges rather than on composition.

The next important parameter required for rectification, as mentioned above, is the minimum reflux. The minimum reflux is the value of reflux for which the number of theoretical trays is infinite. In technical terms, a rectification tower that has the number of trays over a hundred can be taken as having infinite number of trays.

The McCabe-Thiele diagram is used (Fig. 5.14) for the calculation of the minimum reflux.

The x-axis in this diagram shows the composition of the liquid phase in equilibrium with the composition of the vapor phase (y-axis). The straight lines B2 and H2 are balance lines. These lines show the balance between two trays (the composition of the vapor on the n - 1 tray equals the liquid composition on the nth tray). The two points 3 and 2 are called pinch-points; on these points, the compositions of the liquid and vapor do not change any more from tray to tray. That means that an infinite number of trays is required for rectification of such a mixture with this reflux value.

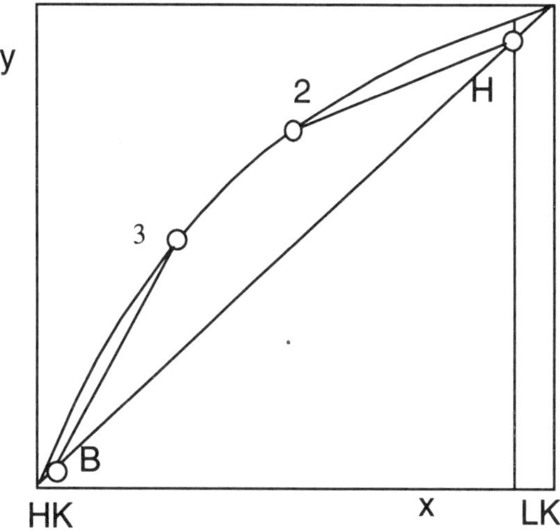

Fig. 5.14: McCabe-Thiele diagram.

The following four operations need to be carried out to calculate the minimum reflux:

1. assume a minimum reflux,
2. carry out the evaluation from tray to tray from B and H,
3. add a small amount of the lightest component to the mixture at the "pinch-point" 3, and evaluate from tray to tray until the head of the tower,

4. repeat as in point 3 for "pinch-point" 2. If the results of the third and fourth points are equal, then the assumed value was correct. If not, then repeat all the operations from the first until fourth point.

Rectification units in the petroleum industry are used to separate the crude oil into fractions for subsequent processing in secondary processes such as catalytic reforming, cracking, alkylation, or coking. In turn, each of these complex secondary-processing units incorporates a fractional distillation tower to separate its own reaction products.

Modern petroleum rectification units operate continuously over long periods of time. For rectification, petroleum is drawn from storage tanks at ambient temperature and pumped at a constant rate through a series of heat exchangers in order to reach a temperature of approximately 120°C. The preheated petroleum passes though an oven (see position O, Fig. 5.15). At this position it is heated to a temperature between 315 and 400°C, depending on the type of petroleum and the quality of the end products desired. A mixture of vapor and nonvaporized oil passes from the oven into the rectification tower. The size of an industrial rectification tower is about 45 meters tall with 20 to 40 fractionating trays. The most common trays used in the petroleum industry are sieve or valve trays. Petroleum vapor rises up through the trays to the top of the tower. It is condensed to a liquid in a water- or air-cooled condenser at the top of the tower. A small amount of gas remains uncondensed (see position 2, Fig. 5.15) and is piped into the refinery fuel-gas system. A pressure control valve on the fuel-gas line maintains rectification tower pressure at the desired value, usually near atmospheric pressure. Part of the condensed liquid, reflux, is pumped back into the top of the column and descends from tray to tray, contacting rising vapors as they pass through the slots in the trays. The liquid progressively absorbs heavier constituents from the vapor and, in turn, releases lighter constituents to the vapor phase.

The intermediate products, or side-streams (see positions 4, 5, Fig. 5.15) are drawn at several points from the tower dependent on the fractions desired. Usually these fractions are gasoline and diesel fractions. Dependent on the production type in the refineries, jet fraction can be drawn as well. In addition, modern petroleum rectification units employ intermediate reflux streams (Fig. 5.6).

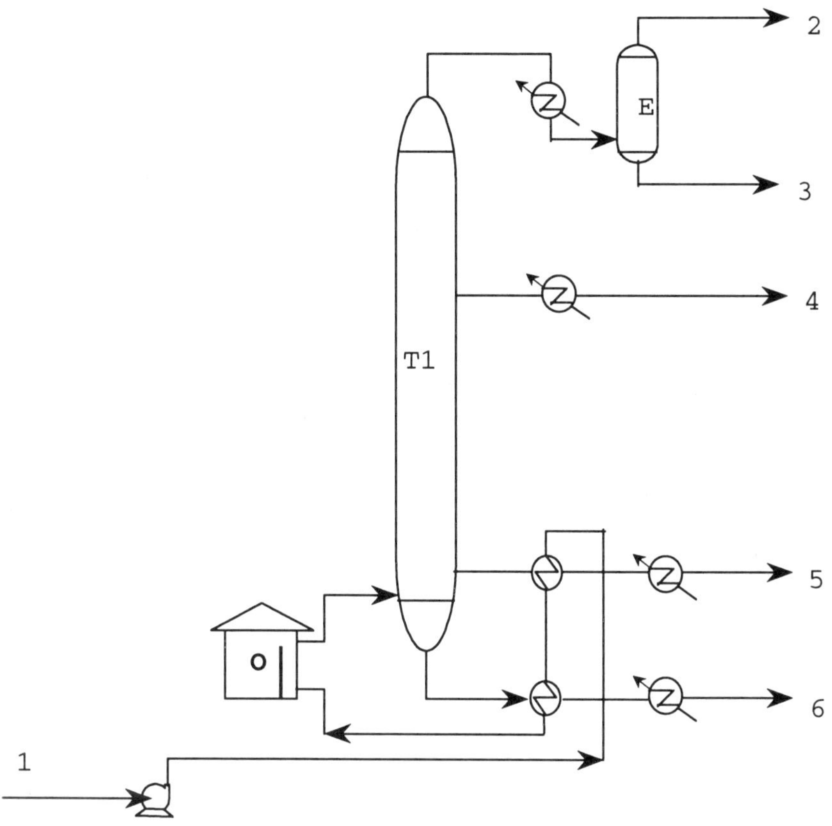

Fig. 5.15: The simple atmospheric rectification:
O – Oven
T1 – Rectification tower
E – Separator
1. Petroleum, 2. Light product, 3. Gasoline, 4. Kerosene, 5. Diesel, 6. Bottom product

Typical boiling ranges for various streams are as follows: light straight-run naphtha (head), 20-95°C; heavy naphtha (side-stream), 90-165°C; crude kerosene or jet (side-stream), 150-245°C; light gas oil or diesel (side-stream), 215-315°C. Non-vaporized oil entering the tower flows downward over the trays (below the feed tray) called stripping trays (Fig. 5.6). These act to remove any light constituents remaining in the liquid. In any case, steam is injected into the bottom of the tower in order to reduce the partial pressure of the hydrocarbons; this plays a similar role as a vacuum during vacuum rectification (see section 5.5). The residue that is obtained from the bottom of the tower is suitable for blending into residual fuels. Alternatively, it may be further distilled under vacuum conditions to yield quantities of distilled oils for manufacture into lubricating oils or for use as feedstock in gas oil cracking processes.

In the modern petroleum industry, a more complicated scheme of the rectification unit, the so called "two-step rectification", is used. An illustration of this type of rectification unit is shown in Figure 5.16.

The main difference between the simple rectification and two-step rectification is the presence of the second rectification tower (T2) in the two-step unit. In the latter case, the feed passes to the first rectification tower at a temperature of approximately 200–240°C. Because of the low temperature used, the volatile product from the first tower only forms part of the gasoline fraction. The bottom product from T1 (Fig. 3.16) passes to the oven for heating until it reaches a feed

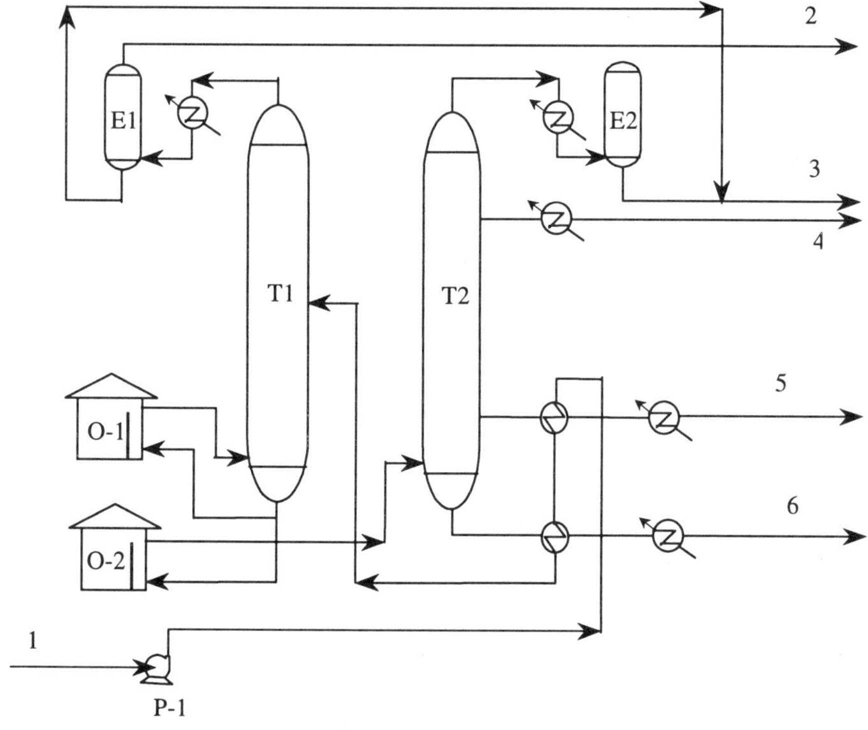

Fig. 5.16: "Two-step" rectification unit:
O-1 and O-2 – Oven
T1 and T2 – Rectification tower
E1 and E2 – Separator
1. Petroleum, 2. Light product, 3. Gasoline, 4. Kerosene, 5. Diesel, 6. Bottom product

temperature of approximately 300–340°C. Further rectification of the petroleum proceeds in the same way as in simple rectification. As compared with the two-step rectification scheme, the main disadvantage of the simple rectification scheme is that if very light (over 15% of gasoline) fractions or incompletely dry petroleum passes to the simple rectification unit, it can cause an increased pressure in the heat exchangers and the tower. This can lead to a decreased efficiency of the rectification unit or, in the worse case, can lead to breaking of heating units.

In both rectification methods, a bottom product is produced that will be separated into fractions in the vacuum distillation unit.

5.2.4 Vacuum Rectification

The principles and main units for vacuum rectification resemble those for atmospheric rectification. The major exceptions are that larger-diameter towers are used to maintain comparable vapor velocities at reduced operating pressures. A vacuum of 50 to 100 mm of Hg absolute is produced by a vacuum pump or steam ejector. The capacity of modern vacuum rectification units is about 3.5 million tons per annum.

The primary advantage of vacuum rectification is that it allows the distilling of heavier materials at lower temperatures than would be required at atmospheric pressure, thus avoiding thermal cracking of the components. The input temperature of the atmospheric residue in the vacuum tower usually does not exceed 425°C. The residue from atmospheric rectification is preheated against distillates and vacuum residue before heating up in the fired heater. From the heater outlet the stream is fed to the vacuum tower. The high specific volume of vapor at low pressure demands large tower diameters, particularly in the upper part. The lower part of the tower (below the feed inlet) is used as a stripping section using steam to reduce the partial pressure of hydrocarbons in the vapor phase.

An example of the one-step vacuum rectification is shown in Figure 5.17. In Figure 5.17, the feed (atmospheric residue) passes through oven O1 to the vacuum tower T1. In this tower, the feed is typically separated into the following fractions: vacuum residue 8, middle oil distillate 4, and light oil distillate 3. The middle oil fraction is fractionated in tower T2 into fraction with narrower boiling temperature ranges. Steam 2 is used to reduce the partial pressure of the oil fraction which, finally, helps in the separation of the light fractions from the oil.

The one-step vacuum rectification unit does not allow for production of oil fractions with desired market quality. This is why one-step vacuum rectification units can be found nowadays only in small refineries. The scheme that allows for the production of oil fractions with a higher quality is the two-step vacuum rectification. An illustration of the two-step vacuum rectification plant is shown in Figure 5.18.

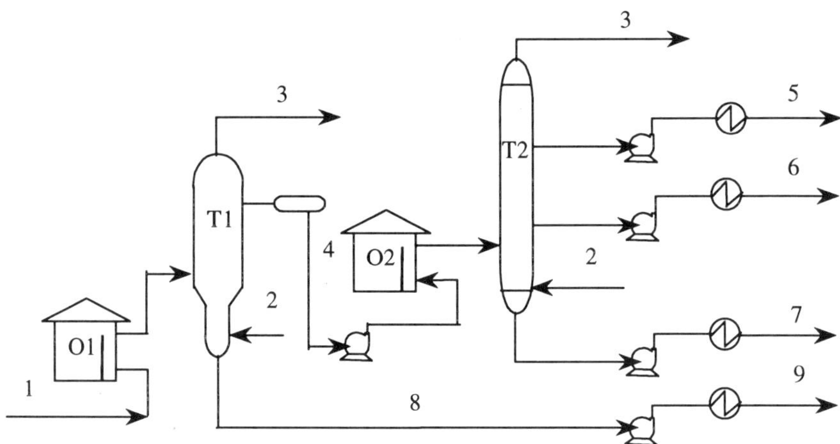

Fig. 5.17: One-step vacuum rectification:
O-1 and O-2 – Oven
T1 and T2 – Rectification tower
1. Residue from the rectification under the atmosphere pressure, 2. Light oil, 3-6. Distillate, 7-9. Vacuum residue

In the two-step rectification unit, the feed 1 in the first tower is distilled to obtain the following fractions: light oil 2, middle oil 3, and partly distilled vacuum residue 6. The vacuum residue from the first tower passes through oven O1 to rectification tower T2, where it is fractionated to narrower fractions. In comparison with the one-step scheme, the two-step scheme requires more energy for production, but the quality of the oil fractions is much higher.

The residue remaining after vacuum rectification is called "Goudron". This may be used for blending to produce road asphalt or residual fuel oil, or it may be used as a feedstock for thermal cracking or coking units. Vacuum rectification units are an essential part of the many processing units required for the production of lubricants.

In modern refineries, atmospheric and vacuum rectification processes are rarely carried out in separate units. Usually, combined atmospheric-vacuum rectification units are used for these processes. An illustration of a scheme for this combined unit used in refineries in Russia is shown in Figure 5.19.

Petroleum 1 for rectification is fed from storage tanks at ambient temperature through a series of heat exchangers in order to attain the temperature of approximately 120°C (see Fig. 5.19). A controlled amount of fresh water is introduced, and the mixture is pumped into an electrical desalting and drying unit (EDDU), where it passes through an electrical field and a salt water phase is separated. If the salt is not removed at this stage, it will be deposited later on the tubes of the ovens or heat exchangers. This can cause plugging and corrosion.

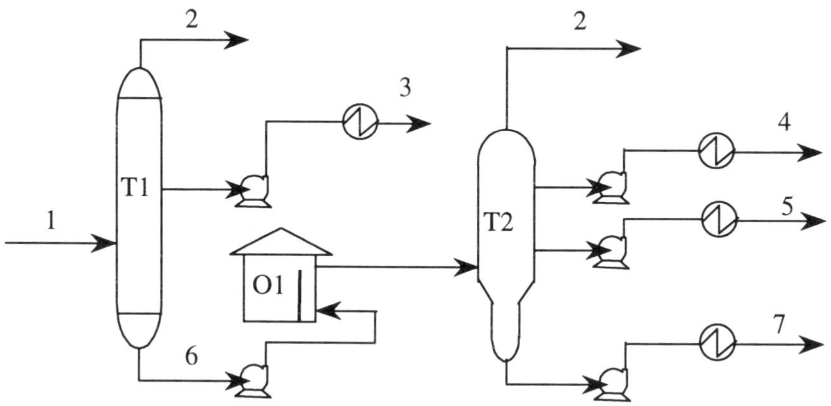

Fig. 5.18: Two-step vacuum rectification:
O-1 – Oven
T1 and T2 – Rectification tower
1. Residue from the rectification under the atmosphere pressure, 2.Light distillates, 3-5. Vacuum distillate, 6-7. Vacuum residue

The desalted crude oil passes through additional heat exchangers to the first atmospheric tower T1. Light gasoline 2 is separated at the top of this tower from a mixture of anti-corrosive additives added to the raw petroleum because of its high sulfur content. The sulfur compounds can be very corrosive to the metallic tower walls and this can lead to destruction of the units. Anti-corrosive additives slow down corrosion of the rectification units. The bottom product from rectification tower 1 passes to oven O2. There, it is heated to a temperature of approximately 350°C. A mixture of vapor and petroleum liquid passes from oven O2 into rectification tower T2. The volatile product obtained in T2 is gasoline fraction. This fraction passes to the next rectification tower T3, where it is fractionated into fractions of light hydrocarbons, gases 3, and residue. The residue from T3 passes to the section for secondary gasoline distillation (SSGD), where the gasoline fraction is fractionated into narrower fractions (4–7) used for the production of fuels and petrochemicals. The side-streams from rectification tower T2 consist of the following fuel fractions: 8 – kerosene fraction with boiling range of 180–230°C, 9 – light diesel fraction (230–280°C), and 10 – heavy diesel fraction (280–350°C). Rectification tower T2 employs intermediate reflux streams, which come from the stripping section of three towers. Stream 15 is fed to the stripping sections of the three towers as well as tower T2 in order to reduce the partial pressure of the light products during rectification.

Petroleum from T2 that is not vaporized passes to oven O4, where it is heated to a temperature of approximately 425°C (the input temperature in vacuum tower T4). After this, the heated atmospheric residue passes to the vacuum tower

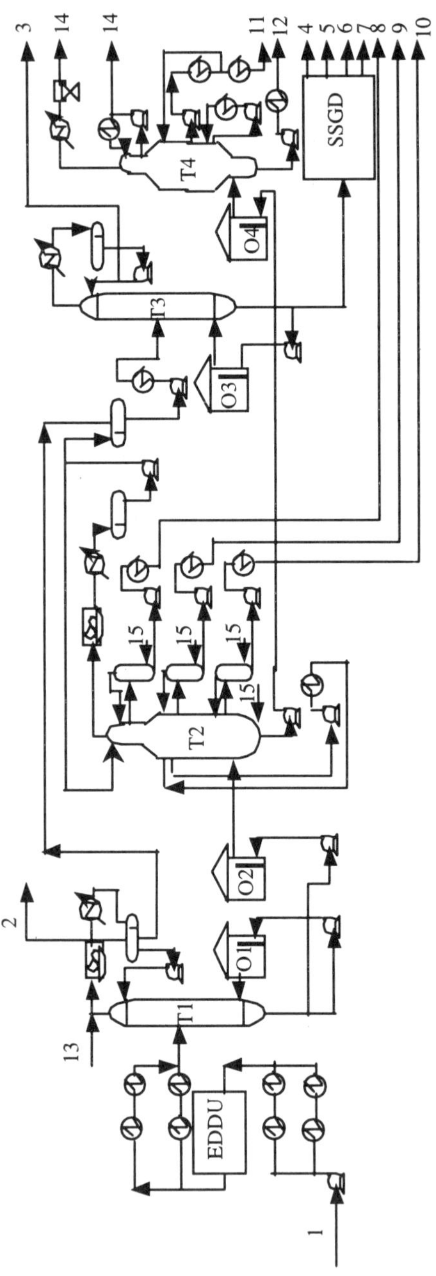

Fig. 5.19: Combined atmospheric vacuum rectification unit:

T4 where it is fractionated into: oil fraction 11, gases, cracking products and the rest of heavy diesel fraction 14, and vacuum residue or goudron 12.

It is seen that in the illustration (Fig. 5.19), a two-step atmospheric rectification unit is combined with a one-step vacuum rectification unit. In modern refineries, however, the combined rectification unit consists of a two-step atmospheric unit and a two-step vacuum rectification unit. At the MIDER refinery in Germany (the most modern refinery in Europe), a combination of a three-step atmospheric and a one-step vacuum rectification unit is used. A scheme for this unit is shown in Figure 5.20.

The unit shown in Figure 5.20 saves about 50,000 tons of fuel oil per year, compared with a one-step rectification process with the same capacity. The process development was based on the objective of avoiding unnecessary overheating of light components. Additionally, it avoids the thermal degrading associated with drawing off of heavy cuts. To this end, crude oil is pre-fractionated in the tower operating with best conditions of 125°C and 2 bar, and then in a second stage at 131°C and 1 bar. The pre-distilled petroleum is then fractionated in the main atmo-

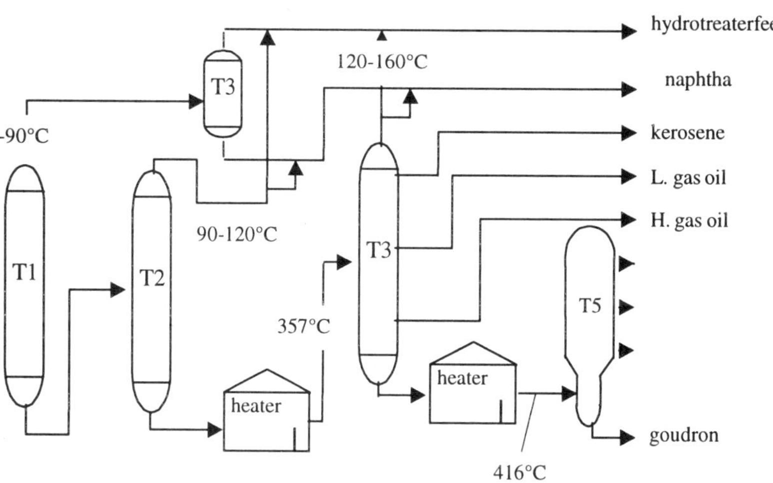

Fig. 5.20: Combined rectification unit:
T1-T5 – Rectification towers

Fig. 5.19 Legend:

O-1–O-4 – Oven
T1-T4 – Rectification tower
EDDU – Electrical desalting and drying unit
SSGD – Section for secondary gasoline distillation
1. Petroleum, 2. Gases, 3. Light gasoline, 4-7. Narrow fractions, 8. Kerosene, 9. Light diesel, 10. Heavy diesel, 11. Oil fraction, 12. Vacuum residue, 13. Fraction 180–230°C, 14. Heavy diesel, 15. Water - steam

spheric tower at 1 bar and 357°C. The bottom product (atmospheric residue) is subsequently heated to 416°C and passed to the vacuum column.

Processing of petroleum under the above-specified conditions results in a vacuum residue of 585°C cut point. The combined rectification yields the following cuts:

- Straight-run gasoline (boiling begins 80°C)
- Naphtha as feedstock for the petrochemical industry (80-95°C)
- Heavy naphtha (96-160°C)
- Kerosene (160-225°C)
- Light and heavy gas oil and vacuum gas oil (225-400°C)
- Medium and heavy vacuum fraction as cat cracker feed (400-585°C)
- Vacuum residue for bitumen production and feed for visbreaker.

5.5 Heat Exchangers and Separators

The distillation tower is the most important equipment used in rectification units. The different types of rectification towers and schemes have been described already in the preceding sections. In addition to towers, there are also other major items of equipment used in the rectification unit. These are essentially units used for heat transfer. Heat exchangers play a very important role not only in rectification units, but also in petroleum processing as a whole. All the heat exchangers used in the petroleum industry can be classified as follows:

- air-coolers
- tube-bundle or shell and tube heat exchanger
- finned heat exchanger
- U-tube heat exchanger
- double-pipe heat exchanger

The first four types of heat exchangers are especially popular in the petroleum industry. The double-pipe heat exchanger is only used for heating high viscosity materials such as vacuum residue or bitumen. Figure 5.21 shows an example of the horizontal tube-bundle heat exchanger.

The example of the heat exchanger shown in Figure 5.21 is used in the low temperature range because of thermal expansion of the tubes. The U-tube heat exchanger and finned heat exchanger were developed for operating at higher temperatures. The last has the largest temperature interval of operation. Figures 5.22 and 5.23 show examples of these two types of heat exchangers, respectively.

From Figures 5.22 and 5.23, it is seen that there is room to accommodate the thermal expansion of the tubes. In the case of the finned heat exchanger, the tubes can expand in an expandable shell, whereas in the case of the U-tube heat exchanger, expansion room is provided because of the existence of free fixed tubes in a firm shell.

The heat exchangers can be installed both vertically and horizontally. The heaters for the rectification-tower bottom-product reflux are special units of interest. There are four types of heat exchanger schemes. These are shown in Figure 5.24.

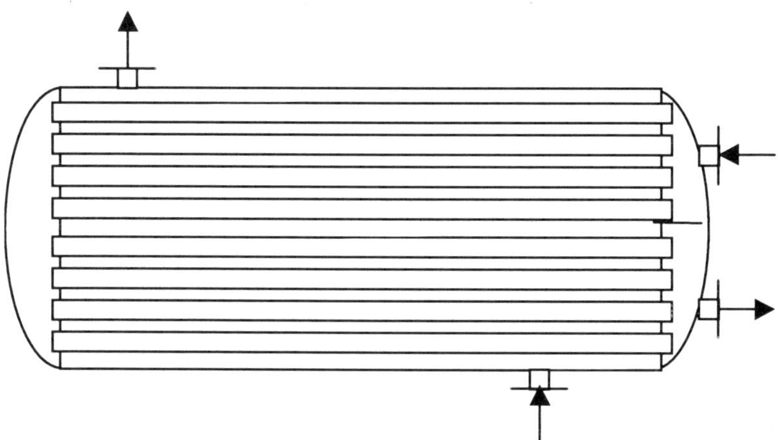

Fig. 5.21: Horizontal tube-bundle (shell and tube) heat exchanger.

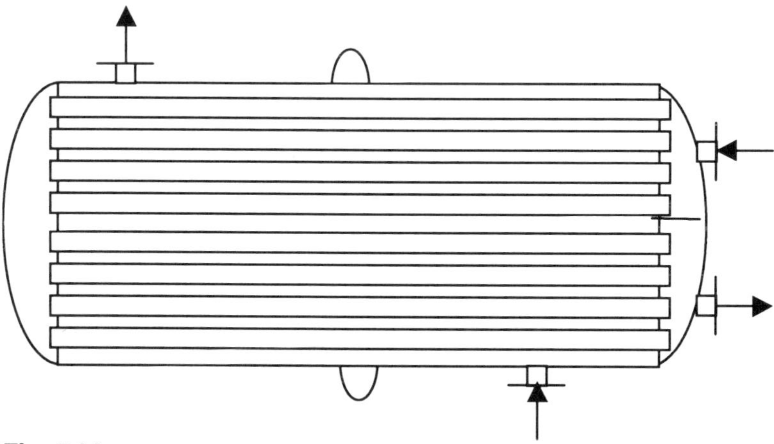

Fig. 5.22: Finned heat exchanger.

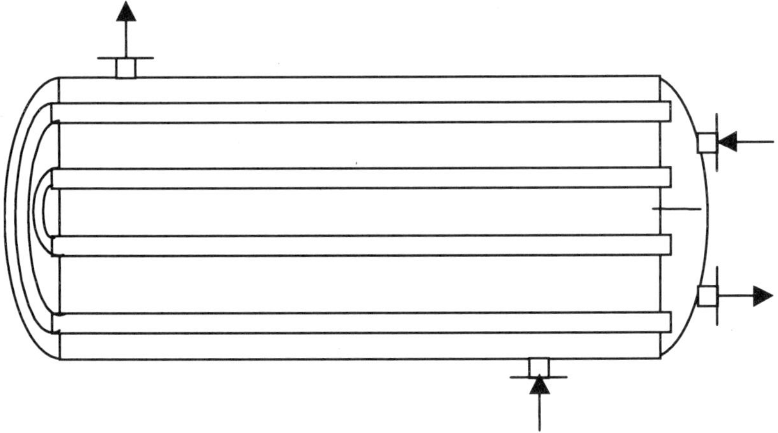

Fig. 5.23: U-tube heat exchanger.

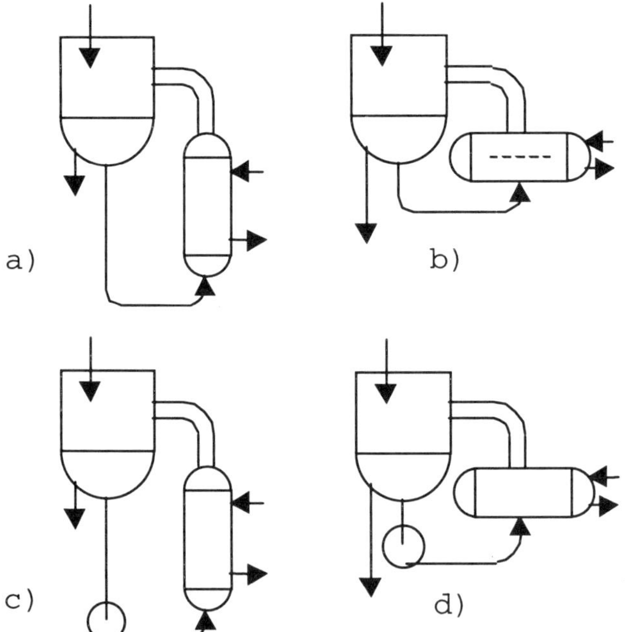

Fig. 5.24: Examples of the bottom product reflux:
a) Natural reflux with vertical heater
b) Natural reflux with horizontal heater
c) Forced reflux with vertical heater
d) Forced reflux with horizontal heater

The schemes with natural reflux are operated based on the principle of natural convection (i.e. circulation based on density difference between cold and heated fluid streams) to create a thermo-siphon. The hotter stream has the lower density. In natural reflux, the heated feed passes to the top of the heat exchanger and then to the bottom of tower by itself while the colder stream drains from the tower bottom to the heater. Forced reflux is based on using pumps for the circulation.

The last major items of equipment in rectification units are the separators. Separators are also used in many other technological schemes in the petroleum industry where there is the need to separate materials based on density difference. Usually, separators are vessels and can be classified as:

- horizontal
- vertical
- spherical

6

Processing of Light and Heavy Distillates

INTRODUCTION

In foregoing sections, we presented only the primary processes for petroleum treatment. It is important to note that all these processes are called "primary processes" because the petroleum only undergoes physical treatment. This means that chemical changes to the feed do not occur. All the processes we will describe in the sections of this chapter are referred to as "secondary processes". This is intended to indicate that chemical changes of the feed occur during these treatments.

6.1 THERMAL CRACKING

With the advent of the mass production of large numbers of gasoline-powered vehicles, the demand for gasoline has grown dramatically. On the other hand, distillation processes produce only a specific amount of gasoline from a given quantity of crude oil of specific characteristics. In 1913, the thermal cracking process was developed. In this process, heavy fuels containing large molecules are broken into smaller ones to produce additional gasoline and distillate fuels by application of both pressure and intense heat.

Thermal cracking is a radical chain process. The chain process contains three main stages: chain start, chain growth and chain termination.

In the "chain start", cracking of the hydrocarbons in the heavy feed proceeds at the weakest bonds of the hydrocarbons. Also, cracking of C-C bonds proceeds first, because the energy requirement for C-C bond breakage in hydrocarbons is always lower than that for C-H bond scission. In a long chain, the weakest C-C bond (C-C bond with the lowest energy requirement for breakage) is at the center of the molecule. Furthermore, the difference in energy of the C-C bond at different locations in a molecule is lower at a higher temperature than at lower temperatures. This means that at moderate temperature (about 400–450°C), cracking of the hydrocarbon chains proceeds symmetrically. At a higher temperature, cracking can proceed almost with the same probability at every C-C bond in the hydrocarbon molecule.

It should be noted that paraffins and olefins are formed during paraffin cracking (see (6.1)).

(6.1)

The olefins formed during thermal cracking are characterized by the fact that the bond in the β-position (the second bond from the double bond) is weaker than the C-C bond in the paraffin chain. The energy of the bond in the paraffin chain is approximately 320 kJ/mol, whereas the energy of the bond in the β-position is 259 kJ/mol and the energy of the C-C bond in the α-position is 371 kJ/mol. This means that the olefins formed during cracking can be cracked more extensively than the initial paraffins.

The side chains of aromatic rings can be cracked very extensively. The energy of the bond in the β-position for these compounds is 273 kJ/mol.

The next step after "chain start" is "chain growth". The same reactions as occurred during "chain start" also occur in this stage of thermal cracking. However, the most important reaction of "chain growth" is the reaction for formation of light stable molecules (HR), from the radical (·R) formed during chain start as well as from the heavier radical from the feed molecule (6.2).

$$\cdot R + CH_3\text{-}CH_2\text{-}CH_2\text{-}CH_2\text{-}CH_3 \rightarrow HR + \cdot CH_2\text{-}CH_2\text{-}CH_2\text{-}CH_2\text{-}CH_3 \quad (6.2)$$

Different radicals have different reactivity for thermal cracking. In the following example, the radicals are presented in order of their reaction abilities:

$$R\text{-}H_2C\cdot > (R)_2\text{-}C\cdot > (R)_3\text{-}C\cdot$$

Reaction (6.2) is one of the possible reactions of "chain growth". Another reaction that can also occur in this stage is the cracking of radicals at the β-bond (6.3).

$$\begin{array}{c} \cdot CH_2\text{-}CH_2\text{-}CH_2\text{-}CH_2\text{-}CH_3 \rightarrow \cdot CH_2\text{-}H_2C\cdot + \cdot CH_2\text{-}CH_2\text{-}CH_3 \\ \downarrow \\ CH_2{=}CH_2 \end{array} \quad (6.3)$$

Another reaction type that occurs during the "chain growth" step of thermal cracking is the addition reaction involving a radical and a double bond (6.4).

$$\cdot CH_3 + CH_2{=}CH\text{-}CH_3 \rightarrow CH_3\text{-}CH_2\text{-}\dot{C}H\text{-}CH_3 \quad (6.4)$$

Chain termination is the last stage of every chain reaction. There are two types of reactions that typically occur in thermal cracking chain termination. One type is radical recombination (6.5).

$$\cdot CH_2\text{-}CH_3 + \cdot CH_2\text{-}CH_2\text{-}CH_3 \rightarrow CH_3\text{-}CH_2\text{-}CH_2\text{-}CH_2\text{-}CH_3 \quad (6.5)$$

The second type is radical disproportionation (6.6).

$$\cdot CH_2\text{-}CH_3 + \cdot CH_2\text{-}CH_2\text{-}CH_3 \rightarrow CH_3\text{-}CH_3 + CH_2{=}CH\text{-}CH_3 \quad (6.6)$$

The greatest problem during thermal cracking arises from reactions involving aromatic feed. Aromatic compounds in the feed have a very high tendency to undergo polycondensation reactions that lead to coke formation. Coke formation decreases the yields of the desired gasoline and diesel fractions. One example of a polycondensation reaction is shown in reaction (6.7).

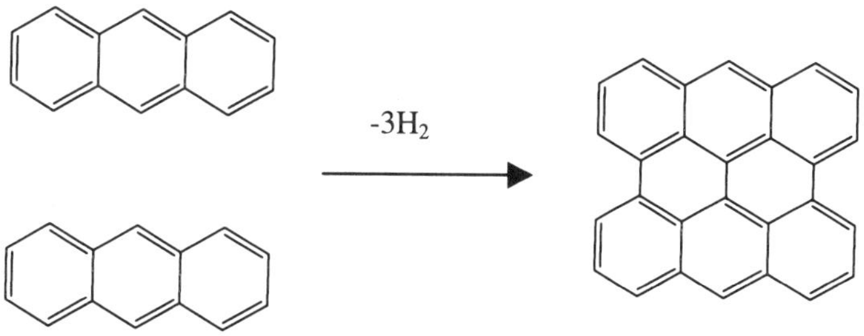

$-3H_2$

(6.7)

Reaction (6.7) is an illustration that molecules of some products from thermal cracking reactions can sometimes be larger than the feed molecules.

It is not only the mechanism that is important in the processing of light and heavy distillates. The kinetics of the chemical process is also important. Chemical kinetics deals with the velocity of material conversion by the chemical reactions.

The smallest step of a chemical reaction is called an elementary reaction. The reaction equation for an elementary reaction j with N components can be presented as in equation (6.8).

$$\sum_{i}^{N} \nu_i \cdot X_{i,j,R} = \sum_{N}^{N} \nu_i \cdot X_{i,j,P} \tag{6.8}$$

where ν_i – stoichiometric coefficients
$X_{i,j,R}$ – reactant i in reaction j
$X_{i,j,P}$ – product i in reaction j

Depending on the sum of the stoichiometric coefficients of the reactants involved in the elementary reaction, one refers to such a reaction as being monomolecular, bimolecular, termolecular, etc. The concentration change caused by one elementary reaction can be described by equation (6.9).

$$d[X_{i,j}] / d[t] = \nu_{i,j} \cdot r_{i,j} \tag{6.9}$$

where $r_{i,j}$ – reaction velocity of reaction j

Reaction velocity is proportional to the concentration of the reactants raised to the power of the reaction order. Reaction order is equal to the stoichiometric coefficient for an elementary reaction. It has been shown in the foregoing discussion that thermal cracking is a very complicated process involving reversible and irreversible reactions. In this case, the reaction velocities for the elementary reactions of the cracking process can be described by equation (6.10).

$$r_j = k_j \cdot \prod_{i}^{N} (X_{i,j,R})^{\nu_{i,j}} - k_{-j} \cdot \prod_{i}^{N} (X_{i,j,P})^{\nu_{i,j}} \tag{6.10}$$

Where k_j – reaction velocity coefficient for the forward reaction
k_{-j} – reaction velocity coefficient for the reverse reaction

The ratio of k_j to k_{-j} is called equilibrium constant. This constant depends on the thermodynamic properties of the system and can be calculated by equation (6.11).

$$K_j = \frac{k_j}{k_{-j}} = \prod_{i}^{N} \frac{(X_{i,j,R})^{\nu_{i,j}}}{(X_{i,j,P})^{\nu_{i,j}}} \tag{6.11}$$

where K_j – equilibrium constant for reaction j

The equilibrium constant calculated in this manner shows the maximum possible degree of conversion of the feed into products. However, as was mentioned earlier, the equilibrium constant can be calculated based on the thermodynamic conditions of the system. This can be done by using equation (6.12)

$$K_{P,j} = \exp(-\Delta G_j/RT) \qquad (6.12)$$

where $K_{p,j}$ – equilibrium constant for reaction j, at temperature T and pressure P
ΔG_j – Gibbs energy difference for reaction j, at temperature T and pressure P
R – gas constant (8.317 J/mol)
T – reaction temperature (K)

The reactions that proceed with changes in volume and pressure are very important. Thermal cracking proceeds at high temperatures, and many products of this process are gaseous. This means that there is a volume increase during the reaction. This dependence can be described by equation (6.13).

$$\Delta G = \Delta G_0 + 19.13 \cdot \log(P \cdot T) \qquad (6.13)$$

where ΔG – Gibbs energy difference for reaction j, at temperature T and pressure P
ΔG_j – Gibbs energy difference for reaction j, at temperature T and by standard pressure
P – pressure in the reaction zone
T – reaction temperature

It is obvious from equations (6.12) and (6.13) that the higher the pressure is, the smaller is the equilibrium constant. This is valid for reactions such as thermal cracking, which proceed with a volume increase. It means that a higher pressure leads to an acceleration of polycondensation, alkylation, hydrogenation and other reactions that proceed with a volume decrease.

The thermodynamic possibility of a reaction can be estimated by the value of the Gibbs energy difference. Gibbs energy is the part of the internal energy of the substance, which can be converted into work. This means that a reaction can go spontaneously only in the case where the difference between the Gibbs energy of products and reactants is negative. This implies that part of the Gibbs energy has been converted into the work needed for making the reaction go.

Gibbs energy contains two values: enthalpy and entropy. Generally, this energy can be presented as in equation (6.14).

$$G = H - T \cdot S \qquad (6.14)$$

where G – Gibbs energy at temperature T
H – enthalpy at temperature T
S – entropy at temperature T

T – temperature

The next equation (6.15), which is a derivation from equation (6.14), is used for the calculation of the difference of the Gibbs energy. This equation is known as Gibbs-Helmholtz equation.

$$\Delta G = \Delta H - T \cdot \Delta S \qquad (6.15)$$

where ΔG – difference in Gibbs energy at temperature T
ΔH – difference in enthalpy at temperature T
ΔS – difference in entropy at temperature T
T – temperature

The difference in enthalpy in this equation represents the total energy change that takes place between the system and its environment. The multiplication of temperature and the term involving the difference in entropy represents the energy used to take care of the intermolecular activity. This is wasted energy and has to be subtracted from the total energy. An analogy can be drawn here between a mechanical engine and a chemical reaction. When a mechanical engine performs useful work, not all of the energy output of the engine goes towards work. Some of the energy output is wasted to overcome friction of the moving parts. The energy to overcome friction is wasted energy and must be subtracted from the total energy output in order to obtain useful energy that is capable of performing a task or work. Similarly, the difference in Gibbs energy represents useful energy of a chemical reaction.

The difference in the Gibbs energy for many reactions can be calculated based on empirical tables that are available in references 7-10 at the end of part III of this book.

However, most of the calculations of Gibbs energy are based on the rule that Gibbs energy has a linear dependence on temperature. Usually, this dependence is as presented in equation (6.16).

$$\Delta G = A + B \cdot T \qquad (6.16)$$

where A and B – coefficients specific for the chemical reaction
T – temperature

More about thermodynamic evaluations can be found in references 11 and 12 at the end of part III of this book.

Thermodynamics can only show how much of the reactants can be converted into the product. Usually, thermal cracking reactions are relatively slow reactions. This is why thermodynamic evaluation alone is often not enough to guide the industrial thermal cracking process since this would suggest carrying out the reactions at a low temperature to obtain a high conversion. Reactions at a low temperature can be quite slow and impossible to carry out at industrial scale. Thus, often it becomes necessary to increase the temperature of the process in order to increase the velocity of the reaction, despite that this may not lead to the maximum possible degree of conversion of the reactants. It is well known that increase of the temperature by 10°C leads to an increase in the reaction velocity of 2–4 times. This dependence can be described by the Arrhenius equation (6.17).

$$k = k_0 \cdot \exp(-E_A/RT) \quad (6.17)$$

where
k – reaction velocity constant
k_0 – frequency factor
E_A – activation energy
R – gas constant (8.317 J/mol)
T – reaction temperature (K)

Activation energy in equation (6.17) shows the minimal energy level which molecules must have to be able to react. The physical significance of this value can be explained based on Boltzmann law, which is presented in equation (6.18).

$$N_E = N \cdot \exp(-E/RT) \quad (6.18)$$

where
N_E – number of molecules with energy higher than E
N – total number of molecules
E – energy
R – gas constant (8.317 J/mol)
T – reaction temperature (K)

The Arrhenius equation can be developed based on the Boltzmann's law. The physical significance can be better explained based on a comparison of equation (6.17) and (6.18). For example, the frequency factor can be compared with the number of particles which are available in the reaction zone and can potentially react to form the product. Reaction velocity constant can be compared with the number of molecules that has energy higher than the activation energy.

The graphical presentation of Boltzmann's law can show the temperature influence on reaction velocity. This is presented in Figure 6.1. The diagram of the distribution of the kinetic energy on the molecules (Figure 6.1) shows that more molecules with a high energy level are available at the higher temperature. If one assumes that only the collisions of molecules with a certain energy (i.e. activation energy) can lead to reaction, then the number of these molecules increases with a temperature increase, and thus the increase in the reaction velocity. This implies that only the molecules with energy higher than the activation energy for the reaction can react to form reaction products. The effect of the reaction rate increase with increasing temperature is due to the increase in the number of molecules with sufficiently high energy.

Every real chemical reaction leads to formation of product over many steps. It also leads to formation of compounds that exist only for a short time as well. This was shown in reaction (6.2). This chemical reaction can be represented by an overall general equation (6.19).

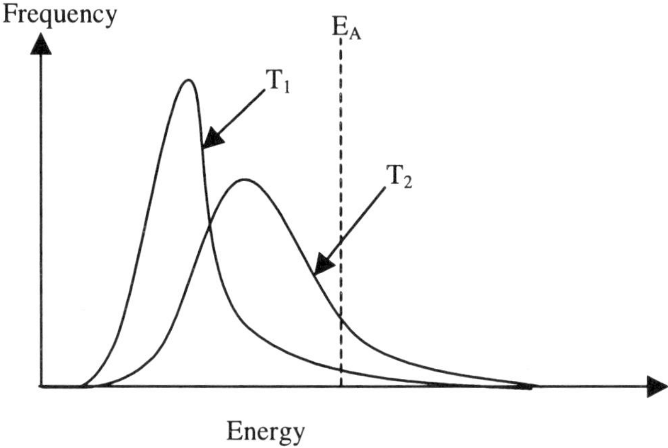

Fig. 6.1: Graphical presentation of Boltzmann's law:
T_1 first temperature, T_2 second temperature, $T_1 < T_2$, E_A activation energy

$$A + B \rightarrow A \cdots B \rightarrow P \tag{6.19}$$

where A and B – reactants
A···B – transition complex
P – products

The transition complex in equation (6.19) is derived from the collision theory. The main idea in this theory is that two molecules form a transition complex by collision, which can exist only for a short time. Then, the activation energy is the difference in the energy levels of the reactants and the transition complex. This is presented in Figure 6.2.

From Figure 6.2, it is obvious that the formation of a transition complex is possible only if a higher energy than the activation energy is released by the collision of two molecules. This implies that the kinetic energy of the movement of molecule, as well as the rotational energy and vibrational energy of the reactants must be higher than the activation energy to enable the formation of the transition complex.

From equation (6.17) and Figure 6.1, it is obvious that temperature has a large influence on every type of reaction. Desired reactions during thermal cracking are reactions that lead to breakage of hydrocarbons. These types of reactions are referred to as monomolecular reactions. This means that only one molecule of reactant is needed to react to the product. The reaction velocity of such a reaction can hardly be influenced by pressure. On the other hand, bi-molecular reactions can be greatly influenced by pressure. These include polycondensation, alkylation, hydration, and recombination. Most polymolecular reactions are undesired in thermal cracking. However, most modern thermal cracking plants operate at high pressures. The reason to carry out the industrial process under pressure is to prevent further cracking of the light cracking products to gas. It should be noted that the desired product during thermal cracking is the light liquid fraction.

One additional kinetic theory that deserves to be mentioned is the theory based on the Eyring equation. The essence in this theory is captured by the following phrase credited to Henry Eyring in 1945: "... a molecular system ... passes ... from one state of equilibrium to another ... by means of all possible intermediate paths, but the path that is most economical in energy will be more often traveled." The equation of the chemical reaction (6.19), which was used for the interpretation of the kinetic theory based on the Arrhenius equation, is modified in order to explain the Eyring equation, as shown in equation (6.20).

$$A + B \leftrightarrow A \cdots B \rightarrow P \tag{6.20}$$

where A and B – reactants
A···B – transition complex
P – products

It is obvious that the difference between equation (6.19) and (6.20) is that the formation of the transition complex A···B is an equilibrium reaction. This means, for example, that by reaction (6.20) shown for thermal cracking, the transition complex is an unstable compound that reacts immediately to products or back to reactants. The reaction velocity of the transition complex formation can be calculated by equation (6.21).

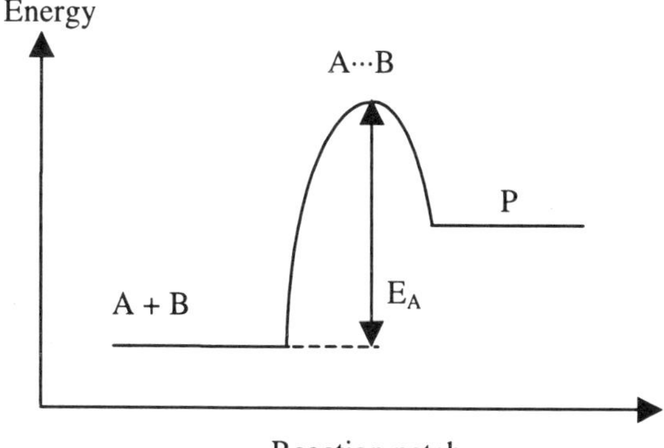

Fig. 6.2: Definition of activation energy.

$$\frac{d[A\cdots B]}{dt} = k_1 \cdot [A] \cdot [B] - k_{-1} \cdot [A\cdots B] - k_2 \cdot [A\cdots B] \tag{6.21}$$

where $[A]$ and $[B]$ – concentration of reactants
$[A\cdots B]$ – concentration of transition complex
k_1 – reaction velocity constant for the formation of transition complex from A and B
k_{-1} – reaction velocity constant for the consumption of the transition complex to form A and B
k_2 – reaction velocity constant for the formation of product P from the transition complex

The transition complex is an unstable compound that is independent of its reaction velocity. It is also permanently in equilibrium with reactants A and B. It means that the terms $k_1 \cdot [A] \cdot [B]$ and $k_{-1} \cdot [A\cdots B]$ in equation (6.21) are equal to each other. Hence, equation (6.21) can be rewritten more simply as equation (6.22).

$$\frac{d[A\cdots B]}{dt} = - k_2 \cdot [A\cdots B] \tag{6.22}$$

The reaction velocity constant k_2 can be calculated based on statistical thermodynamics. This is presented in equation (6.23).

$$k_2 = k_b \cdot T / h \tag{6.23}$$

where k_2 – reaction velocity constant
k_b – Boltzmann constant ($1.381 \cdot 10^{-23}$ J · K^{-1})
T – reaction temperature
h – Planck constant ($6.626 \cdot 10^{-34}$ J · s)

Because k_2 depends only on temperature, and this is a compound independent value, it is often called the universal constant of transition complex formation. The value of k_2 at room temperature for all types of reactions is approximately $6 \cdot 10^{-12}$ sec^{-1}.

The concentration of the transition complex at certain reaction times can be determined by the kinetic equation. It can also be determined by other means. It was shown earlier that the formation of the transition complex from reactants is an equilibrium reaction. This implies that the concentration of the transition complex can be determined based on the mass law for the equilibrium reactions as presented in equation (6.24).

$$[A\cdots B] = K^{=} \cdot [A] \cdot [B] \tag{6.24}$$

where $K^{=}$ – equilibrium constant

Equation (6.22) can be rewritten based on equations (6.23) and (6.24) to the form in equation (6.25).

$$\frac{d[A\cdots B]}{dt} = -K^{=} \cdot k_b \cdot T \cdot [A\cdots B] / h \tag{6.25}$$

From a comparison of the classical kinetic equation (6.10), it is obvious that equation (6.25) can be written in a more simple form as equation (6.26) where the Eyring reaction velocity constant is calculated using equation (6.27).

$$\frac{d[A\cdots B]}{dt} = -k \cdot [A\cdots B] \tag{6.26}$$

where k – Eyring reaction velocity constant

$$k = K^{=} \cdot k_b \cdot T / h \tag{6.27}$$

It was shown earlier in equation (6.12) that the equilibrium constant depends on the thermodynamic characteristics of the reaction system. This dependency can be rewritten as in equation (6.28) for the transition complex.

$$\Delta G^{=} = -R \cdot T \cdot \ln(K^{=}) \tag{6.28}$$

Where $\Delta G^{=}$ – activation Gibbs energy at temperature T
R – gas constant (8.317 J/mol)
T – reaction temperature (K)
$K^{=}$ – equilibrium constant

The basic form of the equation for the calculating of the Gibbs energy was given by equation (6.14). However this equation must be rewritten by equation (6.29) to calculate the activation Gibbs energy.

$$\Delta G^{=} = \Delta H^{=} - T \cdot \Delta S^{=} \tag{6.29}$$

where $\Delta G^{=}$ – activation Gibbs energy at temperature T
$\Delta H^{=}$ – activation enthalpy at temperature T
$\Delta S^{=}$ – activation entropy at temperature T
T – reaction temperature

The activation enthalpy in the case presented in equation (6.26) can be described as the enthalpy difference between the reactants and the transition complex. This is presented in Figure 6.3.

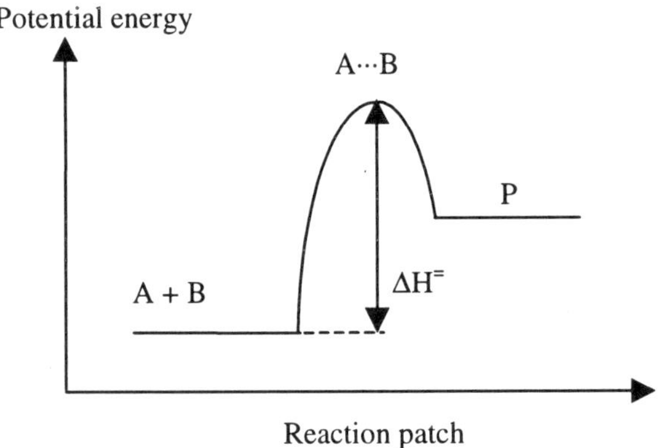

Fig. 6.3: Definition of activation enthalpy.

The meaning of the activation entropy can be explained in the same way as was done for the activation enthalpy. Activation entropy is presented as the difference between entropy of reactants and entropy of transition complex.

The thermodynamic meaning of activation Gibbs energy is the thermodynamic possibility of reaction from reactants to transition complex. However, the value of the activation Gibbs energy can be interpreted as a measure of the stability of the transition complex. The less the activation Gibbs energy is, the more stable is the transition complex formed during the reaction. The reaction velocity depends on the concentration of the transition complex, which increases with its increasing stability. There are three important consequences regarding the reaction system which depend on the activation Gibbs energy:

$\Delta G^{=} > 0$ – reaction is impossible

$\Delta G^{=} = 0$ – equilibrium reaction, there is no change of reactants concentrations

$\Delta G^{=} < 0$ – reaction happens spontaneously

The equilibrium constant $K^{=}$ can be calculated based on equations (6.28) and (6.29) as shown in equation (6.30).

$$\ln(K^{=}) = -\Delta H^{=} / R \cdot T + \Delta S^{=} / R \tag{6.30}$$

The Eyring equation can be written based on equations (6.27) and (6.30) as shown in equation (6.31).

$$k = (k_b \cdot T / h) \cdot \exp(-\Delta H^{=}/R \cdot T) \cdot \exp(\Delta S^{=}/R) \tag{6.31}$$

Equation (6.31) must be rewritten as a linear equation, as it is shown by equation (6.32), for the practical determination of the activation parameters.

$$\ln(k) = \ln(k_b \cdot T / h) - \Delta H^{\neq}/R \cdot T + \Delta S^{\neq}/R \tag{6.32}$$

However, equation (6.32) must be rewritten as shown in equation (6.33).

$$\ln(k/T) = \ln(k_b / h) - \Delta H^{\neq}/R \cdot T + \Delta S^{\neq}/R \tag{6.33}$$

All the thermodynamic parameters can be determined graphically as shown in Figure 6.4.

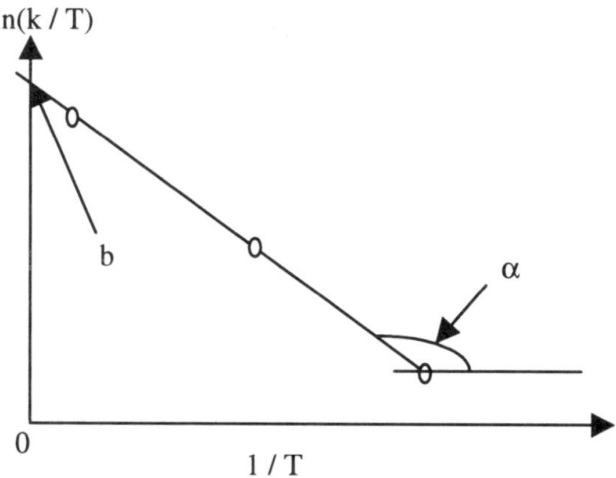

Fig. 6.4: Graphical determination of the thermodynamic parameters of the Eyring equation.

Activation enthalpy is determined from the value of the angle, α, as given by equation (6.34).

$$\Delta H^{\neq} = R \cdot \tan(\alpha) \tag{6.34}$$

Activation entropy is determined from the point of intersection, b, by equation (6.35).

$$b = \ln(k_b / h) + \Delta S^{\neq} / R \tag{6.35}$$

However, the use of the Eyring equation instead of Arrhenius equations must be done very accurately. If it is impossible to draw the line over the experimental points as used in the diagram shown in Figure 6.4, the use of the Eyring equation for calculating the kinetic parameters should be discontinued. The high value of the error in the Eyring equation method is often caused by the fact that in real reaction systems, especially complicated systems such as the cracking of crude oil, many reactions take place at the same time. This leads to the inappropriateness of using this theory by crude oil chemists in many cases. However, the Eyring equation gives much more information about the mechanism of the reactions occurring in the system in comparison to the Arrhenius equation. As was shown earlier, both theories have their advantages and disadvantages. Hence, the application of any of the theories must be decided in every case based on the complexity of the reactions occurring within the reaction system and the importance of studying the reaction mechanism for the system.

More about kinetic evaluation of chemical reactions and the different types of chemical reaction can be found in references 13-16 at the end of part III of this book.

The kinetic evaluation of the chemical reactions can be carried out by much more complicated models than has been presented. The evaluation with such models is often possible by numerical methods alone. There are many programs and special software packages in the market which can be used for kinetic evaluation. References 17-20 show a few programs and methods which can be used for the evaluation of cracking processes.

Thermal cracking units provide a severe treatment to the feed and, often, convert up to 50% of the incoming feed to naphtha and light diesel oils.

To obtain a more severe treatment of the feed during thermal cracking, a two-stage cracking unit is employed. A schematic of such a unit is shown in Figure 6.1.

In the scheme (Fig. 6.5), the feed (heavy residue from atmospheric rectification) passes to the bottom part of tower T3 and to the upper part of the low-pressure evaporator T4. The feed in T4 mixes with the heavy gas oil vapors and then passes to T3. The feed from the bottom of T3 passes to oven O1 for the heavy feed.

Tower T3 has a blind tray in its middle section. The vapor from the bottom part of the tower can pass through this tray to the upper section of the tower, but the liquid from the upper section cannot pass to the bottom section. The liquid from the upper section of tower T3 passes to the deep cracking oven O2.

The cracked products from both the ovens are directed to the reaction tower T1. The mixture of vapor and liquid from T1 passes to the high-pressure evaporator T2. In T2 the cracking residue is separated from the vapor. This residue passes to the low-pressure evaporator T4, and is partly evaporated. The evaporator T4 has a blind tray in the middle section similar to that of tower T3. The vapor product (heavy gas oil) is partly condensed in the upper section of T4 and mixed with the fresh feed. The non-condensed part leaves the cracking unit as kerosene – gas oil fraction 2.

The volatile product from T2 passes to the rectification tower T3. The gas 1 and gasoline fractions leave the upper section of tower T3. The gasoline fraction passes to the rectification tower T5 for further fractionation into fractions of light gasoline 3 and gasoline fraction 4.

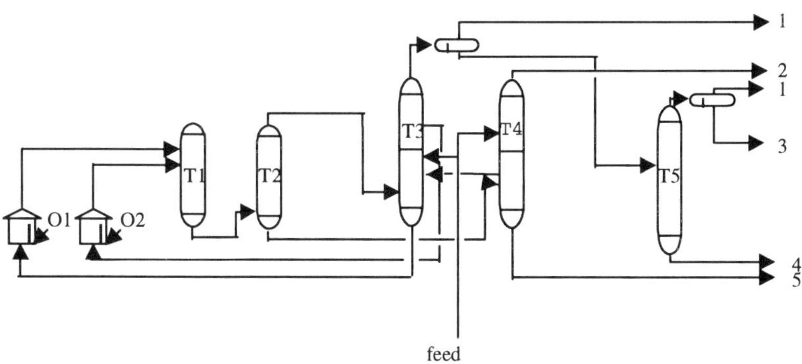

Fig. 6.5: Two-stage thermal cracking unit:
O1 and O2 – Oven
T1 – Reaction tower
T2 – High pressure evaporator
T3 – Low pressure evaporator
T4 – Rectification tower
T5 – Rectification tower
1. Gas, 2. Kerosene – gas oil fraction, 3. Light gasoline, 4. Gasoline, 5. Residue

The last product of the thermal cracking unit is the cracking residue 5, the bottom product of the low-pressure evaporator T4.

Thermal cracking processes can only convert up to 50% of the feed, as stated previously. Catalytic cracking was developed in order to improve the conversion level of the heavy feed.

6.2 CATALYTIC CRACKING

The use of thermal cracking units to convert gas oils into naphtha began in 1913. These units produced small quantities of unstable naphtha and large amounts of by-product coke. While they succeeded in providing a small increase in gasoline yields, it was the commercialization of the fluid catalytic cracking process in 1942 that really established the foundation for modern petroleum refining. The process not only provided a highly efficient means of converting high-boiling gas oils into naphtha to meet the rising demand for high-octane gasoline, but it also represented a breakthrough in catalyst technology.

The use of a catalyst in the cracking reaction increases the yield of high-quality products under much less severe operating conditions than in thermal cracking. Several complex reactions are involved, but the principal mechanism by which long-chain hydrocarbons are cracked into lighter products can be explained by the carbonium ion theory. According to this theory, the catalysts can be classified into two groups as follows:

- catalysis on metals
- catalysis on acid catalysts

Before we look at the mechanism of catalytic cracking on catalysts, it is appropriate to give a definition of a catalyst. Catalysis is derived from the Greek word "katalysis" meaning destruction or weakening. A catalyst is a substance that changes the rate of a chemical reaction but remains chemically unchanged at the end of the reaction.

All the catalytic processes can be divided into heterogeneous and homogenous catalytic processes. All the reactions take place on the surface of the catalysts during heterogeneous catalysis. Thus, for this type of catalysis, it is especially important to select catalysts that have as large a surface area as possible. This means that porosity of the catalyst must be as high as possible since high porosity is generally responsible for a large surface area.

The reaction mechanism in a heterogeneous catalytic process is more complicated than the mechanism in a non-catalytic reaction. In the first place, there is the influence of many physical stages of the catalysis on the reaction itself and on the reaction velocity. The main stages of the catalytic reaction can be represented as follows:

1. diffusion of reactants through the liquid or gas film on the surface of the catalyst particle,
2. diffusion of reactants into the pores of the catalyst,
3. adsorption of the reactants on the inner surface of the catalyst pores,
4. chemical reaction on the inner catalyst surface,
5. desorption of products from the inner catalyst surface,
6. diffusion from the inner catalyst surface through the pores to the external catalyst surface,
7. diffusion of products through the liquid or gas film on the external surface of the catalyst particle into the reactor zone.

All these reaction steps are presented graphically in Figure 6. 6. From Figure 6.6, it is obvious that catalysis is a very complicated physical and chemical process. Also, each of the seven catalysis steps shown can proceed with a different velocity. The slowest stage is called the rate limiting step of the process, because this limits the overall velocity or rate of the process. The general velocity of the chemical catalytic reaction has the same value as the velocity of the slowest step of the catalytic process. This is why the reaction velocity of a catalytic reaction is called the effective reaction velocity.

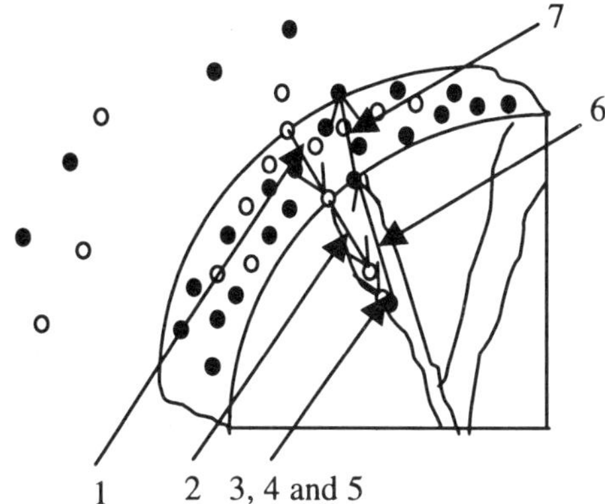

Fig. 6.6: Model of heterogeneous catalysis.
Black points – products
White points – reactants

Depending on the velocity of each of the various steps of catalysis, it is usual to divide it into the following three regimes of catalysis:

1. Kinetic regime – Reaction velocity is the slowest step of the catalytic process. The surface of the catalysts is used fully and most efficiently in this type of catalysis because the pore surface area and the external surface areas are used equally;
2. Pore diffusion regime – Diffusion in the pores is the slowest step in this type of catalysis. Reaction takes place mainly on the external surface of the catalysts,
3. Film diffusion regime – Diffusion through the film on the surface of the catalysts is the slowest step of the catalytic process. Reaction proceeds only on the outside surface of the catalyst.

Table 6.1 shows a comparison of the most important process constants for different regimes of catalysis.

According to the modern view on catalysis, a catalytic reaction can be represented as in equation (6.36).

$$R + X \rightarrow R\cdots X \rightarrow P\cdots X \rightarrow P + X \qquad (6.36)$$

where R – reactant
X – catalyst
R···X – transition complex of reactant with catalyst
P···X – transition complex of product with catalyst
P – product

Table 6.1: Comparison of some process parameters for different catalysis regimes.

	Reaction velocity constant	Mass transfer coefficient	Pore diffusion coefficient
Kinetic regime	small	large	large
Pore diffusion regime	medium	large	small
Film diffusion regime	large	small	very small

From equation (6.36), it is clear that this reaction has more than one activation energy. There are at least two activation energies. It is remarkable that each of the activation energies is smaller than the activation energy of the reaction without catalyst. Figure 6.7 shows a comparison of a non-catalytic reaction and a catalytic reaction.

It is obvious from Figure 6.7 that every chemical reaction can proceed and react from reactants to products only after it has been able to go over the reaction barrier, which is equal to the activation energy. It is to be seen that the non-catalytic reaction only has one activation energy, E_A. The reaction path of the catalytic reaction proceeds over three energetic barriers: E_{A1}, E_{A2} and E_{A3}. However, the adsorption energy of the reactants on catalyst, the activation energy of the catalytic reaction and the desorption energy of products from catalyst are each much smaller than the activation energy of the non-catalytic reaction. Consequently, a catalytic reaction proceeds faster than a non-catalytic reaction. Catalysts can be compared to a leader that leads the reaction over more energetically effective states than happens by non-catalytic reactions.

There are three theories of catalysis that are used in modern science to explain the mechanism of catalytic reactions. These are:

- geometric theory
- electron theory
- chemical theory

The main idea of the geometric theory of catalysis is the assumption that the activity of the catalyst depends on the geometrical form of the crystal grid of the catalysts. Catalysts with crystal grid similar in form to the reactant molecules are usually more active than catalysts with crystal grids different from the reactant molecules.

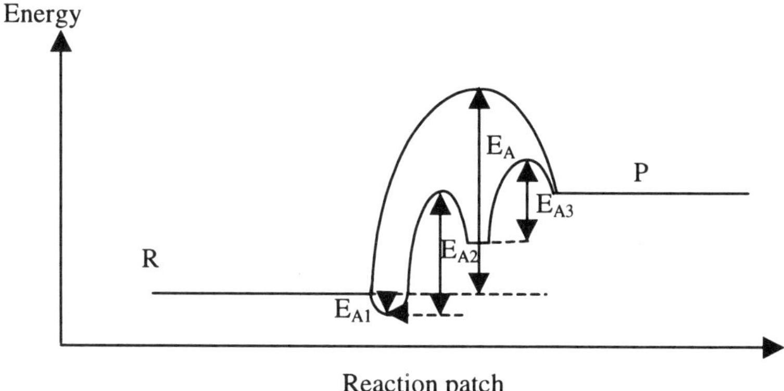

Fig. 6.7: Definition of activation energy during a catalytic process:
E_A – activation energy of the non-catalytic reaction of R to P
E_{A1} – adsorption energy of reactants on catalyst
E_{A2} – activation energy of catalytic reaction of R to P
E_{A3} – desorption energy of products from catalyst

Electron theory is based on the property of the electrons of catalyst to transfer electrical charge from catalysts to reactants. According to this theory, this means that the catalyst's activity is due to their ability to transfer electrical charge to the reactants thereby weakening the bonds in reactant molecules.

The last theory is based on the assumption that catalyst is a chemical substance which can form a chemical substance or transition complex with one reactant, and then the other reactant reacts with this substance with the regeneration of the catalyst and the formation of product. The activation energy of such reaction is smaller than a direct reaction of reactants to products.

More about catalysis and modeling of reactions during catalytic cracking can be found in references 21-27 at the end of part III of this book.

In the mechanism involving catalysis on metals during catalytic cracking, the catalyst promotes the removal of a negatively charged hydride ion from a paraffin compound. The mechanism in this type of catalysis is illustrated in Figure 6.8.

As a result of the interaction between the catalyst and the hydrocarbon molecule, one part of the molecule is bonded with the catalyst via a strong double-electron bond. Another part of the molecule is bonded with the catalyst via a weak one-electron bond. The two surface compounds formed are unstable and are thus very reactive. This explains the high velocity of catalytic reactions. Typical metal catalysts are Fe, Co, Ni, Ru, Rh, Re, Ir and Pt. The activity of the metal catalysts is explained by the non-saturated d-shell (or d-level) orbital that acts as a free radical during thermal cracking.

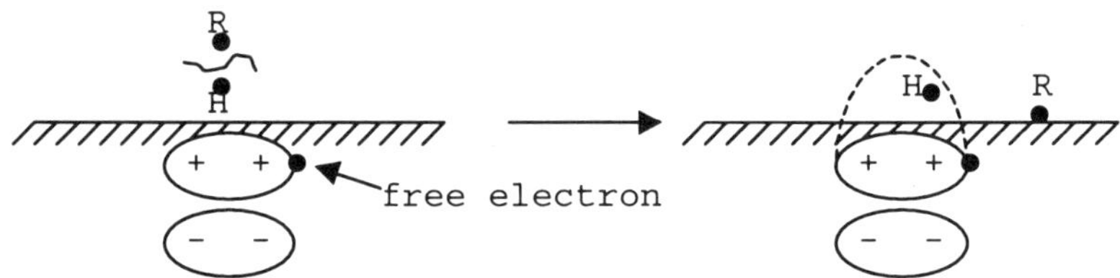

Fig. 6.8: Catalysis on metals.

Another type of catalysis is that on acid catalysts. The mechanism in this type of catalysis is based on the addition of a positively charged proton (H^+) to an olefin compound. This results in the formation of a carbonium ion, a positively charged molecule that has only a very short life as an intermediate compound. It transfers the positive charge through the hydrocarbon. The reaction of the acid catalysts (HX) and olefin is depicted in the reaction equation (6.37).

$$HX + CH_3\text{-}CH{=}CH\text{-}R \rightarrow X^- + CH_3\text{-}CH_2\text{-}CH^+\text{-}R \quad (6.37)$$

Hydrocarbon ion transfer continues as hydrocarbon compounds come into contact with active sites on the surface of the catalyst that promote the continued addition of protons or removal of hydride ions. The result is a weakening of carbon-carbon bonds in many of the hydrocarbon molecules and a consequent cracking into smaller compounds.

The main reactions of the carbon ion is the cracking of the C-C-bonds. This reaction proceeds in the same as was shown for thermal cracking.

Olefins crack more readily than paraffins since double C=C bonds are more friable under reaction conditions. Iso-paraffins and naphthenes are cracked more readily than normal paraffins, which in turn are cracked more readily than aromatics. In fact, aromatic ring compounds are very resistant to cracking, since they readily deactivate fluid cracking catalysts by blocking the active sites of the catalyst (see the next part). The reactions postulated for olefin compounds apply principally to intermediate products within the reactor system, since the olefin content of catalytic cracking feedstock is usually very low.

The most important difference between catalytic cracking and thermal cracking is in the reaction velocity of certain reactions that occur during the cracking. For example, the reactions involving the cracking of paraffins proceed with a reaction velocity a thousand times larger than for thermal cracking [4]. The following reactions of cyclic saturated hydrocarbons proceed by catalytic cracking with noticeable velocity [4]:

Ring destruction with the formation of olefins is shown in reaction (6.38):

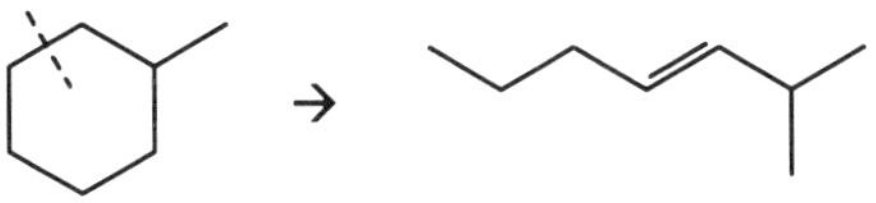

(6.38)

Isomerization of rings is shown by reaction (6.39):

(6.39)

Migration of alkyl side chain is shown by reaction (6.40):

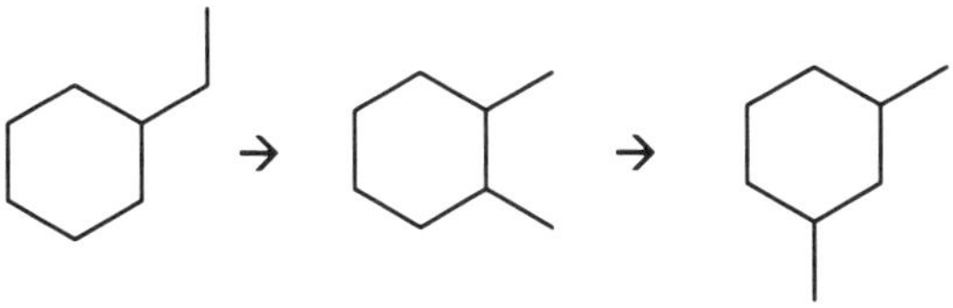

(6.40)

And the typical aromatization reaction is presented by equation (6.41):

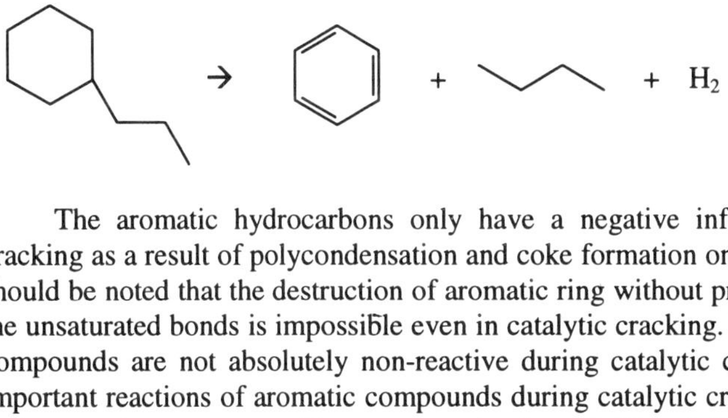

$$+ \quad H_2 \qquad (6.41)$$

The aromatic hydrocarbons only have a negative influence in catalytic cracking as a result of polycondensation and coke formation on catalyst surface. It should be noted that the destruction of aromatic ring without previously saturating the unsaturated bonds is impossible even in catalytic cracking. However, aromatic compounds are not absolutely non-reactive during catalytic cracking. The most important reactions of aromatic compounds during catalytic cracking are isomerization reactions represented by reaction equations (6.42) and (6.43).

The isomerization reactions shown in equation (6.42) can proceed selectively depending on the catalyst used in the process. Aluminosilicates are typical for selective isomerization of aromatic compounds, because they have well defined pore sizes depending on the type of aluminosilicate. The mechanism of selective isomerization can be explained based on Figure 6.9.

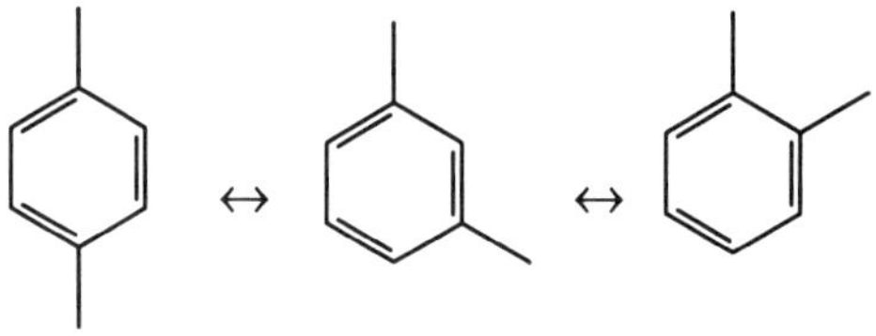

$$\leftrightarrow \quad \leftrightarrow \tag{6.42}$$

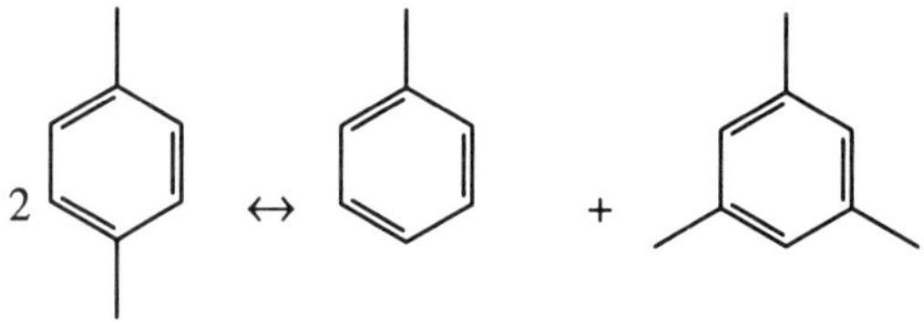

$$2 \quad \leftrightarrow \quad + \tag{6.43}$$

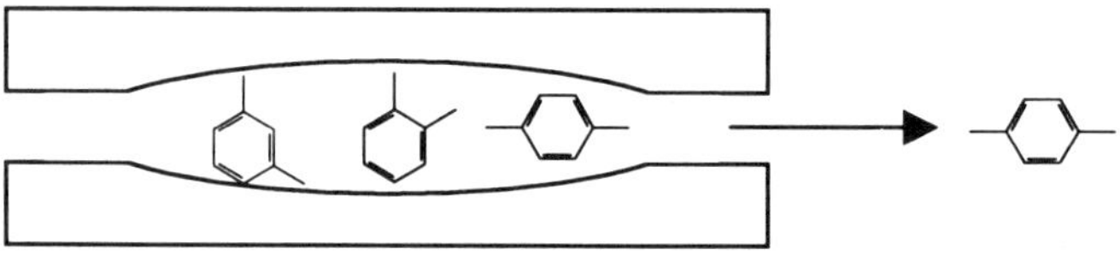

Fig. 6.9: Model of selective isomerization during catalysis cracking on aluminosilicate.

From Figure 6.9, it is obvious that a strictly defined size of catalyst pores allows only the molecules with size less than the pore size to leave the catalyst pore. This leads to the selective isomerization of aromatic compounds as shown in the example in Figure 6.9.

The catalysts used during catalytic cracking or any other catalytic processes have two main properties that determine the choice for the right catalyst for the process. These are

- catalyst activity
- catalyst selectivity

Catalyst activity shows how intensively the catalysts promote the reaction. Selectivity shows the ability to promote the desired reaction and inhibit any other reaction that leads to by-products.

Typical modern catalytic cracking reactors operate at 480-550°C and at relatively low pressures. At first, natural silica-alumina clays were used as catalysts, but by the mid-1970s, zeolites and molecular sieve-based catalysts were used. Zeolite catalysts give more selective yields of products, while reducing the formation of gas and coke.

There are three basic functions in the catalytic cracking process:

- Reaction: Feedstock reacts with the catalyst and cracks into different hydrocarbons.
- Regeneration: The catalyst is reactivated by burning off the coke.
- Fractionation: The cracked hydrocarbon stream is separated into various products.

A modern catalytic cracking unit employs a finely divided solid catalyst that has properties analogous to a fluid when it is agitated by air or oil vapors. The principles of operation of the cracking unit are shown in Figure 6.10. In Figure 6.10, a reactor and regenerator are located side by side. The oil feed is vaporized when it meets the hot catalyst at the feed-injection point, and the vapor flows upward through the riser reactor at a high velocity, providing a fluidizing effect for the catalyst particles. Catalytic reaction occurs exclusively in the riser reactor.

Coke, a byproduct of cracking, is deposited on the catalyst particles. Since these deposits impair reaction efficiency, the catalyst must be continuously put through the regenerator, where the carbon is burned with a current of air. The high temperature of the regeneration process (675-785°C) heats the catalyst to the desired reaction temperature for re-contacting fresh feed in the unit.

In order to regenerate the activity of the catalyst completely, a small amount of fresh catalyst is added to the system from time to time. Figure 6.11 depicts typical reactor-regenerator sections of a catalytic cracking unit used in petroleum refineries.

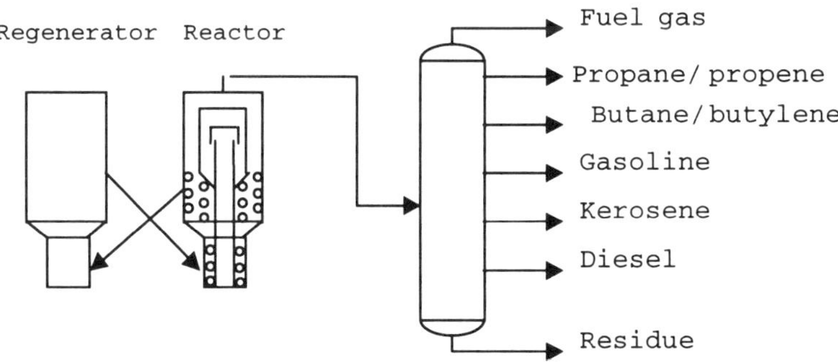

Fig. 6.10: Catalytic cracking.

The three types of catalytic cracking processes in operation in modern refineries are:

- Fluid catalytic cracking (FCC)
- Moving-bed catalytic cracking
- Thermofor catalytic cracking (TCC)

The catalytic cracking process is very flexible, and operating parameters can be adjusted to meet changing product demand. In addition to cracking, catalytic activities include dehydrogenation, hydrogenation, and isomerization.

The most common process is fluid catalytic cracking (FCC) in which the oil is cracked in the presence of a finely divided catalyst maintained in a fluidized state by the oil vapors. The fluid cracker consists of a catalyst section and a fractionating section that operate together as an integrated processing unit. The catalyst section contains the reactor and regenerator. The fluid catalyst is continuously circulated between the reactor and the regenerator using air, oil vapors, and steam as the conveying media (see Fig. 6.11).

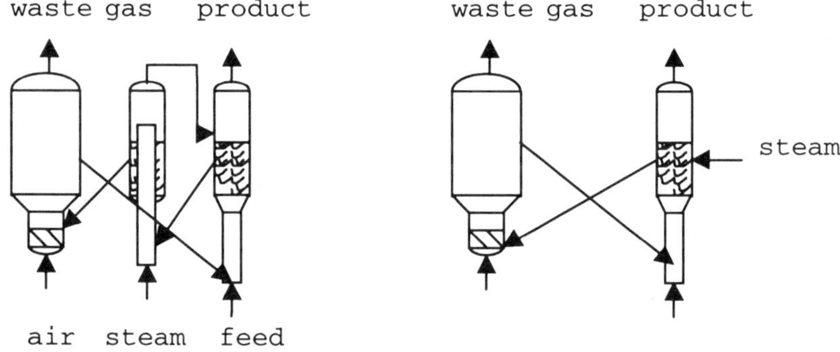

a) Complex two-step catalyst regeneration
b) Simple, direct regeneration of catalyst

Fig. 6.11: The reactor-regenerator sections.

A typical FCC process involves mixing a preheated hydrocarbon charge with hot, regenerated catalyst as it enters the riser leading to the reactor. The charge is combined with a recycle stream within the riser, vaporized, and raised to reactor temperature (500-550°C) by the hot catalyst.

The moving-bed catalytic cracking process is similar to the FCC process. The catalyst is in the form of pellets that are moved continuously to the top of the unit by conveyor or pneumatic lift tubes to a storage hopper, then flow downward by gravity through the reactor, and finally to a regenerator. The regenerator and hopper are isolated from the reactor by steam seals. The cracked product is separated into recycled gas, oil, clarified oil, distillate, naphtha, and wet gas.

In a typical thermofor catalytic cracking unit, the preheated feedstock flows by gravity through the catalytic reactor bed. The vapors are separated from the catalyst and sent to a fractionating tower. The spent catalyst is regenerated, cooled, and recycled. The flue gas from regeneration is sent to a carbon monoxide boiler for heat recovery.

In a recent investigation involving the promotion of catalytic cracking, it was shown that ozonolysis of heavy oil fractions and catalytic upgrading of fuel distillates are good methods for increasing the degree of oil refining as well as improving the quality of the oil product.

A low temperature cracking (350°C) of the ozonated crude oil raw material initiated by ozonides and sulfoxides leads to a significant increase of distillate fraction yield and a significant decrease in the sulfur content of the products.

More about catalytic cracking can be found in many references [28-32] at the end of part III of this book.

6.3 VISBREAKING

The demand for light fractions grows all the time, at the same time the requirement for heavy residual fuel oils declines. Furthermore, many of the new sources of crude petroleum have yielded heavier crude oils with higher natural yields of residual fuels. As a result, refiners have become even more dependent on the conversion of residue components into lighter products.

In 1930, large volumes of residue were being processed in visbreaking units. This simple process unit basically consists of a large oven that heats the feedstock to a temperature in the range of 450 to 500°C. The residence time in the furnace is carefully limited to prevent much of coke formation from taking place and clogging the oven tubes. The heated feed is then charged to a reaction chamber, which is kept at a pressure high enough to permit cracking of large molecules, but restricts coke formation. The process fluid from the reaction chamber is cooled to inhibit further cracking and then charged to a rectification tower for separation into components.

Visbreaking units typically convert about 15% of the feedstock to naphtha and diesel oils and produce a lower-viscosity residual fuel.

More details about the chemistry of visbreaking are given in chapter 8.

6.4 COKING

Coking is a severe thermal cracking process. Coke formation decreases the feed conversion during thermal cracking, and the cracking reactor has to be frequently cleaned because of the coke formed on the reactor walls. This negative influence of coke formation reaction can be decreased if a process is used in which coke is one of the products required. Coking is such a process. Cokers (the coking units) produce no liquid cracking residue, but yield approximately 30% of coke. Much of the low-sulfur coke from the coking process is employed to produce electrodes for electrolytic smelting of aluminum. Most of the lower-quality coke is burned as fuel in admixture with coal. Coker economics usually favor the conversion of residue into light products even if there is no market for coke.

All the existing coking units can be classified into three main types:

- Discontinuous coking
- Delayed coking
- Continuous coking

Discontinuous coking is the oldest coking process. A simple metal cube is the equipment for this process. In the first step, the feed is loaded in the cube reactor. Then the reactor is heated. Light product is formed when a temperature of approximately 300°C is reached. Then the temperature of the vapor phase increases quickly from 300°C to approximately 445–460°C. After this temperature, there is a slow down in the increase of the vapor temperature. This means that no more volatile products are being formed. At this stage, coke formation is finished.

In the next step, the temperature of the reactor is approximately 460°C. Such a high temperature is required for drying the coke (i.e. removing the volatile product). After about three hours of drying, steam is injected into the reactor. The cube reactor is cooled by steam until the temperature reaches approximately 250–200°C. Air is then used for further cooling. At the temperature of approximately 200–150°C, the coke is evacuated from the cube reactor. The evacuation is done using mechanical equipment only. At the present time, discontinuous coking is used very rarely and at old refineries.

The next coking process is delayed coking. This process is a quasi-continuous process that involves two main steps:

- Coking
- Decoking

In the coking step, the feed is heated to approximately 475 to 520ºC in an oven with a very low residence time and is discharged into the bottom of a large vessel called, a "coke drum" for extensive but controlled cracking. The coke drum is not heated. That means that the feed is heated only in the oven. The cracked volatile product rises to the top of the drum and is drawn off. It is then charged to the product fractionator for separation into naphtha, diesel oils, and heavy gas oils for further processing in the catalytic cracking unit. The heavier product remains and, because of the retained heat, cracks ultimately to coke. Once the coke drum is filled with solid coke, it is removed and replaced by another coke drum.

Decoking is a routine daily occurrence accomplished by a high-pressure water jet. First, the top and bottom heads of the coke drum are removed. Next, a hole is drilled in the coke from the top to the bottom of the drum. Then, a rotating stem is lowered through the hole, spraying a water jet sideways. The high-pressure jet cuts the coke into lumps, which fall out to the bottom of the drum for subsequent loading into trucks or railcars for shipment to customers. Typically, coke drums operate on 24-hour cycles, filling with coke over one 24-hour period followed by cooling, decoking, and reheating over the next 24 hours.

The final coking process is continuous coking. During this coking, the feed is heated in the reactor by contact with a hot medium. Coke formation proceeds on the surface of the medium particles. After coking in the reactor, the medium with coke on the surface passes to the regenerator. Coke is combusted in the regenerator, and the heat generated by coke combustion is used for heating the medium. The heated medium (usually ceramic particles) passes back to the coker.

More details concerning the coking chemistry and the equipment for coking are given in chapter 8.

6.5 HYDROPROCESSING

It was shown in previous sections that it is impossible to convert a hundred percent of the crude oil residue to light fractions by using both thermal and catalytic cracking processes alone. The main reason for this is that cracking reactions need to be accompanied by hydrogen transfer reactions in order to stabilize the product. It is obvious that light fractions such as gasoline or diesel fractions are more hydrogen rich than coke and residue by-products of thermal or catalytic cracking processes. This means that hydrogen transfer proceeds from heavy fractions to light cracking products during the cracking processes. However, the complete conversion of cracking feed to light fractions is impossible because of the shortage of hydrogen in the feed. Also, heteroatom compounds present in the feed tend to form coke on the catalysts. The elimination of heteroatom compounds requires hydrogen, which thereby limits hydrogen transfer to the light fractions.

Catalytic hydroprocessing is a hydrogenation process used to remove heteroatom compounds (i.e., compounds containing nitrogen, sulfur, oxygen, and/or metals) from liquid petroleum fractions. These compounds adversely influence equipment and catalysts in the refinery and the quality of the finished product, especially on the ecological properties of the product. Hydroprocessing units are installed prior to units for processes such as catalytic reforming so that the expensive platinum catalyst is not contaminated by untreated feedstock. Hydroprocessing is also used prior to catalytic cracking to reduce sulfur and improve product yields, and to upgrade middle-distillate petroleum fractions into finished kerosene, diesel fuel, and heating fuel oils. In addition, hydrotreating converts olefins and aromatics to saturated compounds.

All the reactions that take place during hydroprocessing can be classified into four large groups:

- Reactions of sulfur containing compounds
- Reactions of nitrogen containing compounds
- Reactions of oxygen containing compounds
- Reactions of hydrocarbons

The hydroprocessing of sulfur containing compounds proceeds with the formation of hydrogen sulfide. An example of this reaction is shown in the reaction equation (6.44).

$$2R\text{-}SH + 3H_2 \rightarrow 2R\text{-}H + 2H_2S \tag{6.64}$$

Sulfides and disulfides are hydrogenated in two steps. The reactions for the hydrogenation of sulfides and disulfides are shown in the reaction equations (6.45) and (6.46).

$$2RSR + 2H_2 \rightarrow 2R\text{-}SH + 2RH + 3H_2 \rightarrow 2R\text{-}H + 2RH + 2H_2S \tag{6.45}$$

$$RSSR + H_2 \rightarrow 2R\text{-}SH + 3H_2 \rightarrow 2R\text{-}H + 2H_2S \tag{6.46}$$

The hydrogenation of the cyclic sulfide proceeds with ring destruction

$$\text{S} \quad + H_2 \rightarrow CH_3\text{-}CH_2\text{-}CH_2\text{-}CH_3 + H_2S \tag{6.47}$$

The next important class of heteroatom containing compounds are those that contain nitrogen. The hydrogenation of these compounds proceeds with the formation of ammonia (6.48).

$$C_6H_5NH_2 + H_2 \rightarrow C_6H_6 + NH_3 \tag{6.48}$$

The hydrogenation of cyclic nitrogen containing compounds proceeds very slowly and with a very low conversion. This reaction proceeds in four steps shown in the reaction equation (6.49).

$$\text{N} \quad + 2H_2 \quad \rightarrow \quad \text{N} \quad + H_2 \quad \rightarrow \quad CH_3\text{-}CH_2\text{-}CH_2\text{-}CH_2\text{-}NH_2 +$$

$$+ H_2 \quad \rightarrow \quad CH_3\text{-}CH_2\text{-}CH_2\text{-}CH_3 + NH_3 \tag{6.49}$$

During the hydrogenation of oxygen containing compounds, hydrocarbons and water are formed. An example of this reaction is shown in the reaction equation (6.50).

$$R\text{-}COOH + 2H_2 \rightarrow R\text{-}CH_3 + 2H_2O \tag{6.50}$$

The hydrogenation of hydrocarbons involves the saturation of the non-saturated bonds (6.51), which are formed during all the cracking processes shown in the previous sections.

$$CH_3\text{-}CH\text{=}CH\text{-}R + H_2 \rightarrow CH_3\text{-}CH_2\text{-}CH_2\text{-}R \qquad (6.51)$$

The velocity of this reaction (6.51) as well as that for most hydrogenation reactions is very slow without using a catalyst. Moreover, reaction (6.51) is an exothermic reaction, meaning that there is a thermodynamic limitation in that the conversion in this reaction decreases with increasing temperature of the process (see section 6.1). Hydrogenation of many hydrocarbons is already impossible at temperatures over 400°C. In section 6.1, it was shown that pressure has a large influence on reactions that proceed with a volume change. It is obvious from all the hydroprcessing reactions shown that this process proceeds with a volume decrease, meaning that hydroprocessing must be carried out under a hydrogen pressure that is as high as possible.

As mentioned already, the majority of hydroprocessing reactions are exothermic reactions. However, the hydroprocessing of light fractions from the catalytic cracking unit can have a relatively small exothermic heat effect, approximately 80 kJ/kg. On the other hand, the hydroprocessing of more unsaturated and heteroatom rich feed can have an exothermic heat effect of up to 500 kJ/kg. It has therefore become essential that most modern hydroprocessing plants have heat exchangers to take out the heat during the process.

Modern hydroprocessing is carried out with the use of catalysts. The typical catalysts for this process are oxides and sulfides of nickel, cobalt, molybdenum, and vanadium. Recently, new catalysts for hydrotreatment have been developed based on metal cluster compounds.

In a typical catalytic hydroproceesing unit (Fig. 6.12), the feedstock 1 is deaerated and mixed with hydrogen 2. This is preheated in a fired heater O to the temperature of 350-400°C and then charged through a fixed-bed catalytic reactor R. In the reactor, the sulfur and nitrogen compounds in the feedstock are converted into H_2S and NH_3. The reaction products leave the reactor and, after separation in the hot separator T1 and cooling to a low temperature, enter the liquid/gas separators T2 in which wash water 3 is used to remove the ammonia and hydrogen sulfide. The sour water 4 leaves the separator in the bottom section of the equipment. The overhead gas from the cold separator T2 is charged to the high-pressure scrubber T3 to remove hydrogen sulfide from the recycled gas. It is chemically absorbed in an amine solution 5. The purified recycle gas is mixed with fresh hydrogen 2 to compensate for the losses in the hydrogenation reactions. The recycled gas compressor routes the gas back to the reactor loop. The liquid product streams are sent to a rectification tower T4 where the volatile product is hydrotreated naphtha.

The bottom product from the T4 is sent to the vacuum rectification tower T5 for further rectification. The bottom product from vacuum tower T5 is the middle distillate.

Hydrotreating processes differ depending upon the feed utilized and the catalysts used. Typical catalysts for hydroproceesing were described in the preceding section.

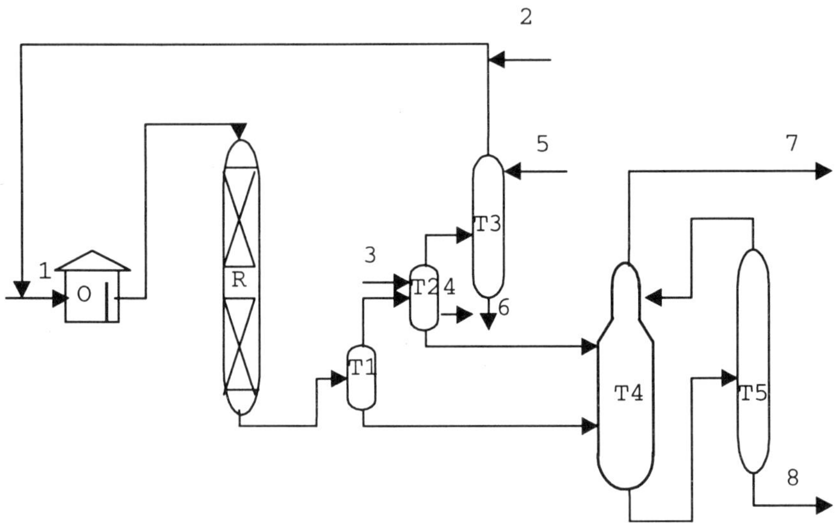

Fig. 6.12: Hydroprocessing unit.
O – Oven
R – Reactor
T1 – Hot separator
T2 – Liquid/gas separators
T3 – High-pressure scrubber
T4 – Rectification tower
T5 – Vacuum rectification tower
1. Feed, 2. Hydrogen, 3. Wash water, 4. Sour water, 5. Amine solution, 6. Hydrogen sulfide rich amine solution, 7. Gases 8. Hydrogenated product

Depending on the feed used, hydroprocessing can be used for the treatment of middle fractions, for example, to improve the burning characteristics of distillates, such as kerosene. Hydrotreatment of a kerosene fraction can convert aromatics into naphthenes.

The hydroprocessing of heavy oil fractions, for example, lubricating oil hydroprocessing, uses a catalytic treatment of the oil with hydrogen to improve product quality. The objectives in mild hydroprocessing include saturation of olefins and reduction of the acid nature of the oil. Mild oil hydrotreating also may be used following solvent processing.

Hydrotreating also can be employed to improve the quality of pyrolysis gasoline, a by-product from the manufacture of ethylene. Traditionally, the outlet for this gasoline has been motor gasoline blending, a suitable route in view of its high octane number. However, only small portions of untreated pyrolysis gasoline can be used for blending owing to its unacceptable odor, color, and gum-forming tendencies. The quality of pyrolysis gasoline, which is high in diolefin content, can be satisfactorily improved by hydroprocessing, whereby conversion of diolefins into mono-olefins provides an acceptable product for motor gasoline blending.

Just like any catalytic process in the crude oil industry, the big problem in hydroprocessing is the deactivation of the catalyst during the process. There two main mechanisms of catalyst deactivation during hydroprocessing:

1. deactivation by coke formation on catalyst surface
2. deactivation by heavy metals deposition on catalyst active centers

Deactivation by coke formation can take place by many mechanisms. For example, coke or similar compounds such as polyaromatic compounds can be adsorbed on the surface of the catalyst as shown in Figure 6.13.

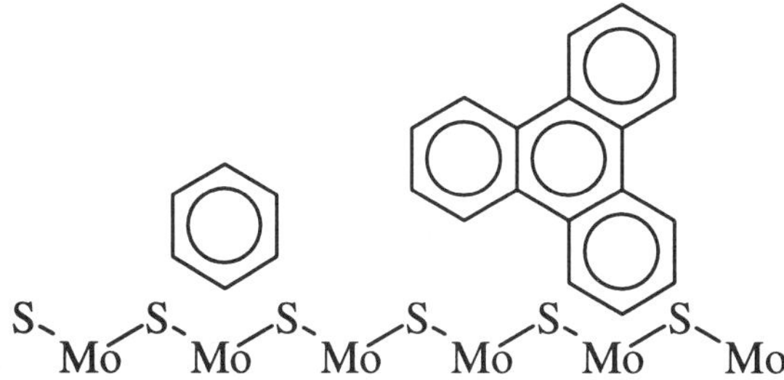

Fig. 6.13: Catalyst deactivation by absorption of coke or similar compounds.

Coke formation on the catalyst proceeds in same way as was shown in section 6.1 for thermal cracking. However, the presence of the catalyst changes the mechanism of the polycondesation reaction. Coke formation in all catalytic processes proceeds by the ion mechanism and not by the radical chain mechanism applicable for thermal processes. One example of a possible pathway for coke formation is shown in reaction (6.52).

Catalyst can also be deactivated by adsorption of asphaltenes during the hydroprocessing of heavy crude oil residues.

All heavy crude oil residues have heavy metals such as Ni, V or Fe in their structrure. These metals are bonded as organometalic compounds. At high temperatures and for hydrogenation reactions, these compounds are cracked and heavy metals are deposited on the catalyst surface. These metals can also react with hydrogen sulfur from the gas phase to form metal sulfides. The deposition of sulfides of iron, vanadium or nickel leads to irreversible poisoning of the catalyst. This is the difference between catalyst deactivation by metals and deactivation by coke; the former leads to an irreversible loss of the catalyst activity.

H H

$$+ H^+ \rightarrow [\quad]^+ + \quad \rightarrow [\quad]^+ \rightarrow$$

$$\rightarrow \quad + n \cdot H_2 \qquad (6.52)$$

Hydroprocessing is one of the most important processes in crude oil processing. However, as a result of the continuous changes in crude oil quality in the world market, it has become necessery to modify the hydroprocessing technology. Reports of new catalysts or new types of hydroprocessing appear every year in the literature. References 33 and 34 at the end of part III of this book show the most modern developments in this area and give references to some articles reporting about the newest technologies in the area of hydroprocessing of crude oil.

6.6 REFORMING

Reforming is the most widespread process for rearranging the hydrocarbon molecules from petroleum fractions. The initial process, thermal reforming, was developed in the 1920s. Thermal reforming employed temperatures of 510-565°C at moderate pressures to obtain gasoline with octane numbers of 70 to 80 from heavy naphtha with octane numbers of less than 40. The product yield, although of a higher octane level, included olefins, diolefins, and aromatic compounds. It was therefore inherently unstable in storage and tended to form heavy polymers and gums, which caused combustion problems.

By 1950, a reforming process was introduced that employed a catalyst to improve the yield of the most desirable gasoline components, while minimizing the formation of unwanted heavy products and coke. In catalytic reforming, as in thermal reforming, a naphtha-type material serves as the feedstock, but the reactions are carried out in the presence of hydrogen, which inhibits the formation of unstable unsaturated compounds that polymerize into higher-boiling fractions.

The main reaction that occurs during catalytic reforming is the aromatization reaction. However, the aromatization reaction involves many simple reactions that are presented in the following set of reactions (6.53):

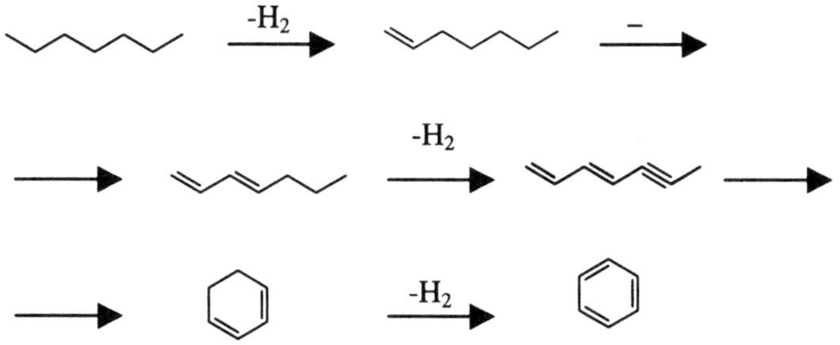

$$\xrightarrow{-H_2} \quad \xrightarrow{-} \quad \xrightarrow{-H_2} \quad \xrightarrow{-H_2} \tag{6.53}$$

From the reaction (6.53) line 1, it can be seen that aromatization proceeds initially from paraffin through olefins, diolefins and triolefins. From this example, it is obvious that less unsaturated compounds are formed during catalytic reforming. They change their structure under the influence of the catalysts from unsaturated and chemically unstable olefins to stable aromatic compounds (see (6.53)).

Chrome oxide, copper chromide and aluminum-molybdenum (MoO_3/AlO_3) were the first catalysts used for the reforming process. This catalyst promotes all the reactions that occur during reforming. This catalyst thus belongs to the catalysts with very low selectivity. During the reforming, as with every other catalytic process in crude oil refining, coke forming occurs, leading to a very fast deactivation of the catalyst. During reforming, the most frequent reactions that occur are cyclization reactions that can proceed further until aromatics are formed. However, recent investigations have shown that cyclization is the last step in these processes. In the case of the reaction of naphthene side-chains, cyclization can actually be the last reaction (6.54)

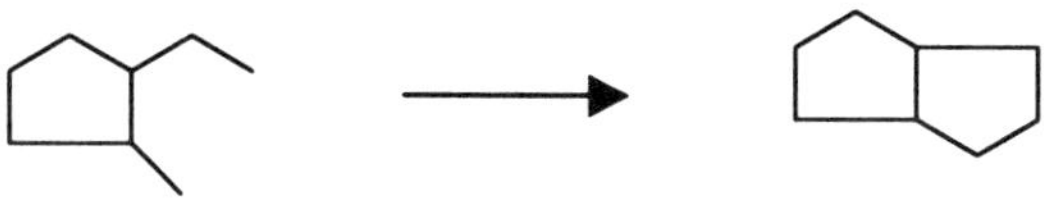

$$\longrightarrow \tag{6.54}$$

From both reactions (6.53) and (6.54), it can be seen that reforming proceeds with the formation of a high amount of hydrogen. However, reforming is carried out under a hydrogen atmosphere in order to reduce coke formation. Because hydrogen is one of the reforming products, the high hydrogen content in the reaction zone slows down the reforming reactions as well. This is why it is so important to choose the hydrogen pressure in the reforming reactor so that the reforming reactions will proceed fast enough for industrial applications.

In modern refineries, platinum is the most widespread active catalyst; it is distributed on the surface of an aluminum oxide carrier, the process is called "platforming". Small amounts of rhenium, chlorine, and fluorine act as catalyst promoters. In spite of the high cost of platinum, the process is economical, because of the long life of the catalyst and the high quality and yield of the products obtained. Platforming is carried out under the atmosphere of a hydrogen containing gas (hydrogen content up to 95%). The principal reactions during platforming are the same as during the described reforming: the breaking down of long-chain hydrocarbons into smaller saturated and unsaturated chains, the formation of iso-paraffins. Formation of ring compounds (cyclization) also takes place, and the naphthenes are then dehydrogenated into aromatic compounds. The hydrogen liberated in this process forms a valuable by-product of catalytic reforming. The desirable end products are iso-paraffins, aromatics and naphthenes, having high octane numbers. The main reactions of hydrocarbons during reforming can be shown by the scheme given in Figure 6.14.

There are many different commercial catalytic reforming processes used nowadays in the petroleum industry:

- Platforming
- Powerforming
- Ultraforming
- Thermofor catalytic reforming

A catalytic reformer comprises a reactor section and a product-recovery section (see Fig. 6.15). The feed preparation section is more or less standard. Here, the feedstock is prepared to the desired specification by combination of hydrotreatment and distillation.

In a typical reforming unit, the naphtha charge after the heater is first passed over a catalyst bed in the presence of hydrogen to remove any sulfur impurities. The desulfurized feed is then mixed with hydrogen and heated to a temperature of 500-540°C in the heater. The gaseous mixture passes downward through catalyst pellets in a series of three to five reforming reactors (Figure 6.15 reactor includes the reactor section of three or five reactors and catalyst regenerator).

After leaving the final reactor, the product is condensed to a liquid and passed to a rectification tower, where the light hydrocarbons produced in the reactors and the hydrogen rich gases are removed by rectification. The reformated product is then available for blending into gasoline without further treatment. The hydrogen leaving the product separator is compressed and returned to the reactor system.

The reformed gasoline has an octane number between 90 and 100. At the higher octane number level, product yields are smaller, and more frequent catalyst regeneration is required. During the course of the reforming process, small amounts of coke are deposited on the catalyst, causing a gradual deterioration of the product yield pattern. Some units are semi-regenerative facilities. That is, they must be removed from service periodically to burn off the coke and rejuvenate the catalyst system. However, increased demand for high octane fuels has also led to the development of continuous regeneration systems, which avoid periodic unit shutdowns and maximize the yield of high octane reformates.

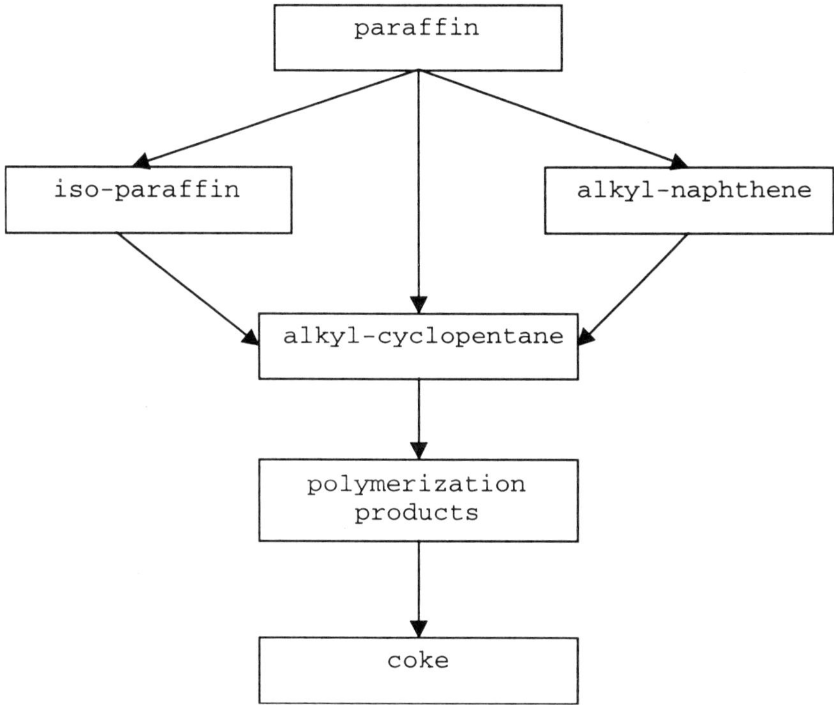

Fig. 6.14: A scheme of the reaction mechanism for catalytic reforming.

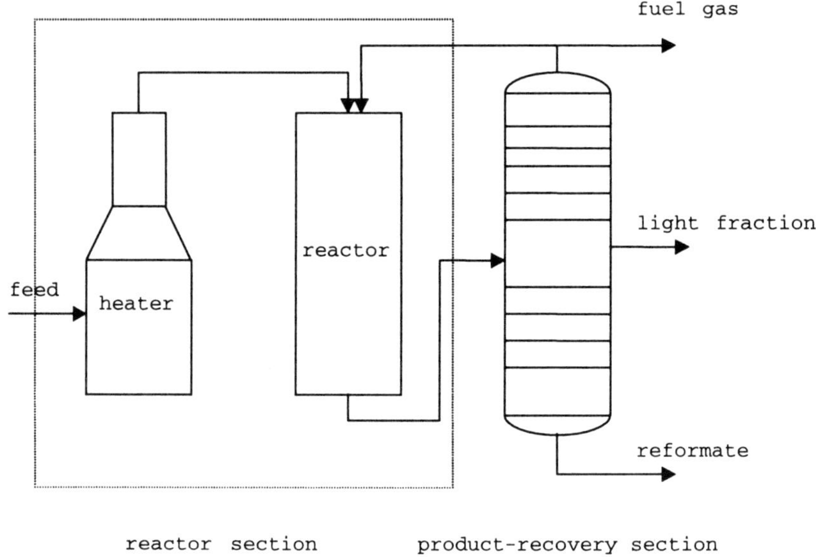

Fig. 6.15: Flow scheme of a reforming unit.

Continuous regeneration employs a moving bed of catalyst particles that is gradually withdrawn from the reactor system and passed through a regenerator vessel, where the carbon is removed and the catalyst rejuvenated for reintroduction to the reactor system.

More about the reforming process can be found in references 4, 35, 36 at the end of part III of this book.

6.7 ISOMERIZATION

The steadily increasing demands for premium gasoline requires higher capacities for isomerization, reforming and alkylation. In addition, the more stringent legislation on aromatics in reformulated gasoline is an important incentive for refineries to saturate benzene and remove naphthenes in isomerization units. The first widespread use of isomerization was during the Second World War, because of the great demand for aviation gasoline, and afterward the quantities of isobutane available for alkylation feedstock were insufficient. This deficiency was remedied by isomerization of the more abundant normal butane into isobutane. The isomerization catalyst is aluminum chloride supported on alumina and promoted by hydrogen chloride gas.

The main reaction of the isomerization process is isomerization (6.55).

(6.55)

The real reaction mechanism on the surface of the catalyst particle can be represented by a more complicated multistage reaction (6.56):

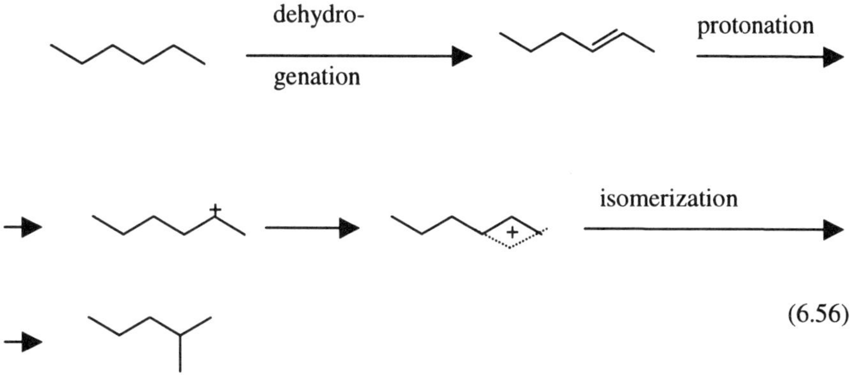

dehydro-
genation
protonation
isomerization

(6.56)

Commercial processes have also been developed for the isomerization of low-octane normal pentane and normal hexane to the corresponding higher-octane isoparaffins. Here the catalyst is usually enhanced with platinum. As in catalytic reforming, the reactions are carried out in the presence of hydrogen. Hydrogen is neither produced nor consumed in the process but is employed to inhibit the undesirable side reactions of coke formation. Molecular sieve extraction and distillation usually follow the reactor step. Though this process is an attractive way to exclude low-octane components from the gasoline blending pool, it does not produce a final product of sufficiently high octane to contribute much to the manufacture of unleaded gasoline.

Figure 6.16 shows a typical isomerization unit.

The feed after hydrotreating and drying in absorber 1 passes to reactor 2. Reactor 2 is a fixed bed reactor. Together with the feed in reactor 2 the promoter (HCl) is within 10^{-4}% of the feed. The liquid product from reactor 3 is separated from the propane fraction in separator 3. The octane number of the product from this unit is approximately 90, according to the research octane method.

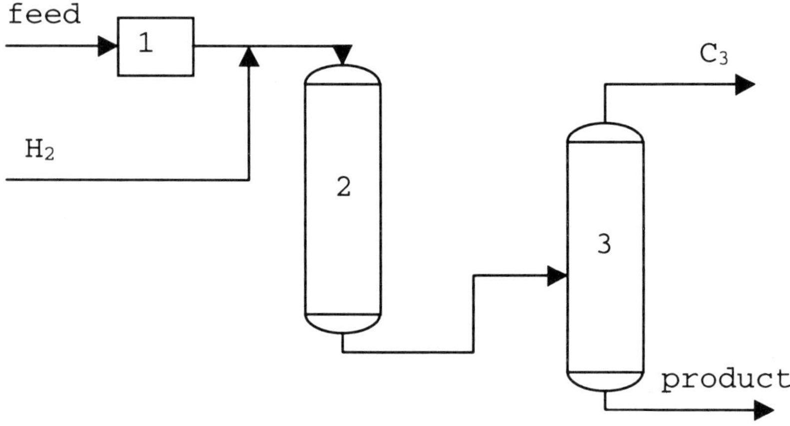

Fig. 6.16: The isomerization unit.
1. Absorber
2. Reactor
3. Separator

6.8 ALKYLATION

The light gaseous hydrocarbons produced by catalytic cracking are highly unsaturated and are usually converted into high-octane gasoline components in the alkylation process. In alkylation, the light olefins (propylene and butylene) are induced to combine, or polymerize, into molecules of two or three times their original molecular weight. The catalysts employed for this are of an acid nature. High pressures are required at temperatures ranging from 175 to 230°C. Alkylated gasolines derived from propylene and butylene have octane numbers above 90 and, with the addition of lead additives, above 100.

Alkylation reactions also produce a longer chain molecule by the combination of two smaller molecules, one being an olefin and the other an iso-paraffin (usually isobutane).

The main reaction of the alkylation process is shown in the reaction equation (6.57).

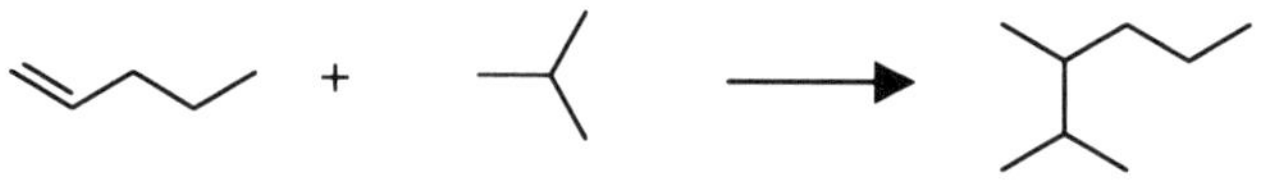

(6.57)

Actually, reaction (e6.57) proceeds on the surface of the catalyst according to a more complicated mechanism involving three stages.

In the first stage, the olefin reacts with the catalyst, for example sulfuric acid (6.58).

$$+ \ H_2SO_4 \longrightarrow \quad + \ ^{+}HSO_4 \qquad (6.58)$$

Next hydrocarbon ion reacts with iso-butane (6.59).

(6.59)

In the second stage, the butyl ion reacts with the olefin (6.60).

(6.60)

The third and last stage is most complicated. First of all, the migration of the hydrogen atom proceeds at this stage (6.61).

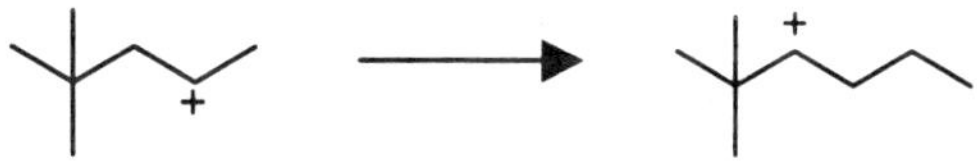

(6.61)

Next, isomerization reaction proceeds very quickly at the third stage of alkylation (6.62).

(6.62)

All the carbon ions formed at this stage finally react with iso-butane in the first stage with the formation of stable molecules and new carbon ions.

All the alkylation units used in the industry can by classified into different groups according to the catalyst used:

- sulfuric acid alkylation
- hydrofluoric acid

In sulfuric acid alkylation, concentrated sulfuric acid of 98 percent purity serves as the catalyst for the reaction that is carried out at 2 to 7°C. Refrigeration is necessary because of the heat generated by the reaction. The octane numbers of the alkylates produced range from 85 to 95. An example of sulfuric acid alkylation is shown in Figure 6.17.

The feed (olefins and iso-butane) passes to reactor 1 (Fig. 6.17) where this is mixed with the sulfuric acid from vessel 2. The temperature in the reactor is approximately ± 1°C. The product from the reactor passes to separator 3 and then to distillation tower 4. The product is fractionated into alkylate (desired product) and n-butane in tower 4. In tower 6, the iso-butane is separated from the propane and passes as recycled iso-butane back to the reactor.

Hydrofluoric acid is also used as a catalyst for many alkylation units. The chemical reactions are similar to those in the sulfuric acid process, but it is possible to use higher temperatures (between 24 and 46°C), thus avoiding the need for refrigeration. Recovery of hydrofluoric acid is accomplished by distillation. An example of the hydrofluoric acid alkylation unit is shown in Figure 6.18.

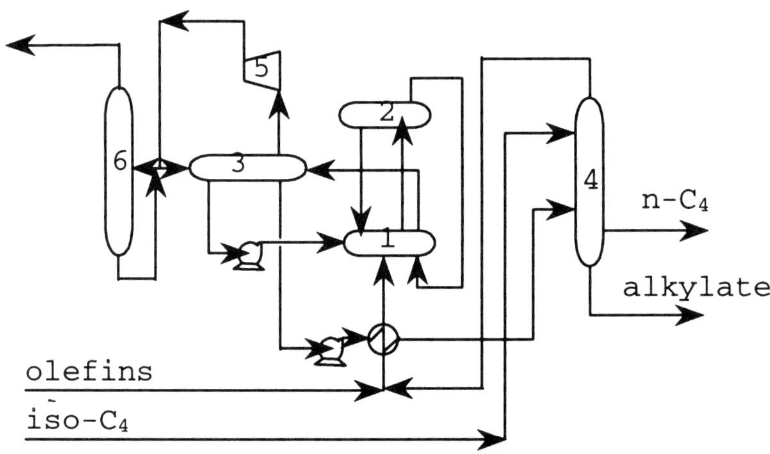

Fig. 6.17: Sulfuric acid alkylation unit.

1. Reactor, 2. Vessel, 3. Separator, 4. Distillation tower, 5. Compressor, 6. Gas separation tower

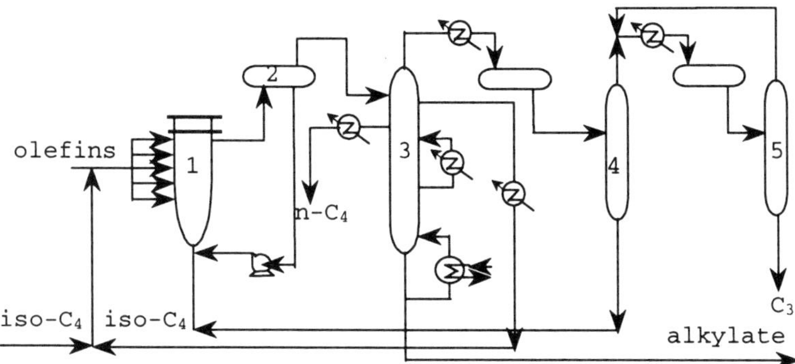

Fig. 6.18: Hydrofluoric acid alkylation unit.
1. Reactor, 2. Separator, 3. Distillation tower, 4-5. Separation towers

The feed and recycled iso-butane pass to reactor 1 (Fig. 6.18). The product from the reactor passes to separator 2, where this is separated to acid and the product. The product is then distilled into an alkylate (desired product) fraction and the volatile product containing the hydrofluoric acid vapor and the rest of the iso-butane. This volatile product passes to the next separation tower 4 where this is separated into propane with the hydrogen sulfide from the product and the bottom product directed to reactor 1. The propane with the hydrogen sulfide passes to the separation tower 5, the bottom product of which is pure propane.

6.9 BLENDING

We have concluded the discussion on the main petroleum treatment processes. However, after petroleum is treated using all the processes described in chapters 5 and 6, there is still a last operation to be performed. This is the blending of the product, from different units to produce market products. These are then delivered to gas stations and car services.

A very interesting technology for blending diesel fuels has been developed by Total-Fina-Elf company. Not only are diesel fractions from different units blended, but also the diesel fuel blend and a small amount of water are mixed together. The name of the new generation of diesel fuel is "Aquazole". It exists as an emulsion of water in diesel fuel.

It is well known that water vapor can be a burning catalyst. In "Aquazole", this property of water has been used for the first time for burning motor fuels. One of the positive effects achieved with this fuel is the improvement of the ecological properties of the diesel fuel. It has been shown that by using this fuel, up to 30% less NO_x formation and up to 50% less emission of particulates have been achieved. Also, the blending of diesel fuel with water leads to a more economical use of diesel fuel. Accordingly, the fuel requirements for trucks have been reduced by up to 4%.

In the example of "Aquazole", it has been shown that even the simplest process such as blending that is carried out in the crude oil refinery can have a large influence on the quality of the crude oil products and the economics of the process.

7

Environmental Issues Facing the Refining Industry

7.1 INTRODUCTION

Many ecological problems that arise in the crude oil industry were already discussed in chapter one of this book. Thus, the objective of this chapter is to show that in order to develop an appropriate technology for crude oil treatment, not only are economically rational decisions required, but also ecologically acceptable decisions are needed. Nowadays it has been very important to change the world-view on environmental issues, such that everyone understands that our environment is a very sensitive system that must be protected.

The first steps that need to be taken in the direction of environmental development of the crude oil industry were already done more than thirty years ago. In January 1, 1970 in the USA, a new ecological law called the National Environmental Policy Act (NEPA) was accepted. The main objectives of this law are to ensure that the necessity to estimate all the possible influences of any industrial activity on the environment is carried out, and to decrease the negative influence of the industry on the biosphere. The highlights of this law are given as follows. An analysis of the influence of the industrial activity on the environment must be done before starting the activity. Also, the analysis must be done not only by specialists from the industry, but also by independent specialists from the area where this industrial activity is to be carried out.

This new ecological system based on NEPA, called Environmental Assessment or Environmental Impact Assessment, was started in the USA. The next country that started using an ecological law in order to change industrial policy was Canada. Then, many European countries started their ecological policies because of the clear necessity for environmental protection. Now almost all the countries around the world use environmental laws in their industrial policies.

This chapter thus deals with the ecological consequences of crude oil treatment and the use of crude oil products. It also deals with the methods currently in place to improve the ecological outlook of the crude oil Industry.

7.2 METHODS OF CLEANING CRUDE OIL CONTAMINATED WATER AND SOIL

It is well known that it is almost impossible to transport, store and treat crude oil without spills and losses. Crude oil contaminates both water and soil through the above mentioned avenues. It is difficult to prevent spills resulting from failure or damage on pipelines. It is also impossible to install control devices for controlling the ecological properties of water and the soil along the length of all pipelines. The soil suffers the most ecological damage in the damage areas of pipelines.

Crude oil spills from pipelines lead to irreversible changes of the soil properties. The most affected soil properties by crude oil losses from pipelines are filtration, physical and mechanical properties. These properties of the soil are important for maintaining the ecological equilibrium in the damaged area.

Natural filtration of petroleum through the soil creates the so-called chromatographic effect resulting in crude oil differentiation and fractionation; heavy components containing asphaltenes, resins and cyclic compounds are accumulated in upper soil layers, while the light compounds penetrate deeper into the bottom mineral layers of the contaminated soil. The possibility of self-cleaning in different soils is different and depends on the contaminated area. For example, accumulation of crude oil contaminants in clay soils can be more extensive than in sandy soils. Damage on pipeline leads to an irreversible change of the ecological system in the contaminated area. This is why it is necessary to remove crude oil from the soil as fast as possible.

The cleaning of the soil is usually done in two steps:

- mechanical step
- biological step

The first operation during the mechanical step of soil cleaning after the damage on a pipeline is to minimize the size of the affected area (i.e. the area of contact between crude oil and the soil should be made as small as possible). This can be done by leading the petroleum spill to natural containments. This is the simplest way but is not reasonable all the time. It is not always possible to find natural containment near the place of pipeline damage. Then, such an operation as described above can actually result in the contamination of a larger soil area while trying to lead crude oil to a natural containment. Thus, if it is impossible to find a nearby natural containment, then it is necessary to build a dam around the place of damage in order to prevent the spread of crude oil to other areas.

Special pumps are used for removing crude oil from the surface of the soil. Then, the surface of the soil must be collected and utilized or disposed of. In very rare cases where the crude oil contamination penetrates very deep into the soil layer, typically over ten meters, it becomes necessary to completely burn the soil layer. After burning up the contaminated soil, it is replaced with fresh soil because the ecological system of the burnt soil had been completely destroyed by burning.

Natural adsorbents such as peat or sand are used to simplify the removal of crude oil from the soil surface. Example of the dam and control system on pipeline place of damage is presented in Figure 7.1.

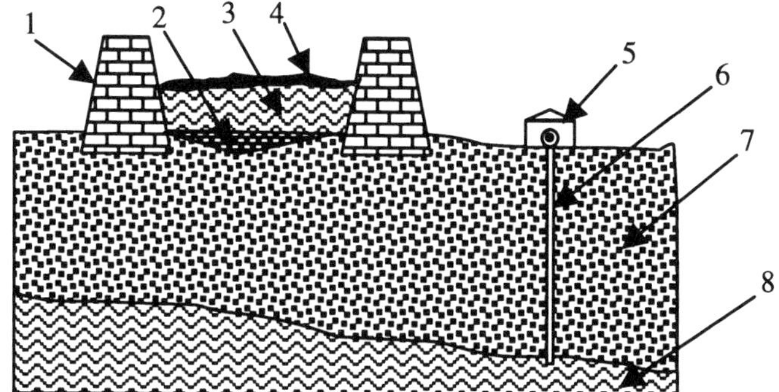

Fig. 7.1: Place of pipeline damage during the restoration of the soil.
1. Dam
2. Crude oil and natural adsorbent layer
3. Water
4. Crude oil
5. Station of ground water sample extraction
6. Borehole to ground water
7. Soil layer
8. Ground water

Water can enter the dam area in Figure 7.1 from rain or snow coming into the damage area. Crude oil film, 4, is collected by special pumps from the dam. The upper layer of the soil together with crude oil adsorbed on natural adsorbent is removed from the damage area by a dredger and either disposed of or put into use. Often, in order to ensure that crude oil is successfully removed from the damage area, ground water analysis is performed in analyzing stations installed near the dam. It is important to control the quality of ground water in the damage area because if the crude oil contamination reaches the ground water it can be spread further to areas around the damage area and contaminate a wider area than was at the time of pipeline damage [38].

Biological self-restoration of the soil after crude oil contamination is a very long process. Usually, it takes over twenty years before the ecological system of the contaminated area reaches a new ecological equilibrium. However, the process can be accelerated simply by loosening the soil. This action minimizes the oxygen deficiency in the soil caused by formation of crude oil film on the surface.

Soil and water contamination can occur not only by damage of pipelines. A majority of crude oil treatment processes need water either as cooling medium or heating agent in heat exchangers, or simply as a reagent. These avenues lead to oil-water contacts, which are responsible for formation of waste water that must be cleaned before it is released back to the environment. There are many different methods used for cleaning waste water. These can be classified as follows:

- mechanical method
- chemical method
- physical chemical method
- biological method

The types of contamination that emanate from the oil industry and can affect natural water can be classified as follows:

- mechanical contamination: characteristic of this type of contamination is the presence of a high concentration of insoluble solid particles in water,
- chemical contamination: characteristic of this type of contamination is the presence of organic or inorganic pollutants in water; this type of water contamination is especially typical of the crude oil industry,
- bacteriological or biological contamination are represented by contamination by bacteria or foreign biological material in certain ecological systems; this type of contamination is atypical of the crude oil industry, and can only occur through improper selection and use of biological cleaning methods,
- radioactive pollutions, this type of pollution is not usually present in the crude oil industry,
- thermal pollution, this type of pollution is present by releasing a warmer process water into the colder natural pools, ponds, or other surface waters; although the potential is there, this type of pollution occurs very rarely in the crude oil industry.

The choice of method and technological scheme for water cleaning in industry depends on many factors. However, the most important factor is the level of cleanliness of the water as required by legislation in the area. Most modern industry processes work with a closed water loop or circuit during production. In this case, water must be only as clean as is needed for the process. In order to have universal applicability, the old plants must be replaced with new and more ecological based processes. For ecologically based processes, water in the circuit only needs to be cleaned to as high a degree as is technically desired and economically feasible.

The main objective in the mechanical method of water cleaning is to remove insoluble or solid contaminants from waste water by settling and filtration. Mechanical solid contaminants are separated by grids or fine filters depending on the size of the contaminant particles. Liquid contamination from crude oil or its products must be separated from the water in special reservoirs by settling. The mechanical cleaning allows the separation from industrial waste water up to 95% of contaminants. A schematic of the mechanical cleaning system is shown in Figure 7.2.

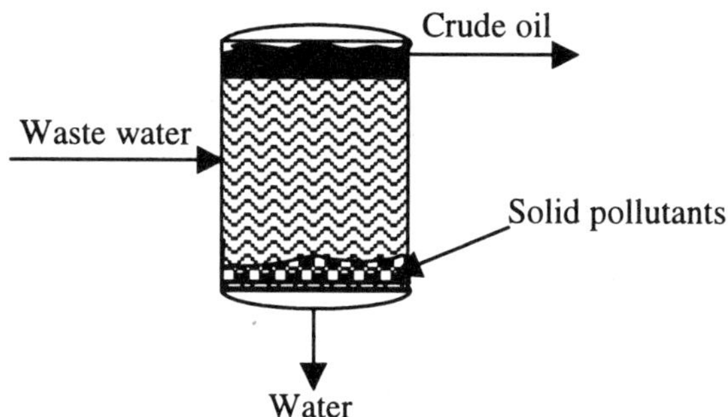

Fig. 7.2: Principle of mechanical water cleaning.

The principle in mechanical water cleaning by settling is separation with regard to density differences of the mixture constituents (see Figure 7.2). The lightest constituents such as crude oil are collected at the top of the separating container and the heaviest at the bottom of the separator.

The main idea in the chemical method of water cleaning is the addition of special chemicals that react with the contaminants to form insoluble heavy residues. These residues can be easily separated from the waste water by settling using the principles discussed for mechanical cleaning.

Fine colloidal or soluble inorganic contaminants are removed from water by physical-chemical methods of waste water cleaning. The main methods used are coagulation, oxidation, sorption, extraction and electrolysis methods.

Biological cleaning methods are based on the principles of natural cleaning of the waste water. However, only waste water cleaned by mechanical and chemical or physical-chemical cleaning can be treated by this method. There are many different methods of biological water cleaning. However, the most used methods are the biological filter and biological pool methods. In the first method, the biological material used for cleaning is deposited on a carrier material as a thin film. The waste water is filtered through this material and the contaminants are destroyed biologically by special types of bacteria. The main objective in the biological pool method is that the waste water should be cleaned in the same way as for self-restoration in natural ecological water systems such as lakes. This means that the waste water is dumped in a special pool in which optimal biological conditions are created for the fastest biological self-restoration of the water.

In modern refineries, it is very rare to use only one of the water cleaning methods. Usually, complicated schemes including many cleaning steps are used. For example, mechanical cleaning is usually used as the first step for waste water cleaning in almost all modern crude oil refineries. The chemical cleaning is used only in special cases if it is necessary to remove any contaminant which cannot by removed by the physical-chemical method. Biological cleaning of waste water is used only in the case where the water must leave the industrial water circuit and be returned to natural lakes or pools.

The physical-chemical methods of waste water cleaning are the most important and most popular as they are included in almost all industrial technological schemes of waste water treatment. Oil–water emulsion destruction by coagulation of the disperse phase is the most popular method used as a second step for water cleaning in water treatment processes in modern crude oil refineries. The objective in this method is to destroy the stable oil–water emulsion by addition of inorganic electrolytes. NaCl, H_2SO_4, $FeSO_4$, $Fe_2(SO_4)_3$, $FeCl_3$, CaO, $Al_2(SO_4)_3$ can be used as inorganic electrolytes. These electrolytes can be used separately or in combination with each other depending on the type of contaminants. These electrolytes, when added, result in a decrease of the electrokinetic potential of oil–water emulsion and the destruction of the structural-mechanical barrier preventing the desired coagulation of oil drops in water. It is important to note that some electrolytes are able to recharge the oil droplets in the emulsion with the formation of an unstable colloidal system. This type of emulsion is called opposite emulsion. Thus, it is very important to determine the optimum concentration of electrolyte needed for application for successful waste water treatment. The use of many electrolytes allows considerable flexibility in this regard, and also leads to increased cleaning efficiency. Currently, aluminum sulfate is the most popular electrolyte used for waste water cleaning.

The coagulation intensity during cleaning by physical-chemical methods can be increased by additional influence of factors such as electromagnetic field, temperature or ultrasound. Ultrasound is used only in special cases because of the deficit of knowledge of such processes. On the other hand, the influence of temperature and electromagnetic field has been well investigated and widely used in industrial waste water treatment plants. The use of such technologies in the crude oil industry is relatively rare because of the high cost of a process that uses an electromagnetic field. However, the use of such a technology can undoubtedly improve the quality of water leaving the industry.

Very often, special filter systems are used for the separation of crude oil from water as a form of waste water cleaning. The principle in such filtration plants is not just ordinary filtration but is adsorptive filtration. The filter in such a plant consists of a layer of adsorbents. Adsorbents usually used are oxidative adsorbents or activated natural fibers.

A general scheme of a waste water treatment plant is presented in the example shown in Figure 7.3. The scheme presented in Figure 7.3 is a very general presentation of the possible water cleaning system in the refinery. Depending on the amount of water and the type of contaminants, it can look different than what is shown in Figure 7.3. However, the pool–collectors and collectors for technical water, which were not described previously, are always part of every scheme of the water circuit in refinery. The need for these elements is based on the possibility that the formation waste water or need for technical water in the refinery can be larger or smaller than the performance of the capacity of the water cleaning plant. It must be noted that the cleanliness of the technical water from the waste water cleaning plant is not sufficient for it to be released into natural pools. However, it can be used in the refinery water circuit again.

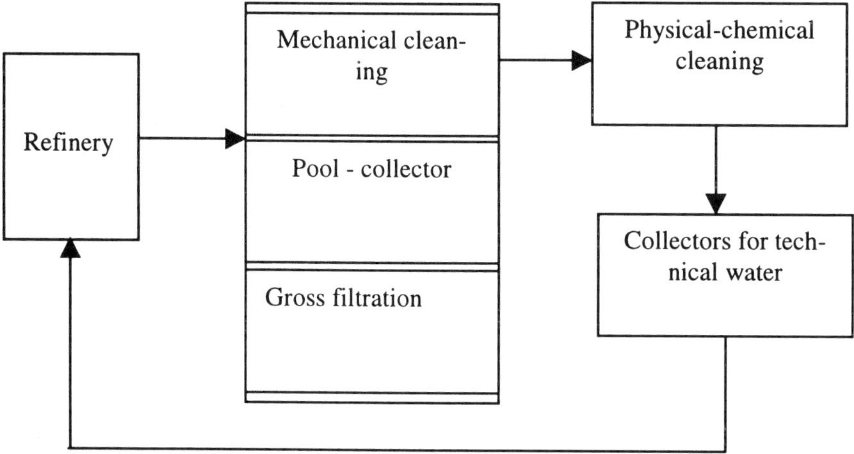

Fig. 7.3: A general scheme of industrial water circuit with water cleaning.

7.3 METHODS OF AIR AND GAS CLEANING USED IN THE CRUDE OIL INDUSTRY

The atmosphere of our planet is a very sensitive system, such that even little changes in the concentrations of the same gases present in the atmosphere can strongly change the ecological equilibrium not only in the area of formation of the pollutants, but also in areas many kilometers around. Using methods of preventing the damage done by gas pollution as was shown for soil and water contamination in section 7.2 is impossible because of the very high speed of spreading in the environment for gaseous pollutants. This is why it is important to control the quality of the waste air or gas leaving the industrial circuit.

All the types of gas pollutants can be divided into:

- Mechanical particles (solid and liquid) in the form of aerosols
- Gaseous and vapor pollutants

The first group consists of solid particles of inorganic and also liquid droplets (e.g. fog) all of which could be in the form of aerosols. Dust is a stable colloidal system containing more large particles than smokes and fogs. The concentration in dusts in terms of number of particles in 1 cm^3 is small in comparison with smokes and fogs. The second group, gaseous and vapor pollutants contained in industrial waste gas, is much more extensive than the first group. Acids, halogens and halogen containing substances, gaseous oxides, aldehides, ketones, spirits, hydrocarbons and many other components of industrial waste gases are part of this group.

The complete removal of pollutants from industrial gases is often impossible or uneconomical. This is the reason for the development of tables, which limit the concentration of dangerous pollutants. If the waste gas contains more than one pollutant, a new limit of concentrations for the gases must be calculated by equation (7.1)

$$C_1/C_{L1} + C_2/C_{L2} + \ldots + C_n/C_{Ln} = 1 \tag{7.1}$$

where C_1, C_2, C_n – real concentration of gases 1, 2 …,n
C_{L1}, C_{L2}, C_{Ln} – standardized limit of concentrations of gases 1, 2 …,n

If it is impossible to achieve the limit of concentration by cleaning methods, sometimes the dilution of the pollutants in the higher layers of the atmosphere is applied. This reduces emission of gases in the bottom layers of the atmosphere by venting or shooting the waste gases through a tall pipe. The pollutants are thereby dispersed in the top layers of the atmosphere. The theoretical definition of concentration of pollutants in the bottom layers of the atmosphere depends on the height of the pipe and other factors connected with the laws of turbulent diffusion of pollutants in the atmosphere. There is no accurate method for such an evaluation. Consequently, empiric equations such as equation (7.2) are used such calculations.

$$C_{HL} = \frac{C_L H^2 \sqrt[3]{V\Delta\Delta}}{AFm} \tag{7.2}$$

Where C_L – standardized limit of concentration of the gas in the bottom layers of the atmosphere
C_{HL} – calculated limit of concentration of the gas in the higher layers of the atmosphere
H – height of the pipe [m]
V – volume velocity of the gas stream from the pipe [m^3/s]
Δt – temperature difference between waste gas and atmosphere [K]
A – coefficient specific for the pollutant and atmosphere
F – sedimentation coefficient for the pollutants in the atmosphere
m – coefficient of the output conditions of the gas from pipe

The last coefficient in equation (7.2) must be calculated by equation (7.3).

$$m = \frac{1.5 * 10^3 \omega^2 D}{H^2 \Delta t} \qquad (7.3)$$

where ω – velocity of gas output from the pipe [m/s]
D – diameter of the pipe [m]
H – height of the pipe [m]
Δt – temperature difference between waste gas and atmosphere [K]

However, the method using the dilution of the pollutants in the higher layers of the atmosphere does not really save the environment from the contaminants. It only moves the pollutants from the area of formation to other areas. Consequently, it is better not to use this method if it is possible.

All the methods for gas cleaning can be classified into the following three classes:

- mechanical cleaning
- electrostatic coagulation method
- cleaning using ultrasound

The mechanical methods can be classified into dry and humid methods. The dry methods can be different such as in:

- gravitational settling
- centrifugal settling
- filtration

Gravitational settling is based on settling of the pollutant particles by gravity through movement of dusty gas at low speed with no change in the direction of flow. The process is carried out in pipe settlers and settling chambers. A set of horizontal trays at a distance of 40-100 mm from each other is installed in settling chambers, which split the gas flow into flat jets. This is done to reduce the height of particles settling in settling chambers. Productivity of settling chambers can be calculated by equation (7.4).

$$P = A \cdot w \qquad (7.4)$$

where S – area of horizontal section of the chamber
w – speed of particles settling under the influence of gravity [m/s]

Gravitational settling method is effective only for pollutant particle size larger than 50-100 microns. This method is suitable only for preliminary cleaning of gases.

Centrifugal methods of gas cleaning are based on the action of centrifugal force that arises from rotation of flow of dirty gas in the cleaning chamber or by rotation of parts of the chamber. Cyclones of various types are used as centrifugal chambers. Cyclones are the most frequently applied units for the cleaning of aerosols (solid particles in gas) in the industry. Cyclones are characterized by their high productivity in comparison to gravitational settling units. The size of the particle that can be separated by this method can be approximately calculated by empirical equation (7.5).

$$d = \sqrt{\frac{9\mu D}{2N\ \omega(\rho_p - \rho_g)}} \tag{7.5}$$

where
μ – viscosity of the gas
D – diameter of the end pipe at the cyclone
N – effective number of gas rotations in the cyclone
ω – gas velocity at the input into the cyclone
ρ_P – particle density
ρ_g – gas density

The next method for dry gas cleaning is filtration. Filtration is based on passage of dirty gas through various filtering fabrics such as cotton, wool, chemical fibers, fiber glass, etc. or through other filtering materials such as ceramics, metal ceramics, porous filters from plastic, etc. Special fibrous materials such as fiber glass or wool are often used for gas cleaning. Fabric filters are applied for gas cleaning at the temperature of waste gas of 60-65°C. Depending on colloidal structure of initial waste gas, up to a 85-99% degree of cleaning can be achieve by this method. The hydraulic resistance of the filter is about 1000 Pa.

The use of fiber filters allows the achievement of a degree of cleaning up to 99.9%. Using fiber glass as a filter material allows the cleaning of the waste gas at a temperature up to 275°C.

Filtration is the most frequently used method for fine cleaning of waste gases in industry. However, the energy needed for this method is higher than the energy required for settling methods. Thus, this method should only be used in the case where a high degree of cleaning is really needed.

Humid cleaning of waste gases is the next group of cleaning methods for waste gases. This is based on washing of gas by a liquid (usually by water). The contact surface area of liquid with particles in the waste gas stream should be as large as possible, and the mixing of waste gas with liquid should also be as intensive as possible. This is a universal method of particles of dust, smoke and fog from industrial gases. It is the most widespread method used as a final stage of mechanical cleaning. There are many different devices that are developed to use the humid method of gas cleaning as a rational cleaning method.

Towers with special packings are used and they are characterized by their simplicity of design and operation, stability in work, small hydraulic resistance (300-800 Pa), and a rather small need of energy. In such devices, it is possible to clean gases with initial gas pollution up to 6 g/m^3. The efficiency of one step of cleaning for this device for waste gas with pollutant particle size over five microns is usually between 70% and 80%. The disadvantage of this method is that the void volume of the packed tower is quickly blocked by dust, especially at a high initial dust content.

Humid cyclones are used for cleaning large volumes of gas. They have small hydraulic resistance - 400-850 Pa. The efficiency of gas cleaning for waste gas with pollutant particle size of 2-5 microns is approximately 50%.

Foam devices are used for cleaning waste gas from pollutants with polydisperse colloidal structure. Intensive foam formation by this method is created by linear speed of gas between 1 and 4 m/s. Foam cleaners have high efficiency of gas cleaning and small hydraulic resistance (app. 600 Pa). For pollutants with particle size over 5 microns, up to a 99% degree of cleaning can be achieved for foam gas cleaners.

The scrubber is the most effective device for waste gas cleaning. However, it consumes a large amount of energy for it to work. The speed of the waste gas into a narrowing pipe can be up to 100-200 m/s, and in some special devices, it can be up to 1200 m/s. At such a speed, the waste gas is highly dispersed into smallest drops. This leads to an intensive collision of pollutant particles with drops of liquid, and catching of these particles under the action of inertial forces. Scrubbers are universally small-sized units ensuring cleaning of fogs up to 99-100%, waste gases with pollutant particles with size between 0.01 and 0.35 microns up to 50-85%, and waste gases with pollutant particles with size between 0.5-2 microns up to 97%. The degree of cleaning for waste gases with particle size between 0.3 and 10 microns can be calculated by equation (7.6).

$$\eta = 1 - e^{-KL\sqrt{\varphi}} \qquad (7.6)$$

Where
- η – degree of cleaning
- K – coefficient
- L – volume of liquid introduced into the gas [dm^3/m^3]
- ϕ – inertia coefficient

One example of the scrubber is presented in Figure 7.4.

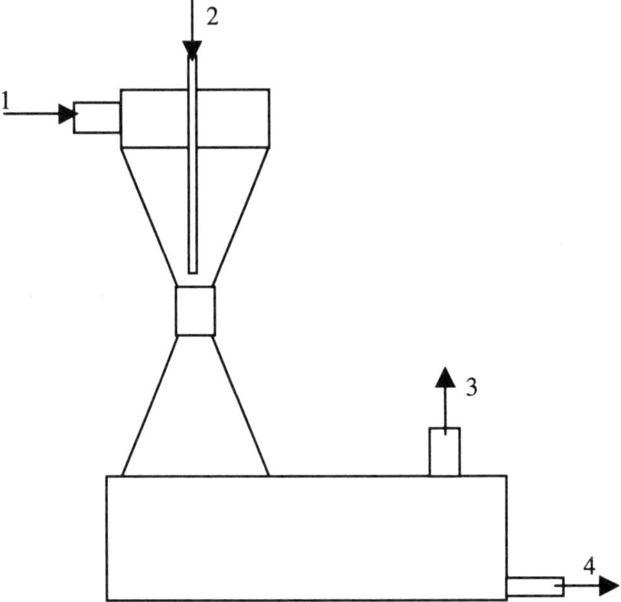

Fig. 7.4: Example of a scrubber.
1 – waste gas,
2 – washing liquid,
3 – clean gas,
4 – liquid with pollutants.

The biggest disadvantage of all humid methods of gas cleaning is the formation of a large amount of liquid residue, which must be utilized or disposed of. If in the general scheme of the cleaning plant in the refinery, there is no plant for utilizing such residues, this method will only lead to transfer of pollutants from gas into water.

The electrostatic cleaning of gases serves as a universal method which is suitable for cleaning all types of waste gases that contain solid or liquid pollutants, including fogs of acids. The method is based on the ionization of pollutant particles in the waste gas by passing the gas through an electrical field of a high voltage created by electrodes. Settling of the particles occurs in settling electrodes. The industrial electrofilters consist of a number of plates or pipes through which the waste gas passes. Between the settling electrodes are hung wire electrodes, which have voltages of 25-100 kV. The theoretical equation for calculating the degree of cleaning of waste gas in tubular electrofilters is presented in equations (7.7) and (7.8).

$$\eta = 1 - e^{-y} \tag{7.7}$$

$$y = \frac{2ul}{r\omega} \tag{7.8}$$

where η –degree of cleaning
y – coefficient calculated by equation (7.8)
u – particles velocity
l – length of electrode
r – radius of settling electrode
ω – velocity of waste gas

The degree of cleaning that can be achieved by this method is up to 99.9%, depending on the type of pollutants in the waste gas.

Cleaning using ultrasound is a relatively new method of gas cleaning and used very rarely at industrial scale. The degree of cleaning that can be achieved by this method is approximately the same as was shown for electrostatic settling methods.

Waste gases do not only contain solid or liquid particles as pollutants. They also contain gaseous and vapor contaminants. These pollutants are also dangerous to the environment. These pollutants cannot be separated from the waste gas by any of the methods described before. The methods for waste gas cleaning from gaseous and vapor pollutants can be classified into three major classes:

- absorptive methods
- adsorptive methods
- catalytic methods

The absorption of pollutants from waste gases by liquids is applied in the oil industry for the extraction of gases such as sulfur dioxide, hydrogen sulfide and other sulfur containing compounds, nitrogen oxides, acids (HCl, HF, H_2SO_4), various organic compounds such as phenol, and volatile solvents.

All the absorptive methods are based on selective solubility of the gas components in the washing liquid. There are two different methods of absorption:

- physical absorption
- chemical absorption

All the absorptive cleaning methods used in modern industry are continuous methods. The washing liquid is usually regenerated and used in the cleaning again. Heating or decreasing the pressure is usually used for regenerating the washing liquid. One example of absorptive gas cleaning plant is presented in Figure 7.5.

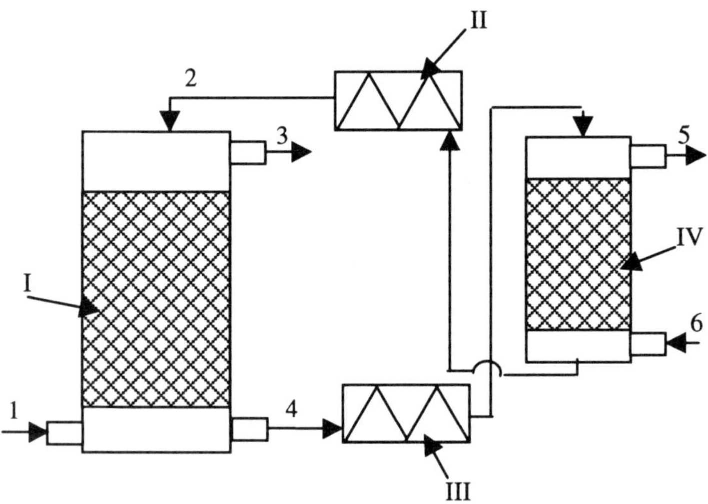

Fig. 7.5: Absorptive gas cleaning unit.
1 – waste gas,
2 – washing liquid,
3 – clean gas,
4 – liquid with pollutants,
5 – gas,
6 – vapor,
I – absorber,
II – cooler,
III – heater,
IV – regenerator.

Absorbents, used in an industry can be estimated by the following parameters:

1. Absorbent capacity, it means solubility of a certain pollutant in an absorber depending on temperature and pressure;
2. Selectivity characterized by different solubility of different gases and speeds of their absorption;
3. Minimal pressure of the absorbent vapor, in order to prevent pollution of cleared gas by absorbent vapors;
4. Cost;
5. Absence of corrosive effects on the material of construction of the equipment.

Water, solutions of ammonia, salts of manganese, oil and so on are used in modern industry as absorbents for gas cleaning.

Adsorptive methods are based on the selective adsorption of pollutants from the waste gas on the surface of an adsorbent. These methods are used in modern industry more often as the most used universal method for cleaning waste gas from toxic pollutants. Adsorbents usually used are activated coal, silica gel, aluminum gel and synthetic zeolite.

Industrial adsorptive cleaning plants usually work continuously. However, the adsorbent is a solid material, which makes it difficult to transport continuously from adsorber to regenerator. In order to circumvent this difficulty, modern adsorptive cleaning plants make use of at least two absorbers. When one adsorbent is working, the other adsorbent is switched to the regenerating regime. In that way, it is possible to carry out the cleaning continuously. One example of an adsorptive cleaner is shown in Figure 7.6.

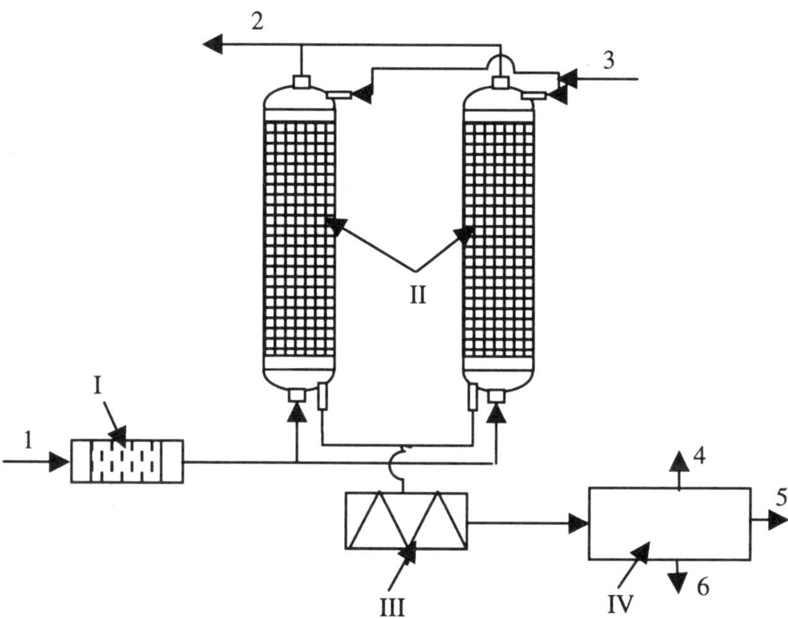

Fig. 7.6: Adsorptive gas cleaning unit.
1 – waste gas,
2 – clean gas,
3 – vapor,
4 – vapor and gas,
5 – condensed pollutant,
6 – condensed vapor,
I –filter,
II – adsorber,
III – cooler,
IV – separator.

The next class of gas cleaning methods is the catalytic cleaning method. The main difference of this method from all the other methods described in this chapter is the fact that with this method pollutants are not separated from the waste gas but transformed into compounds that are not dangerous vis-à-vis the environment. The reactions of pollutants to new products take place in presence of catalysts. This is why this cleaning method is called the catalytic method.

The best example for the use of the catalytic method in the oil industry is the cleaning of waste gas of H_2S and the production of sulfur as a byproduct. The transformation of hydrogen sulfide to sulfur proceeds according to the reaction given in equation (7.9).

$$H_2S + 1/2\ O_2 = H_2O + S \tag{7.9}$$

Catalytic methods of gas cleaning are used very often in modern industry as a result of the possibility of very deep cleaning of the waste gas. Up to 99.9% of pollutant can be separated from the waste gas by this method. However, the disadvantage of this method is the formation of new compounds, which must be useable. In the example shown of waste gas cleaning from the H_2S, the formed by-product can be used in the chemical industry. However, the amount of sulfur formed in crude oil refineries is often so high that it cannot be moved completely to the chemical industry. It then becomes important to find new ways of using sulfur.

The last method of gas cleaning is the thermal method. This method was very intensively used in the past. However, it has been the desire recently to reduce the number of such plants in modern refineries. The main idea of this method is that of burning the waste gases thereby destroying the pollutants.

Using only one of the described methods in the refinery for waste gas cleaning is impossible. Modern refineries have different schemes in their technological structure for gas cleaning, each consisting of many methods. It is only the use of many methods in combination that will allow us to achieve the needed cleanliness of the gas with optimum use of energy and chemical recourses of the refinery.

7.4 CONCLUSION TO PART III

The main processes involved in crude oil refining were shown in part III of this book. Now is the time to show how all these processes can be combined in one crude oil refinery. However, it must be noted that every crude refinery is built uniquely and there are no two crude oil refineries in the world having the same technological scheme of processes used. The technological scheme of the refinery must be chosen based on the properties of crude oil that the refinery is supposed to treat. Nevertheless, it is possible to classify all modern refineries in two groups:

- working by fuel variant
- working by oil variant

This classification is very much simplified, but it can help us to understand the tendencies in the development of the modern crude oil industry. From the names of both these groups, it is obvious towards which type of products they are oriented.

The fuel working crude oil refineries produce only fuels. In these refineries, crude oil is treated with a deep transformation of heavy fractions into light fuel fractions as much as is possible. The most important processes in these refineries are catalytic cracking, hydrotreatment of fuel fractions, hydrocracking and all destructive processes. In these refineries, no lubricants or lubricating oils are produced.

The oil-oriented refineries often have no processes for secondary treatment of heavy fractions of crude oil. Thus, these refineries produce mainly lubricants and lubricating oils, but still produce fuels. The main processes in these types of refineries are processes that improve the quality of lubricating oils such as deparaffinization, hydrotreatment of heavy crude oil fractions, etc.

Depending on the sulfur content in the crude oil, both types of crude oil refineries can have the plants for catalytic waste gas cleaning with the production of sulfur in their technological structure. If more sulfur is produced than the refinery can sell to the chemical industry, it is often reasonable to install plants for sulfuric acid production inside the refinery.

Because of the necessity to decrease the use of thermal methods for gas cleaning, new methods of using organic pollutants from waste gases have been developed. Thus, the installation in many modern refineries of plants for methanol production. The light gas from secondary processes is used for processes in such plants.

In part III of this book, the complexity of modern refinery was shown and the fundamentals of most important processes were explained. The crude oil industry is a very fast developing industry. In order to attempt to keep abreast with new developments, additional information have been given in references at the end of part III of this book. For more detailed discussions, it may be necessary to look at further references recommended in these references. It is almost impossible to show all the most modern developments of crude oil industry in one book. Also, this is not the object of the book. Instead, we propose to prepare the reader for independent search for information in articles, scientific communications, letters, etc. because these contain the most factual information about the newest developments in the field of crude oil treatment.

In the next part (part IV) of this book, the fundamentals of the most modern developments in the area of secondary crude oil treatment such as co-processing are shown. These are based on the latest research results and have not yet been applied at industrial scale so far. These technologies therefore constitute the possible future technologies for crude oil refineries.

Bibliography

1 P Seidel. Schweres Erdöl - ein alternativer Rohstoff zur Erzeugung von Treibstoffen. Expert Verlag, Renningen-Malmsheim 1994.

2 C Jentsch. Erdölverarbeitung, in Ullmans Encyclopaedia der Technischen Chemie. 4. Auflage, Bd. 10, pp. 641-714. Weinheim: Verlag Chemie, 1975.

3 I Pfeiffer, R Saal. J. Phys. Chem., 44, 139, 1940.

4 W N Erih, M G Rasina, M G Rudin. Chimiya i Technologija Nefti i Gaza. Chimiya, Leningrad, 1977.

5 J D Elliot. Delayed Coker Design and Operation: Recent Trends and Innovations. Foster Wheeler USA Corporation, 1996.

6 A F Orlicek, H Poell, H Walenda. Hilfsbuch für Mineralöltechniker. Springer, Wien, 1955.

7 R J Kee, F M Rupley, J A Miller. The CHEMKIN Thermodynamic Data Base, SAND95-8215, 1995.

8 L V Gurvich, I V Veyts, C B Alcock. Thermodynamic Properties of Individual Substances. Hemisphere Publishing Co, 4th edition, 1989.

9 J D Cox, D D Wagman, V A Medvedev. Key Values for Thermodynamics. Hemisphere Publishing Co, 1989.

10 M W Chase Jr, C A Davies, J R Downey Jr, D J Frurip, R A McDonald, A N Syverud. JANAF thermochemical tables. J. Phys. Chem. Ref., 14, 1985.

11 G Kortuem, H Lachmann. Einfuehrung in die Chemische Thermodynamik. 7. Auflage, Verlag Chemie, 1981.

12 J Gmehling, B Kolbe. Thermodynmik. Second Edition, VCH, 1992.

13 R G Gilbert, S C Smith. Theory of Unimolecular and Recombination Reactions. Blackwell Scientific Publications, Oxford, 1990.

14 R Zellner. Bimolecular Reaction Rate Coefficients. W C Gardiner Jr, Springer, New York, 1984.

15 W C Gardiner, J Troe. Rate Coefficients of Thermal Dissociation, Isomerization, and Recombination Reactions. Springer, New York, 1984.

16 W Tsang. Rate Constants for the Decomposition and Formation of Simple Alkanes Over Extended Temperature and Pressure Ranges. J. Combust. and Flame, 78:71-86, 1989.

17 D Golden, A Baldwin. OLCHEM Chemical-Rate-Equation Integrator. Molecular Physics Laboratory, Stanford Research Institute, 1998.

18 W D Hinsberg, F A Houle. Chemical Kinetics Simulator 1.01. IBM Almaden Research Center, San Jose, California, 1996.

19 R G Gilbert, S C Smith, M J T Jordan. UNIMOL Program Suite, Calculation of Rate Coefficients for Unimolecular and Recombination Reactions. 1994.

20 H Briesen, W Marquardt. Adaptive Model Reduction and Simulation of Thermal Cracking of Multicomponent Hydrocarbon Mixtures. J. Comput. Chem. Eng. 24, 1287-1292, 2000.

21 B C Gates, J R Katzer, G C A Schuit. Chemistry of Catalytic Processes. McGraw-Hill, New York, 1979.

22 B C Gates, L Guczi, H Knözinger. Metal Clusters in Catalysis. Elsevier, Amsterdam, 1986.

23 B C Gates. Catalytic Chemistry. Wiley, New York, 1992.

24 P G Smirniotis, E Ruckenstein. Catalytic Cracking of Gas Oil: Effect of the Amount of Zeolite Composite Catalysts. J. Chemical Engineering Communications, Vol. 116, 171-191, 1992.

25 H C Beirnaert, J R Alleman, G B Marin. A fundamental kinetic model for the catalytic cracking of alkanes on a USY-zeolite in the presence of coke formation. Ind. Eng. Chem. Res. 40, 1337-1347, 2001.

26 N V Dewachtere, F Santaella, G F Froment. Application of a single-event kinetic model in the simulation of an industrial riser reactor for the catalytic cracking of vacuum gas oil. Chem. Eng. Sc., 54, 3653-3660, 1999.

27 N V Dewachtere, G F Froment, I Vassalos, N Markatos, N Skandalis. Advanced modeling of Riser-type catalytic cracking reactors. Applied Thermal Engineering, 17, 8-10, 837-844, 1997.

28 P G Smirniotis, E Ruckenstein. Catalytic Cracking of Gas Oil: Effect of the

Amount of Zeolite Composite Catalysts, J. Chemical Engineering Communications, Vol. 116, 171-191, 1992.

29 P G Smirniotis, E. Ruckenstein. Comparison of the Performance of ZSM-5, b-Zeolite, Y, USY and Their Composites in the Catalytic Cracking of n-Octane, 2,2,4-Trimethylpentane and 1-Octene. Industrial & Engineering Chemistry Research, Vol. 33, 800-813, 1994.

30 M L Occelli, H Eckert, M Kalwei, A Wölker, A Auroux. The effects of steam-aging temperature on the properties of an HY zeolite of the type used in FCC preparations. in "Fluid Catalytic Cracking V: Technology for Next Century", M L Occelli, P. OíConnor Eds.; Elsevier, Amsterdam, 2001.

31 A Brait, K Seshan, H Weinstabl, A Ecker, J A Lercher. Evaluation of commercial FCC catalysts for hydrocarbon conversion II. Time-on-stream behavior of n-hexane conversion and comparison of n-hexane conversion to MAT. J Applied Catalysis A: General 169, 315-329, 1998.

32 M Absi-Halabi, J Beshara, H Qabazard, A Stanislaus. Catalysts in petroleum refining and petrochemical industries. Proceedings of the 2nd International Conference on Catalysts in Petroleum Refining and Petrochemical Industries, Kuwait, April 22-26, 303-365, 1995.

33 M Absi-Halabi, J Beshara, H Qabazard, A Stanislaus. Catalysts in petroleum refining and petrochemical industries. Proceedings of the 2nd International Conference on Catalysts in Petroleum Refining and Petrochemical Industries, Kuwait, April 22-26, 99-293, 1995.

34 M C Obala, S S Shih. Catalytic hydroprocessing of petroleum and distillates. Based on the proceedings of the AIChE Spring National Meeting, Houston, Texas March 28 – April 1, 1993, Marcel Dekker, Inc., New York, 1993.

35 W Keim, J Herwig. Einsatz eines Mikro-Kreislaufreaktors zur Bestimmung kinetischer Konstanten von Reforming-Katalysatoren. J. Chem. Soc., Chem. Commun., 1592, 1993.

36 W Keim. Schwefelbeständigkeit von Pt-Ir-L-Zeolith-Reforming-Katalysatoren. IG-Chemie, 30.11. 1993/94.

37 Internet publication: Wenn TotalFinaElf Wasser in sein Dieselöl mischt. http://www.totalfinaelf-service.de/home/Bibliothek/index.htm, 2002.

38 Internet publication: I B Shumilova, N G Maximovich, S M Blinov, L N Kuznezov. Vozmozhnye puti borby s posledstviyami razlivov nefti, Nauka, 2002.

Part IV

HEAVY OIL PROCESSING - CHEMISTRY OF ASPHALTENES

OVERVIEW

From the title given to part IV of this book, it is obvious that in the next two chapters we will be engaged in studies involving the chemistry of the heaviest compounds of crude oil - asphaltenes. The presence of these compounds in crude oil leads to many problems during crude oil treatment. In this part of the book, these problems and ways to circumvent them are discussed.

The present trend in the petroleum industry shows an increasing demand for light products such as gasoline, jet fuel and diesel fuel. In order to meet the market demand, refineries convert a portion of their residual heavy oils into light fractions by destructive processes as was highlighted in chapter 6. This conversion also results in the production of modern heavy fuels, which contain a greater concentration of sulfur, vanadium, and asphaltenes.

Asphaltenes are considered to be part of the "bottom of the barrel". They constitute the non-volatile, high molecular weight fraction of petroleum. In addition, since asphaltenes are non-soluble in heptane, they remain in the solid form in crude oil as well.

The chemistry of asphaltenes is very complicated and it is the least studied field of crude oil chemistry. Because of the complexity of asphaltenes' structure, there is no information about the exact chemical structure of an asphaltene molecule. It is natural that only the average asphaletene molecular is possible as given in the literature. The use of such a chemical structure (i.e. average molecular structure) for the asphaltene molecule is warranted because of the wide molecular weight range and the diversity of chemical groups in the structure of asphaltenes.

The chemistry and understanding of the nature of asphaltenes is as fascinating as it is complicated. Many properties of asphaltenes are still not fully investigated. This, perhaps, is the object of the next generation of crude oil chemists; that is to discover more fully new properties of asphaletenes, and also, to find out ways to convert these compounds to useful crude oil products.

In the next two chapters, the reader will be exposed to the main problems that arise for the crude oil industry due to the presence of asphaltenes in crude oil. The reader will also be exposed to the latest results of investigations of various possibilities for conversion to useful products of asphaltenes and heavy oil.

8

Chemistry of Crude Oil Asphaltenes

8.1 INTRODUCTION

For many years, crude oil has been the cheapest source of liquid fuels in many countries. The balancing between product yield and market demand without manufacturing large quantities of low-commercial-value fractions has long required processes for the conversion of hydrocarbons of one molecular weight range and/or structure into another molecular weight and/or structure. Basic processes for this are the so-called "cracking" processes in which relatively heavy hydrocarbons are broken down (i.e. cracked) into smaller, lower-boiling fractions. In the present market, there is increasingly less high quality crude oil but more bitumen with a very high content of asphaltenes. This is why one has to realize that heavy bitumen and/or vacuum residues from petroleum refineries have value as an alternative feed for the production of liquid fuels. In Canada, for example, 60% of all crude oil sources are in the form of bitumen (tar) sands [1]. But by using almost all the existing cracking processes, coke formation is inevitable. This makes all these processes non-economic; i.e., the present thermal treatment processes cannot completely solve the problem of production of fuel fractions from heavy residues of crude oils.

Reactions that lead to coke formation are polycondensation reactions involving the heaviest and most highly aromatic compounds of crude oil – the asphaltenes. Asphaltene is derived from the word "asphaltu" meaning "to split", and adopted later by the Greeks, signifying "firm", "stable" or "secure". Asphaltenes are present in crude oil in its original state as the so-called "native asphaltenes". Asphaltenes are also found in larger amounts in residual fractions because, there is an increasing trend to extract large portions of light fractions from crude oil by cracking and visbreaking, while asphaltenes stay in the heavy residues [2].

The classic definition of asphaltenes is based on the solution properties of petroleum residuum in various solvents. Broadly speaking, asphaltenes are insoluble in paraffin solvents but soluble in aromatic solvents. Structurally, asphaltenes are condensed polynuclear aromatic ring systems bearing mainly alkyl side-chains. The number of these rings in oil asphaltenes can vary from 6 to 15.

The presence of asphaltene in crude oil causes problem [3] for:

- Oil recovery (in the oil reservoir, in the well, in the pipeline).
- Visbreaking processes (degraded asphaltenes are smaller in size, more aromatic and less soluble in maltene than original asphaltene. This leads to coke formation).
- Blend of mineral oil residue (flocculation of asphaltenes).
- Storage (sludge and plugging due to further oxidation, among other things).
- Preheating (the preheating of fuel oils prior to their burning encourages the precipitation of asphaltenes and coking).
- Combustion (poor combustion causes boiler fouling, poor heat transfer and stack solid emission).

All these problems arise from the sedimentation of asphaltenes. Sedimentation begins when asphaltenes achieve a predetermined size of asphaltene molecules and/or asphaltene particles. The growing of asphaltene molecules is caused by polycondensation reactions. Lowering the solubility of asphaltenes in oil causes polycondensation. This implies that in the case of deep cracking of asphaltene, polycondensation reactions cannot take place.

The objectives of this chapter are twofold. First is to show the fundamentals of the chemistry and process engineering of asphaltenes during thermal treatment for achieving deep asphaltene cracking to increase and/or improve distillable yields of crude oil. The second is to present the current major processes for utilization of heavy oils and residuum fractions.

8.2 PROBLEMS OF CRUDE OIL RESIDUE TREATMENT WITH RESPECT TO ASPHALTENES

The presence of asphaltene in crude oil causes problems for oil treatment as follows [4]:

- Asphaltenes formed as a result of thermal treatment are smaller in size, but more aromatic and less soluble in maltene than the original asphaltene. This leads to coke formation on the pipe walls resulting in very high heat consumption since coke has a very low thermal conductivity. Coke formation on pipe walls can also result in blocked pipes and this will lead to a high pressure drop in the pipes.
- Catalytic treatment: deactivation of catalysts is caused by coke formation, and also catalysts are poisoned by the heavy metals from asphaltenes
- In blending of mineral oil residue asphaltenes can be flocculated
- Storage (sludge and plugging due to further oxidation, among other things)

- Preheating (the preheating of fuel oils prior to their combustion encourages the precipitation of asphaltenes and coking).

8.2.1 Coke Formation and Reduction of Heavy Metals

Generally the mechanism of coke formation can by described be polycondensation reactions of asphaltenes or asphaltenes with aromatic or unsaturated compounds. In the simplest form all these reactions can be represented by the following reactions:

As + As → coke + gas
As + n · Ar → coke + gas
As + n · olefins → coke + gas

where As: asphaltene molecule
Ar: aromatic compound
gas: hydrogen or hydrocarbon gas.

Polycondensation of asphaltenes proceeds quickly only if asphaltenes are flocculated from the maltenes. It is appropriate to obtain a better understanding of the mechanism for flocculation and, consequently, that of polycondensation. In this context, the original colloidal model of petroleum of Pfeiffer and Saal [5] can be used to explain any characteristics of crude oil residues and asphaltenes. One form of this model is shown in Figure 8.1. According to the present version of this model, asphaltenes associate to form microscopic solids that are dispersed by amphiphilic resins. The resins are attracted to asphaltenes on one end of the resin molecule and to small ring aromatics on the other end. While small ring aromatics act as solvent for asphaltene-resin dispersion, saturates act as a non-solvent. Thus, this model is neither a solution nor a colloidal model, but a hybrid of both. A modern corroboration of this model is that the addition of resins to asphaltenes in toluene reduces its radius of gyration, measured by small angle X-ray scattering (SAXS) [6]. Furthermore, the addition of low concentrations of amphiphilic model compounds greatly increases asphaltene solubility. As most models, this is an oversimplification that enables an approximation of reality within a certain degree of accuracy. Once the limitations are determined by comparing with actual data, directions for improving the model are usually suggested.

With the hybrid model of petroleum, it can be seen that asphaltenes are held in a delicate balance that can be easily upset by the addition of alkanes or the removal or upgrading of resins or small ring aromatics. It has not generally been known, but the blending of crude oil residues with other compounds can change the delicate balance and precipitate or stabilize asphaltenes. When crude oil residues are thermally processed, the side chains are cracked off asphaltenes and resins. This causes resins to combine and form more asphaltenes, and the asphaltenes to become less soluble [7]. This strongly accelerates the polycondensation reactions and leads to coke formation. In addition, if the thermal conversion is carried far enough, the asphaltenes can precipitate as a liquid crystalline phase or carbonaceous mesophases and very quickly combine to form coke [7]. On cooling, the less soluble asphaltenes with fewer resins as natural dispersants can precipitate from the product [8].

mmm
maaam
maRRRam
maRAARam
maRAARam
maRRRam
maaam
mmm

Fig. 8.1: Colloidal dispersion - solution hybrid model of petroleum:
A – asphaltenes
R – resins
a – aromatic compounds
m – maltenes

During the thermal treatment of crude oil vacuum residue, asphaltenes undergo polycondensation reactions that consequently lead to coke formation. As was mentioned earlier, this results in a high pressure drop since the pipes are blocked by coke, and to higher heat consumption since the coke has a very low thermal conductivity. This is why every thermal cracking process has to be stopped periodically in order to clean off the coke. In contrast to thermal cracking, catalytic cracking is affected adversely by deactivation of the catalysts due to coke formation and by catalyst poisoning by reactions with heavy metals. Coking can be rapid but can be controlled by continuous regeneration as in FCC catalysts, or slowed down by subjecting to monthly or yearly maintenance, as in hydrotreating catalysts.

The heavy metals in crude oil residues are agglomerated, first of all, in asphaltenes in the form of porphyrin compounds. One example of this compound is presented in Figure 8.2:

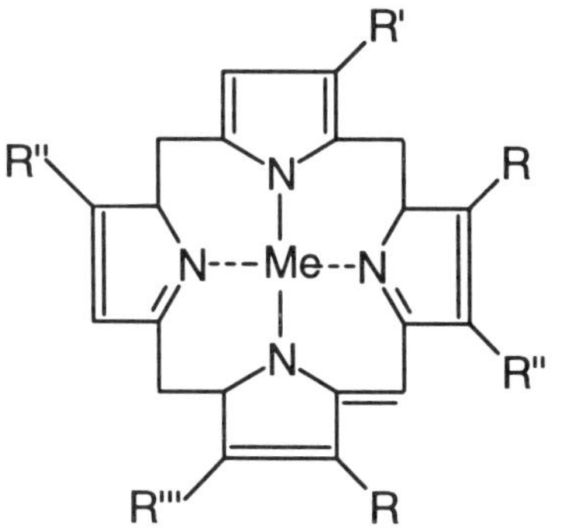

Fig. 8.2: Example of porphirin compound.

Vanadium, nickel and iron generally represent the heavy metal (Me) in the asphaltenes in Figure 8.2. The molecular weight of this type of compound varies between 420 and 520, i.e. from C27N4 – C33N4 [10]. During catalytic treatment of the crude oil residues, the destruction of asphaltenes and the formation of non-bonded heavy metals occur. A possible mechanism of catalyst poisoning will be presented in a later part of section 8.2.

8.2.2 Treatment Possibilities for Crude Oil Residues

Treatment for crude oil residues is intended to minimize the adverse effects of the presence of asphaltenes during thermal processing. Generally, two types of treatment possibilities are available. There are physical and chemical treatments. The physical treatment essentially involves deasphalting whereas the chemical treatment includes thermal processes such as visbreaking and coking.

8.2.2.1 Physical treatment - deasphalting

Feedstocks for cracking processes are now usually the vacuum residue or heavy distillates from vacuum distillation.

Heavy oils and residues contain substantial amounts of asphaltenes, which preclude the use of these residues as fuel oils or lubricating stocks. Subjecting these residues directly to thermal cracking is economically advantageous, since, on the one hand, the final result is the production of lower-boiling fractions; on the other hand, asphaltenes in the residue are regarded as coke-forming constituents and may even promote coke formation from the compounds of the residue.

Furthermore, to avoid catalyst poisoning or reduction in catalyst activity, it is essential that as much nitrogen and heavy metals as possible are removed from the feedstock. It has been shown [9] that a greater part of the heteroatoms (nitrogen, oxygen and sulfur) and the heavy metals are contained in or associated with asphaltenes. It is necessary for the cracking processes that asphaltenes are removed from the cracking feedstock.

There are a number of thermal processes, such a visbreaking and coking, that are directed at upgrading feedstock by the removal of asphaltenes. There is also the method of deasphalting with liquid hydrocarbons or gases such as propane, butane or iso-butane. This is very effective in the preparation of vacuum residues as cracking feedstock.

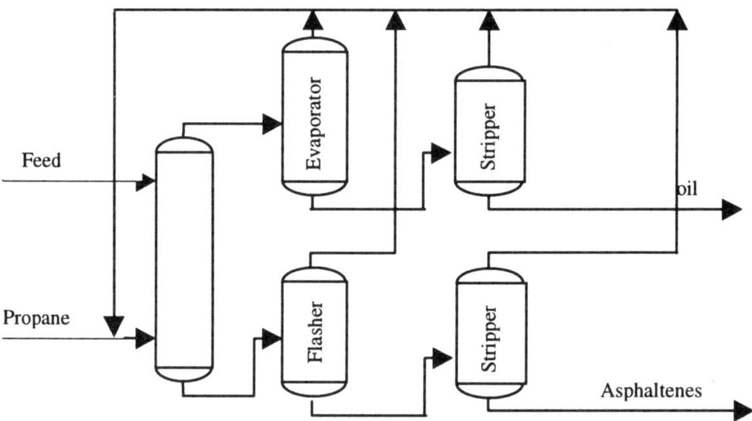

Fig. 8.3: Propane deasphalting.

In general, propane deasphalting (see Fig. 8.3) or modifications of this process are extraction processes in which the desirable oil in the charge is dissolved in liquid propane and the asphaltenes remain insoluble. In practice, a solvent deasphalting unit associated with an oil refinery mixes residual oil produced by a refinery with a light hydrocarbon solvent for producing two liquid product streams. One stream is substantially free of asphaltenes and contains deasphalted oil and solvent, and the other stream contains asphaltene and solvent within which some deasphalted oil is dissolved. These product streams are fed to a solvent recovery section that extracts most of the solvent from the product streams. The resultant solvent-free deasphalted oil is returned to the refinery for conversion to gasoline, jet fuel, lubricating oils; and the resultant solvent-free asphaltene can be combined with a diluent, such as diesel fuel, for conversion to residual fuel. In some units, the solvent recovery section includes a supercritical solvent recovery section that removes a large percentage of solvent from the product streams, followed by an evaporative solvent recovery section that removes the balance of the solvent. In other units, only the evaporative solvent recovery section is used. In both cases, the output of the evaporative solvent recovery section is deasphalted oil product and asphaltene product having acceptable levels of the solvent. In an evaporative solvent recovery section, each of the liquid product streams of deasphalted oil and solvent or asphaltene and solvent is first flashed to produce a vaporized solvent stream and a reduced solvent liquid product stream. Each of the reduced solvent liquid product streams so produced are then subjected to serial flashing and/or stripping until the final product stream is free of solvent to the desired degree. The vaporized solvent produced in this manner is condensed and reused.

In order to reduce the amount of heat lost as a result of the condensation of the vaporized solvent, the temperature at which flashing operations are carried out is kept as low as possible. Thus, the flash drums to which the solvent containing product streams are applied operate to produce solvent vapor at about 100°C. Heat contained in these vapors is of such low quality that economic recovery is not practical. As a consequence, such heat is extracted from the solvent by air or water cooling, and is lost to the environment.

The solubility of propane decreases with increasing temperature and is markedly influenced by pressure. Generally, the extraction temperatures can range from 50 to 90°C with a pressure of 25 to 40 bar. Propane:oil volume ratios ranging from 4:1 to 10:1 are normally used. Butane alone may also be employed as solvent, but is usually mixed with propane.

8.2.2.2 Chemical treatments

As the availability of lighter crude oil sources diminishes, refiners are being forced to deal with heavier crude oil feedstocks. This comes at a time when exploring for oil and gas formations is becoming increasingly more expensive while there is an increasing demand for refined products, particularly transportation fuels, such as gasoline and diesel fuel. Refiners are faced with the need for finding conversion processes to convert the heavy crude oil feedstocks and the various crude oil residues (residue) that result from the normal refining processes to more useful and profitable lighter products while minimizing the production of heavy fuel oils and coke.

Existing processes for converting heavy crude feedstocks and residue to useful, lighter products include fluid catalytic cracking, residue catalytic hydrocracking, thermal cracking, delayed coking, and fluidized bed coking. Visbreaking and coking were presented in chapter 6 as processes for treating residue fractions. In this section, we present some details concerning their chemistry as well as equipment used to carry out the processes.

8.2.2.3 Visbreaking

Visbreaking as a mild form of thermal cracking significantly lowers the viscosity of heavy crude-oil residue without affecting its boiling point range. Residue from the atmospheric distillation tower is heated to 425-510°C at mild pressure and mildly cracked in a heater. It is then quenched with cool gas oil to control overcracking, and flashed in a distillation tower. Visbreaking is used to reduce the pour point of waxy residues and to reduce the viscosity of residues used for blending with lighter fuel oils. Middle distillates may also be produced, depending on product demand. The thermally cracked residue tar that accumulates in the bottom of the fractionation tower is vacuum flashed in a stripper, and the distillate is recycled.

Figure 8.4 shows the principle of visbreaking. The feed in this visbreaking unit is preheated against steam and visbreaker residue to higher than 300°C. The feed is then charged to the reaction section which consists of the heater and the soaking drum. In the heater, the raw material is further heated to more than 450°C and partially vaporized. The maximum temperature is limited to prevent coke formation. A mild thermal cracking process takes place in the heater. The severity of cracking depends on the temperature - the main process variable. The heater effluent is sent to the soaking drum, where additional residence time is provided to crack to the desired conversion. The soaking drum is adiabatic whereas the cracking reactions are endothermic. Consequently, the fluid outlet temperature is approximately 20°C lower than the inlet temperature. The maximum cracking severity is determined by the stability of the produced fuel oil. The effluent of the reaction section enters the fractionation column at a temperature of more than 400°C. In the fractionator, gas, naphtha, light and heavy gas oil are separated from the residue. Alternatively, the gas oil fraction can be included with the visbreaker residue, removing only gas plus naphtha from the visbreaker effluent. The alternative procedure minimizes the need for lighter cutter stock to meet viscosity specifications.

The main reactions of the crude oil vacuum residue during visbreaking are cracking reactions. These kinds of reactions can be described by the following equation (8.1):

$$C_nH_{2n+2} \rightarrow C_{(n-x)}H_{2(n-x)+2} + C_xH_{2x} \tag{8.1}$$

The basic structural changes in asphaltene composition during visbreaking are as follows:

- aromatic carbon content increases.
- the number of aromatic and naphthenic rings in the average molecule slightly decreases.
- the number of side chains and their lengths also decrease.

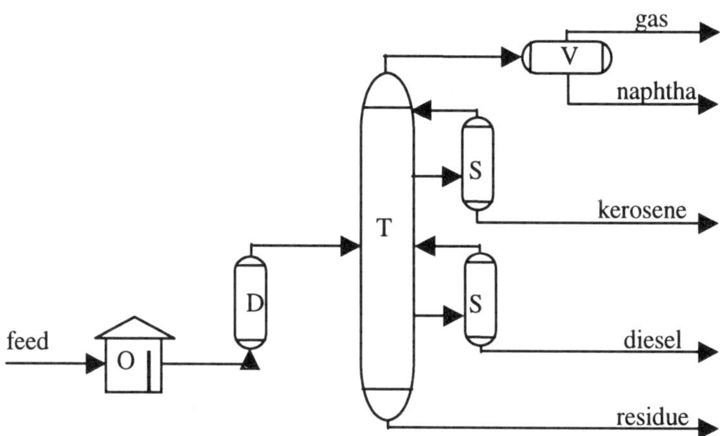

Fig. 8.4: Visbreaking unit.
O – oven
D – soaking drum
T – rectification tower
S – stripper
V – separator

Two contrasting processes that relate to asphaltenes take place during visbreaking:

1) cracking of asphaltenes that results in a decrease in both their content and molecular weight, but increase in their aromaticity in the oil
2) asphaltenes formation from the polar components of the feed

These processes may be expressed as resulting in a decrease and an increase in asphaltenes content, respectively. Reactions of condensation of asphaltene precursors (resins, etc.) prevail for light stocks. For heavier stocks, cracking of asphaltenes plays a more significant role. Thus, any increase in asphaltene content takes place mainly due to concentrating them in the visbreaking residue as a result of distillation of the visbroken product.

8.2.2.4 Coking

Delayed coking is a thermal cracking process used in refineries to upgrade and convert crude oil residue known as vacuum tower bottom product (i.e. the bottoms fraction from a vacuum rectification tower) into liquid and gas product streams leaving behind a solid concentrated carbon material, coke. The vacuum towers referred to are generally used to further fractionate virgin atmospheric-reduced crude oil (see chapter 5). Typically, the vacuum tower bottom product from such vacuum rectification towers generally includes all the fractions boiling above a selected temperature. Usually, this temperature can be as low as 450°C. But often, it is set at 510°C.

A fired heater with horizontal tubes is used in the coking process to reach thermal cracking temperatures of 485 to 505°C. By using short residence times in the furnace tubes, coking of the feed material is thereby "delayed" until it reaches large coking drums downstream of the heater. Three physical structures of coke: shot, sponge, or needle coke can be produced by delayed coking. These physical structures and the chemical properties of the coke determine its end use, which includes being burned as fuel; calcined for use in the aluminum, chemical, or steel industries; or gasified to produce steam, electricity, or gas feedstocks for the petrochemical industry. The technology of coking can be broken into three stages:

1) The feed undergoes partial vaporization and mild cracking as it passes through a coking oven.
2) The vapors undergo cracking as they pass through the coke drum to fractionation facilities downstream where products of gas, naphtha, jet fuel and gas oil are separated. The coke remains in the drum.
3) The heavy hydrocarbon liquid trapped in the coke drum is subjected to successive cracking and polycondensation until it is converted to volatile product and coke.

Delayed coking is the only main process in a modern petroleum refinery that is a batch-continuous process. The flow through the tube oven is continuous. The feed stream is switched between two drums. One drum is filling with coke while the other drum is being steam-stripped, cooled, decoked and warmed up (see chapter 6). The overhead vapors from the coke drums flow to a rectification unit. The rectification tower has a reservoir in the bottom where the fresh feed is combined with condensed product vapors (recycle) to make up the feed to the coker heater.

According to Elliott [10], delayed coker design objectives are:

- High in-tube velocities resulting in maximum inside heat transfer coefficient
- Minimum residence time in the oven
- A constantly rising temperature gradient
- Optimum flux rate with minimum practicable maldistribution based on peripheral tube surface
- Symmetrical piping and coil arrangement within the furnace enclosure
- Multiple steam injection points for each heater pass

In contrast to delayed coking, fluid coking is a continuous process which uses the fluidized solids technique to convert vacuum residue to more valuable products, and coke formed during this kind of coking is a byproduct of the process

[11]. The residue at the temperature 480–570°C is coked by being sprayed into a fluidized bed of hot, fine particles, which permits the coking reactions to be conducted at higher temperatures and shorter contact times than can be employed in delayed coking (see chapter 3). Moreover, these conditions result in decreased yields of coke; better quantities of more valuable liquid product are recovered in the fluid processes.

8.2.3 Coke Forming Reactions During Residue Treatment

A wide variety of reactions occur during the thermal treatment of crude residue, as was shown earlier. However, not all these reactions lead to the formation of coke. In this section, we will look at the reactions that are involved in coke formation for both the catalytic treatment process and the purely thermal treatment process.

8.2.3.1 Catalytic treatment

Hydrocarbons that react on a catalyst surface may not always follow the expected reaction pathway. Some compounds adsorb strongly on the surface and form a coke deposit. This explains why deactivation by coke formation is sometimes called, "self-poisoning". Coke can also be formed by products that are not taking part in the main reactions. A simple form of a possible reaction scheme for coke formation during the catalytic treatment of crude oils is given in reaction equation (8.2). The mechanism of coke deposition is a very complex multi-step reaction sequence. The reaction path consists of adsorption, dehydrogenation, polycondensation, and cyclization of hydrogen-deficient fragments [12]. It is not possible to find a general coking mechanism. Different reaction schemes have been proposed for different systems. Butt and Petersen [13] discuss a carbanion mechanism for formation of higher aromatics from benzene and naphthalene on cracking catalysts. Petersen and Bell [14] propose that coke can be formed from methylcyclohexane on a platforming catalyst via a doubly bonded intermediate that is converted to a six-bonded coke precursor.

$$\text{Olefin}' + \text{Olefin}'' \rightarrow \text{heavier Olefin} - n \cdot H_2 \rightarrow \text{coke} \quad (8.2)$$

Figure 8.5 shows a working metal catalyst surface. On the surface there are reactants, products, coke in the form of polymers, and different hydrocarbon fragments that may react to coke if they are not removed from the surface [15]. The carbonaceous layer is not the only form of coke. On iron surfaces, coke can grow as whiskers with a little metal crystallite.

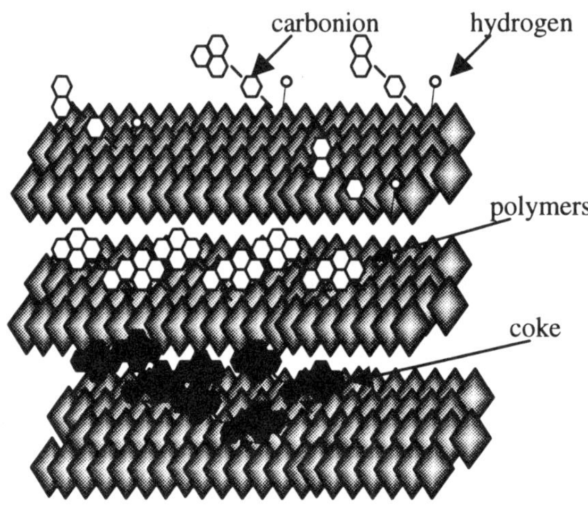

Fig. 8.5: Carbonaceous deposits on the surface of a working metal catalyst.

Coke that accumulates on a catalyst may cause deactivation either by covering active sites or by physically blocking the pores in the catalyst (see Fig. 8.6). When the active sites are covered, the activity can drop rapidly even with relatively slow coking. It is important that the coke accumulation occurs slowly so that the pore mouth is not blocked, cutting off the rest of the pore. When coke is formed on a catalyst, the degree of deactivation can vary greatly for different reactions. The coke may block some sites on the catalyst in preference to others [16].

The effect of coke formation can be reduced either by preventing coke from being formed, or by removing it as soon as it has been produced. Steam is known to reduce the coking rate [17] and it is also possible to remove coke already on the catalyst by adding steam [18] or by burning off the coke from the catalyst.

High hydrogen pressure can prevent coke formation, since hydrogen is a product of polycondensation reactions (coke formation). If alkali and alkaline-earth metals or their oxides are added to the catalyst, these promoters may increase the gasification of coke on the catalyst [19]. Less coke will be formed on a less acidic support [16].

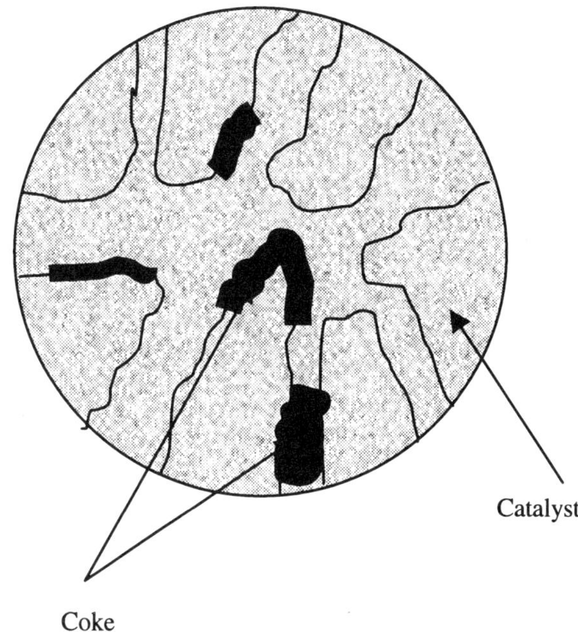

Fig. 8.6: Coke on a supported metal catalyst.

Different catalysts may form coke at different rates. There has been a report [16] of studies conducted on the effect of coke formation on platforming catalysts with different metal dispersions. In the report, it was found that small crystallites had a stronger resistance to coke deposition than larger ones. The explanation for this fact was that cyclopentadiene (cyclopentane in nitrogen was used as feed) adsorbed more strongly on the big crystallites and formed coke. The effect was due to less electronic interaction with the support than for smaller crystallites. Another study [7] on promotion of reforming catalysts with lithium found out that the promoted catalysts became more stable against coke formation.

Catalyst poisoning or diffusion blocking by coke formation can have similar effects on catalyst activity. Assigning a cause to one or the other is an important first step in poisoning studies. Poisons mask active sites or change the selectivity of the catalyst for particular reactions or reaction types by reaction with any compounds of feedstock. Poisons are usually metals or multiple-bond light-gas molecules. In the case of vacuum residue treatment, metals are agglomerated in the asphaltenes. This means catalyst poisoning can be prevented by the removal of asphaltenes from the process. To differentiate between poisoning and diffusion blocking, one can first increase the temperature. Heavy product residue will evaporate off with heating, leaving elemental carbon or poisoning as the culprit.

Substituting an oxidizing gas for the inert gas would remove carbon at high temperature. Such a treatment and the resulting mass and conversion data would show the relative contributions of poisoning, coking, and heavy product retention.

8.2.3.2 Thermal treatment

In contrast to catalytic treatment, coke formation during thermal treatment leads only to pipe blocking and poor thermal conductivity of the reactor walls. The most important difference between coke formation in catalytic and thermal treatment is that a free radical mechanism in coke formation is not possible in the case of thermal processing.

The rate or velocity of coke formation is, first of all, a function of the feedstock characteristics. Seidel [1] reported about the tendencies of different components of heavy vacuum residues to coke formation. The following sequence shows a descending tendency to coke formation during thermal treatment:

asphaltenes > resins > light aromatics > olefins > naphthenes > paraffin

A comparison between the activation energies for coke formation from light aromatics (52–58 kJ/mol) and from asphaltenes and resins (34– 47 kJ/mol) shows that the reaction velocity of coke formation from light aromatics grows faster with increasing temperature than for coke formation from asphaltenes or resins.

Figure 8.7 shows a possible chemical path from paraffins (the most stable compounds against coke formation) to coke [1, 20]. It is evident that direct coke formation from paraffin or naphthenes is impossible. Coke formation can occur only during thermal cracking of paraffinic fractions by the many complicated reactions from paraffins to light and then to heavy aromatics. The utilization of vacuum residue paraffinic compound mixtures in thermal cracking processes can lead to flocculation of the asphaltenes from the maltenes. Flocculated and agglomerated asphaltenes form a melted crystalline mesophase on the reactor walls and this reacts very quickly to coke. Flocculation of asphaltenes is possible from mixtures with aromatic compounds as well. There is a report [12] about the phenomenon of self-association in asphaltene/aromatic mixtures that has been confirmed through measurements of surface tension.

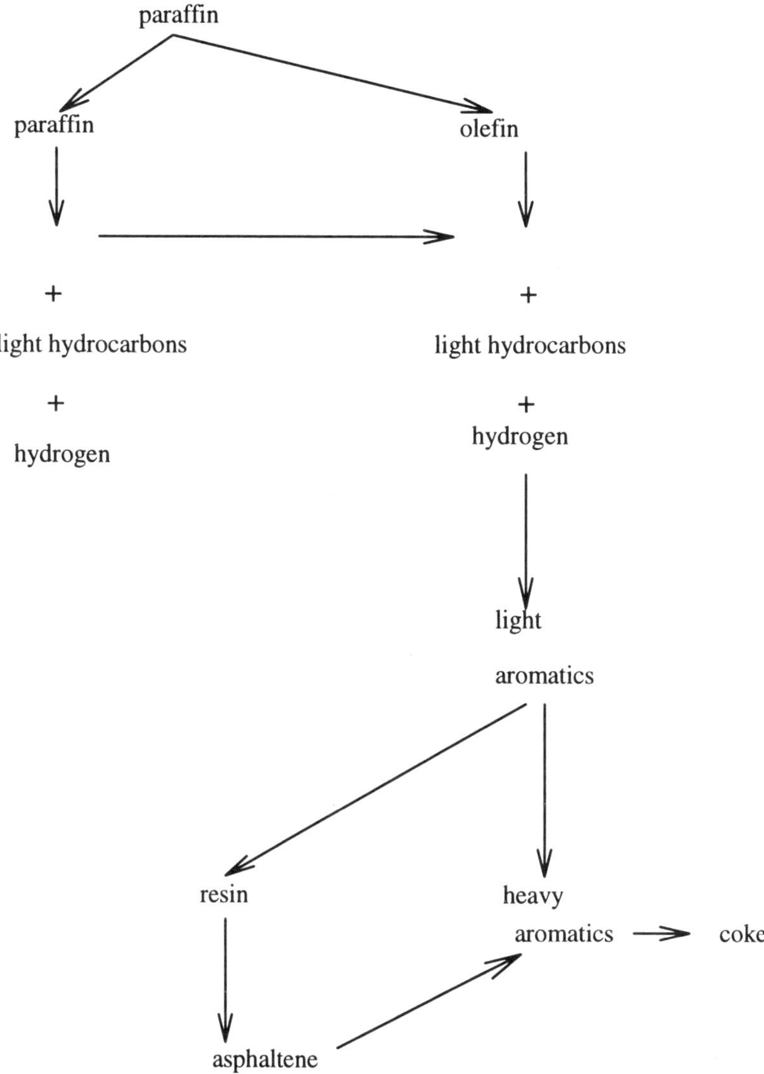

Fig. 8.7: Scheme for coke formation.

It has been shown [21] that, at low concentrations (below the critical micelle formation concentration), asphaltenes in solution are in a molecular state. Above the critical micelle concentration, however, asphaltene micelle formation occurs in a manner similar to that in surfactant systems where surfactant monomers are more uniform in their structure and less polydisperse. Now, it is obvious that coke formation during thermal treatment is a result of displacing the equilibrium colloid solution of asphaltenes in maltenes towards asphaltene flocculation and finally to polycondensation and coke formation.

8.3 METHODS OF ANALYSIS OF CRUDE OIL RESIDUE

8.3.1 Methods and Main Definitions for the Determination of Coke Formation Tendency

Asphaltenes are derived from the root word "asphalt", a "sticky tar-like substance found naturally in petroleum crude oil." Asphaltenes are complex, high molecular weight aromatic compounds suspended within the fuel. They have high melting points and high carbon/hydrogen ratios with low calorific values.

During the refining process, a portion of the asphaltenes is coked to form carbon residue. The actual percentage depends on the refining process. For this reason, the commonly held assumption that asphaltenes in the fuel can be estimated by knowing the Conradson Carbon Residue is erroneous. In actual fact, the asphaltenes vary widely from Conradson Carbon Residue levels and have to be analyzed separately.

The term "carbon residue" is used to designate the carbonaceous residue formed after evaporation and pyrolysis of a petroleum product. The residue is not entirely composed of carbon, but is a coke that can be further changed by pyrolysis. The term is used to indicate the relative tendency of a residue to form carbon deposits that may foul reactor pipes.

There are three tests used to measure carbon residue: Conradson Carbon Residue (CCR), Microcarbon Residue (MCR) and Ramsbottom Carbon Residue (used primarily with distillate fuels). CCR and MCR results are closely correlated, though MCR is considered to be more accurate and is becoming the predominant test procedure in test laboratories. The MCR test is also quicker to run and uses a smaller sample of residue. The standard test method for determining Carbon Residue (Micro Method), ASTM D4530 (ISO 10370:1993) states: "A weighed quantity of sample is placed in a glass vial and heated to 500°C under an inert (nitrogen) atmosphere in a controlled manner for a specific time. The sample undergoes coking reactions and volatiles formed are swept away by the nitrogen. The carbonaceous type residue remaining is presented as a percentage of the original sample as micro carbon residue. The term carbon residue may be misleading as the residue may contain elements other than products from carbon decomposition (such as ash) in many kinds of residue. The ash content of a residue oil is the noncombustible residue found in the residue. These are organometallics from the crude oil, inorganic contaminants or metallic catalyst fines used in the refining process".

Values obtained by the method of Ramsbottom Carbon Residue are not numerically the same as those obtained by ASTM Method D524, or Test for Ramsbottom Carbon Residue of Petroleum Products nor have satisfactory correlations been found between the results of the two methods for all materials which may be tested, because the carbon residue test is applied to a wide variety of petroleum products.

All vacuum residues have carbon residue, which is a result of the refining process or asphaltene presence. The only way to reduce the percentage of carbon residue is to blend the residue with asphaltene free substances.

8.3.2 Analytical Characterization of Heavy Oil Residues and Asphaltenes

8.3.2.1 Solution analysis

The method of solution analysis is based on differences in solubility of the vacuum residue compounds (pseudo-compounds) in different solvents. Figure 8.8 shows an example of the scheme for solution analysis in which the following solvents are used:

- Pentane
- Toluene
- Tetrahydrofuran (THF)

Four fractions of pseudo-compounds are obtained by solution analysis of crude oil residue or its cracked product. During solution analysis of the cracking product from thermal treatment of vacuum residue or mixtures of vacuum residue and plastics (such mixtures were used in our investigation), a first step of solution analysis is soxhlet extraction. In the soxhlet extractor, the liquid/solid product is extracted with fresh warm solvent (THF) that does not contain the extract. This can increase the extraction rate, as the sample is contacting fresh warm solvent. The sample is placed inside a cellulose thimble and placed in the extractor. The extractor is connected to a flask containing the extraction solvent, and a condenser is connected above the extractor. The solvent is boiled, and the extractor has a bypass arm that the vapor passes through to reach the condenser, where it condenses and drips into the sample in the thimble. Once the solvent reaches the top of the siphon arm, the solvent and extract are siphoned back into the lower flask. The solvent reboils, and the cycle is repeated until the sample is completely extracted, and the extract is in the lower flask.

After completion of extraction, the cellulose thimble is placed in a drying-oven at 120°C. After drying, the solid remainder in the cellulose thimble is weighed. The extract contains the liquid product and solvent (THF). THF is distilled from the product and reused in a subsequent analysis.

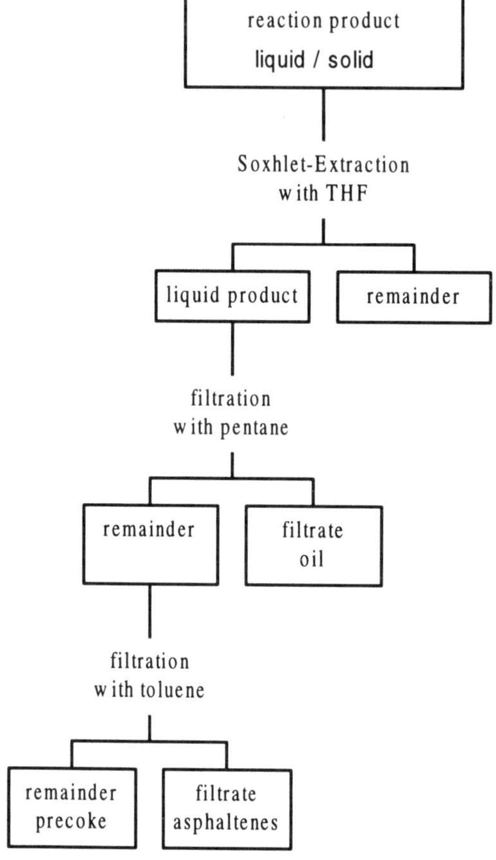

Fig. 8.8: Scheme of solution analysis.

The rest of the THF is removed from the product by drying at room temperature. This can proceed from one through any number of days. Figure 8.9 shows a typical progression of this drying process.

The following balance equation (8.3) is used for the definition of the amount of liquid product:

$$m_{\text{liquid product}} = m_{\text{feedstock}} - m_{\text{remain}} - m_{\text{gas}} \quad (8.3)$$

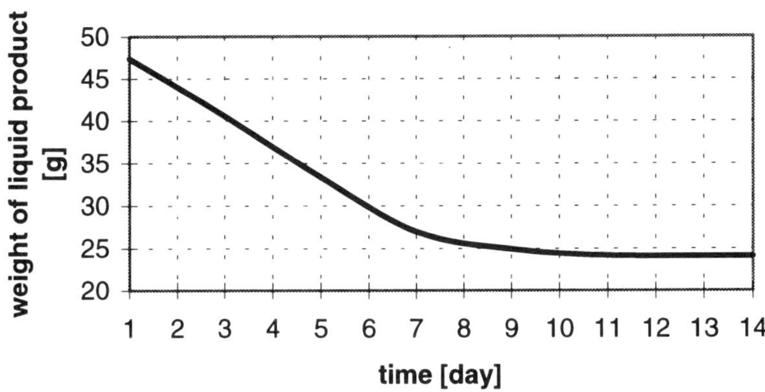

Fig. 8.9: Typical progression of THF reduction in the liquid product.

The drying of the extract has to be complete if equation (8.4) is to be satisfied.

$$m_{extract} \approx m_{liquid\ product} \tag{8.4}$$

After the soxhlet extraction and drying of the extract are completed, the extract is dissolved in pentane first. An ultrasound bath can be used to improve solving. This solution of extract in pentane is filtered. After filtration, the extract is separated into two fractions: pentane soluble fraction or filtrate and the remainder. The pentane soluble fraction is composed of compounds that have similar polarity as pentane or the liquid at the filtration temperature. The remainder contains the asphaltene fraction and precoke. This remainder is dissolved in toluene in the same manner as the extract was solved in pentane, and then filtered. The asphaltene fraction is the filtrate after filtration with toluene, and precoke is the remainder. The solvents (pentane and toluene) are removed from the filtrates by distillation at 40°C and normal pressure for pentane distillation, and 70 mbar for toluene distillation.

The described example of solution analysis was shown for the analysis of crude oil vacuum residue, but this method can be successfully applied for different kinds of feed. In our investigation, we used this method for the analysis of mixtures of vacuum residue and plastics, and pure plastics. In the case of the plastic or product of its cracking analysis, one has to consider that what is left behind after soxhlet extraction is the heavy plastic fraction or non-cracked plastic.

Table 8.1 shows the results of solution analysis for the vacuum residue (Bitumen 200 Elf) and plastics that were used in our investigation:

Table 8.1: Results of solution analysis for the different kinds of feedstock.

Feed	pentane soluble (wt. %)	toluene soluble (wt. %)	THF soluble (wt. %)	remainder (wt. %)
Bitumen B200E	70.70	28.40	0.90	0.00
polystyrene	1.11	51.34	43.45	4.10
Polypropylene	0.00	0.00	0.00	100
Polyethylene	0.00	0.00	0.00	100
PVC	9.70	5.98	45.78	38.54

Based on the results shown in table 8.1, it is obvious that during solution analysis of the cracking product of polyethylene and polypropylene, only non-cracked plastics remain completely in the remainder. In contrast to polyethylene and polypropylene, polystyrene and PVC are over 40 wt. % soluble in THF.

8.3.2.2 Coagulation analysis

Similar to solution analysis, the method of coagulation analysis is based on the different solubilities of vacuum residue compounds (pseudo-compounds) in the different solvents. The following solvents were used for the analysis of feedstock:

- n-heptane
- toluene
- mixture of iso-butanol and cyclohexane (4:1)
- mixture of acetone and chlorinemethylene (1:2)
- iso-butanol

Figure 8.10 shows the general scheme for coagulation analysis that was used for analysis of crude oil vacuum residue in our investigation.

During coagulation analysis, the ratio of the coagulation agent to the substance is 40 to 1 based on recommendations by many authors that this ratio is the optimum for this kind of analysis [1, 22]. The coagulation of the asphaltene fraction in the scheme is done according to the Golde method (DIN 51595, IP 143/57) recommended as the most exact method of asphaltene definition [1, 23, 24].

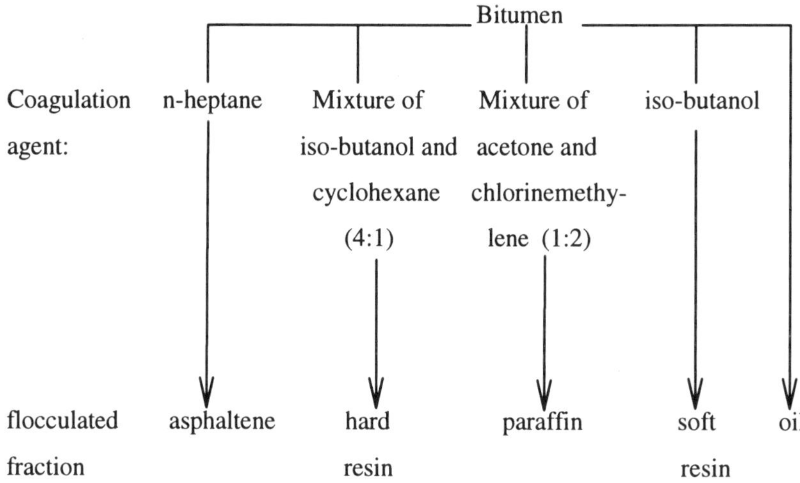

Fig. 8.10: Scheme of coagulation analysis for vacuum residue.

In the first step of the Golde method, the substance is dissolved in n-heptane in the ratio of n-heptane:substance = 1:40. The asphaltene fraction is flocculated from the dissolved sample after 24 hours of coagulation. After this, asphaltenes are filtered from the maltenes. And the rest of maltenes or resins, which stay in the asphaltenes during flocculation with n-heptane at room temperature is extracted by soxhlet extraction (see preceding section). In the next step, the hard resin fraction is coagulated from the maltenes in the same manner as the asphaltenes from the vacuum residue, but during this step, a mixture of iso-butanol and cyclohexane is applied as coagulation agent. In the case of hard resin coagulation, the coagulation agent with the solved substance is not filtered, but removed from the glass with a pipette, and the product (hard resin) is not extracted, but only dried. The coagulation agent is distilled from the rest of the sample, and the rest is analyzed in the next analysis step in the same manner. The analyses proceed until only the oil fraction stays.

An example of the results of this analysis for Bitumen 200 Elf is presented in table 8.2. Coagulation analysis for the vacuum residue was carried out for three samples (1, 2 and 3), which were taken from three different levels of the container; since bitumen is a colloid solution, the result of the coagulation analysis can differ depending on the level in the container. From table 8.2, it is evident that the analyzed vacuum residue represents the colloidal stable solution, because the values of the amount of all fractions are approximately the same for all levels. For each

Table 8.2: Results of coagulation analysis for Bitumen 200 Elf.

	asphaltene	hard resin	paraffin	soft resin	oil
1. a)	11.25	21.03	0.82	14.86	52.03
b)	11.24	22.49	0.83	15.54	49.90
c)	11.25	21.59	0.81	14.86	51.48
mean of 1	11.25	21.70	0.82	15.09	51.13
2. a)	11.17	20.83	0.93	15.20	51.86
b)	11.16	20.19	0.92	15.54	52.18
c)	11.18	19.01	0.94	15.13	53.73
mean of 2	11.17	20.01	0.93	15.30	52.59
3. a)	11.06	19.22	0.94	15.13	53.64
b)	11.07	19.46	0.93	15.20	53.34
c)	11.08	19.41	0.95	15.07	53.49
mean of 3	11.07	19.36	0.94	15.14	53.49
mean	11.16	20.36	0.90	15.18	52.40

sample, coagulation analysis was carried out three times (a, b and c). It is also obvious from table 8.2 that the method of coagulation analysis is very reproducible since the difference between analyses a, b and c is very small.

For the characterization of the product of thermal treatment of vacuum residue and mixtures of vacuum residue and plastics, a simpler method of coagulation analysis was used, since the method described earlier is very complicated to use for the given number of samples. The scheme of coagulation analysis for the product of thermal processing is represented in Figure 8.11:

During the analysis shown in Figure 8.11, the sample is broken down into four fractions:

- Maltene
- Asphaltene
- Precoke
- Remainder

The remainder during this analysis is coke, in the case of thermal treatment of pure vacuum residue, and coke and non-cracked plastics, in the case of the thermal treatment of mixtures of crude oil vacuum residue and plastics.

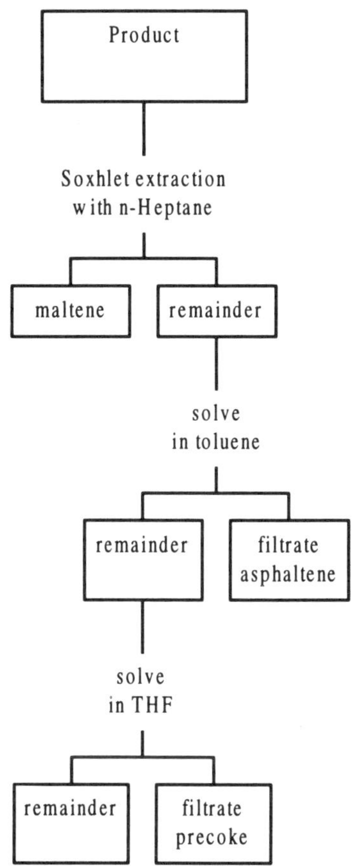

Fig. 8.11: Scheme of the coagulation analysis for the product of thermal treatment.

8.3.2.3 Distillation method

Distillation analysis is based on the fractionation of the sample by boiling temperature. This type of analysis can be carried out only up to a maximum temperature of approximately 350°C, since at higher temperatures cracking of the heavy organic components occurs.

Figure 8.12 shows the composition of the asphaltene fraction after the different solution analyses and distillation. From Figure 8.12, we see that the depth of distillation can be controlled by the level of vacuum. With a higher vacuum, less non-asphaltic compounds stay in the residue.

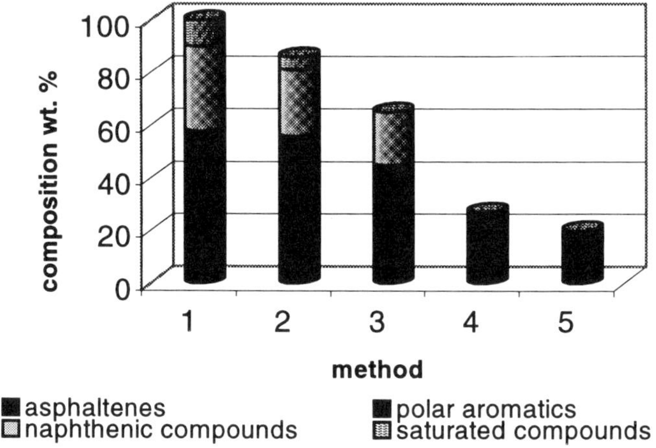

Fig. 8.12: Analysis methods.

1 – vacuum distillation
2 – deep vacuum distillation
3 – flocculation with propane
4 – flocculation with pentane
5 – flocculation with heptane (Golde method)

8.3.2.4 Chromatography

Chromatography is based on the differential adsorption ability of the vacuum residue components on an adsorbent (see chapter 2). As a first step, the solved vacuum residue is adsorbed on the adsorbent and then different components in the sample are desorbed using solvents of various polarities. An example of a scheme for chromatographic analysis (E. D. Radchenko) is shown in Figure 8.13 [1].

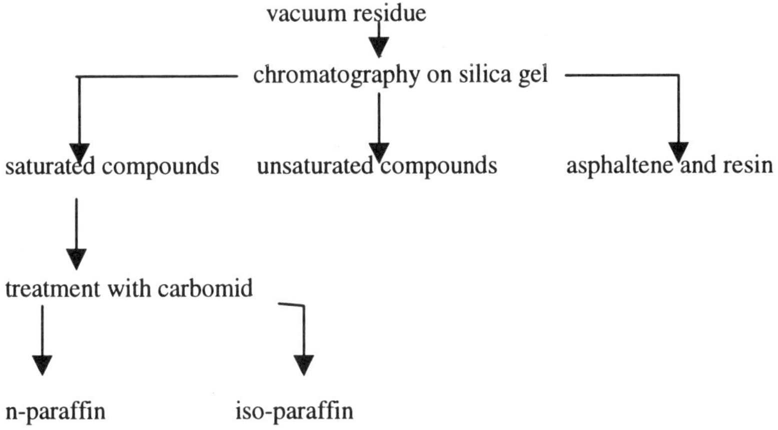

Fig. 8.13: Vacuum residue chromatography on silica gel.

8.3.2.5 ^{13}C – NMR analysis

Nuclear magnetic resonance (NMR) is a phenomenon that occurs when the nuclei of certain atoms are immersed in a static magnetic field and exposed to a second oscillating magnetic field. Some nuclei experience this phenomenon, and others do not, depending on whether they possess a property called, “spin”. Spin is a fundamental property of nature like electrical charge or mass. Spin comes in multiples of 1/2 and can be + or -. Protons, electrons, and neutrons possess spin. Individual unpaired electrons, protons, and neutrons each possess a spin of ½.

When placed in a magnetic field of strength B, a particle with a net spin can absorb a photon of specific frequency. The frequency □ depends on the gyromagnetic ratio, γ, of the particle (see equation (8.5)).

$$\upsilon = B \cdot \gamma \tag{8.5}$$

For ^{13}C, $\gamma = 10.71$ MHz / T

Almost every element in the periodic table has an isotope with a nonzero nuclear spin; the spin for carbon (^{13}C) is ½. In order to understand how atoms with spin behave in a magnetic field, let us consider carbon. Think of the spin of this carbon as a magnetic moment vector, causing the proton to behave like a tiny magnet with a north pole and a south pole. When the carbon is placed in an external magnetic field, the spin vector of the particle aligns itself with the external field, just like a magnet would. There is a low energy configuration or state where the poles are aligned N-S-N-S, and a high-energy state N-N-S-S.

The NMR analysis is a continuous wave experiment. There are two ways of performing this experiment. In the first, a constant frequency continuously probes the energy levels of the molecules while the magnetic field is varied. The energy of this frequency is represented by the blue line in the energy level diagram.

The continuous wave experiment can also be performed with a constant magnetic field and a frequency that is varied. The magnitude of the constant magnetic field is represented by the position of the vertical blue line in the energy level diagram.

Nuclear magnetic resonance spectroscopy is the use of the NMR phenomenon to study physical and chemical properties of matter. As a consequence, NMR spectroscopy finds applications in several areas of science. NMR spectroscopy is routinely used by chemists to study materials. Solid state NMR spectroscopy is used to determine the molecular structure of solids. In our investigation, NMR spectroscopy was used to determine the molecular structure of asphaltene molecules.

Asphaltene and/or coke molecules contain carbon. Unfortunately, the ^{12}C nucleus, which is contained in crude oil residues, does not have a nuclear spin, but the ^{13}C nucleus does, due to the presence of an unpaired neutron. This is why only ^{13}C can be used in C-NMR analyses.

The result of ^{13}C-NMR analyses for the asphaltenes and hard resins in Bitumen 200 Elf is presented in table 8.3. From table 8.3, it is evident that asphaltenes (heavier fraction) have a higher aromaticity in comparison to hard resins and at the same time a lesser amount of paraffin side chains. This shows that asphaltenes have a higher tendency to coke formation in comparison with hard resins (see section 8.2).

Table 8.3: Results of ^{13}C-NMR analysis for asphaltenes and hard resins (% of C in the chemical group to total C in the sample).

	hard resin	asphaltene
$-CH_3$	5.2	3.5
α-CH_3	13.3	9.2
α-CH_2	6.3	4.6
$-(CH_2)_{n>4}-$	14.5	11.8
$-CH_2-_{bridge}$	10.9	8.9
$\rightarrow$CH	7.6	6.9
$C_{protoned\ aromatic}$	27.0	36.2
$C_{alcyl.\ aromatics}$ and $C_{cond.\ aromatics}$	14.5	18.8
$C_{phenol\ Aromatics}$	0.6	0.0
Aromaticity of the fraction	42.1	55.0

8.3.2.6 Ultimate analysis

In our investigation, we used VARIO EL analytical unit for ultimate analysis. In this analysis, the data concerning the amount of carbon, hydrogen, nitrogen and the other heteroatoms are obtained. The applied kind of ultimate analysis is based on oxidation of the sample at 960°C, at which the oxides of carbon, hydrogen and nitrogen are determined. Sulfur determination is very important for crude oil vacuum residues in addition to ultimate analysis, since many authors have reported that feedstock with a higher sulfur content has a stronger tendency to coke formation during thermal or catalytic treatment. The results of ultimate analysis and sulfur determination for feedstock used in our investigation is shown in table 8.4.

The ratio of H / C_{atomic} can be calculated from the obtained data by the following equation (8.6).

$$H / C_{atomic} = 12 \cdot H / C \tag{8.6}$$

where H – amount of hydrogen in the sample (wt. %)
C – amount of carbon in the sample (wt. %)

This value shows the quality of the product or the depth of hydrogenation of the feedstock. The higher the H / C_{atomic} ratio, the more deeply hydrogenated the feedstock is. Feedstock with a H / C_{atomic} ratio higher than two can be used as hydrogen donor during thermal or catalytic treatment of the heavy feed. These values of the H / C_{atomic} ratio are presented in table 8.5.

Table 8.4: Results of ultimate analysis and sulfur determination (wt.%).

<table>
<tr><th>Feedstock</th><th>carbon
C</th><th>hydrogen
H</th><th>nitrogen
N</th><th>sulfur
S</th><th>oxygen
O_{diff}</th></tr>
<tr><td>Bitumen-B200E</td><td>85.24</td><td>10.17</td><td>0.60</td><td>3.25</td><td>0.74</td></tr>
<tr><td>Asphaltene</td><td>84.58</td><td>7.50</td><td>1.50</td><td colspan="2">6.42</td></tr>
<tr><td>Hard resin</td><td>84.46</td><td>8.90</td><td>0.85</td><td colspan="2">5.79</td></tr>
<tr><td>Paraffin</td><td>79.85</td><td>9.70</td><td>0.54</td><td colspan="2">9.91</td></tr>
<tr><td>Soft resin</td><td>84.40</td><td>10.25</td><td>0.52</td><td colspan="2">4.83</td></tr>
<tr><td>Oil</td><td>84.73</td><td>11.21</td><td>0.34</td><td colspan="2">3.72</td></tr>
<tr><td>Polystyrene</td><td>92.29</td><td>7.71</td><td>0.00</td><td>0.00</td><td>0.00</td></tr>
<tr><td>Polypropylene</td><td>85.81</td><td>14.19</td><td>0.00</td><td>0.00</td><td>0.00</td></tr>
<tr><td>Polyethylene</td><td>85.83</td><td>14.14</td><td>0.00</td><td>0.00</td><td>0.03</td></tr>
</table>

Table 8.5: H / C_{atomic} ratio for the feedstock used.

Feedstock	H/C_{atomic}
Bitumen	1.43
Polystyrene	1.00
Polypropylene	1.98
Polyethylene	1.97

From the table 8.5, it is evident that polypropylene and polyethylene have H/C_{atomic} ratio of approximately two, which means that these plastics can be used as hydrogen donors during the co-processing of crude oil vacuum residue and plastics.

8.3.2.7 Molecular weight determination

Vapor pressure osmometry has become the prevalent method for determining asphaltene molecular weights. However, the value of the molecular weight from vapor pressure osmometry must be weighed carefully since, in general, the measured value of molecular weight is a function of temperature and the solvent molecular properties (see next section). The vapor pressure osmometry method is suitable for the determination of the average molecular weight of high molecular substances, which are soluble in different organic solvents.

An example of a vapor pressure osmometer is shown in Figure 8.14. Two thermistors, connected to measure the difference, are suspended in a thermostated measuring cell filled with the saturated vapor of the solvent. The measuring probes, which are first covered with solvent droplets, adapt to the cell temperature.

Thus, there is no temperature difference between them. Exchanging the solvent droplet of one of the probes by a droplet of the solution leads to condensation of solvent vapor due to the lower vapor pressure of the solvent above the solution. Thereby, the released condensation enthalpy increases the temperature of the solution droplet, which simultaneously leads to an increase of the vapor pressure. After reaching the vapor pressure equilibrium of the solution droplet, a relatively stable temperature is obtained at the solution droplet. This temperature is converted by the measuring system to a direct voltage signal and is thereby at the user's disposal as a measuring value. The resulting relative measured value is nearly proportional to the osmolal concentration of the solutions. However, it may be affected by heat loss and the nonideal behavior of the polymer solutions.

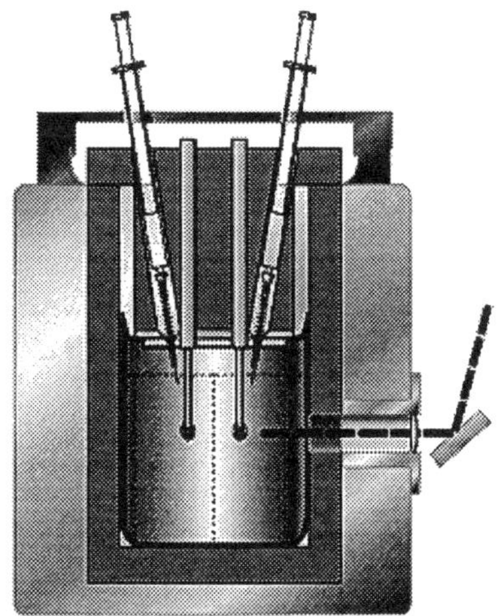

Fig. 8.14: Osmometer.

By calibrating with solutions of known molality and statistical calculations of the measured value, the nonlinear behavior may be eliminated. Figure 8.15 shows an example of the calibration line for benzile, which was used in our investigation. K_{cal} (calibration coefficient) is defined from the calibration line, as is shown in Figure 8.15. The molecular weight is calculated by equation (8.7):

$$M = K_{cal} / K \tag{8.7}$$

K in this equation (8.7) is found in the same manner as K_{cal} from the diagram for the investigated sample. Figure 8.16 shows an example of this diagram for asphaltenes from Bitumen 200 Elf.

Defined molecular weights for the different fractions from Bitumen 200 Elf are presented in table 8.6.

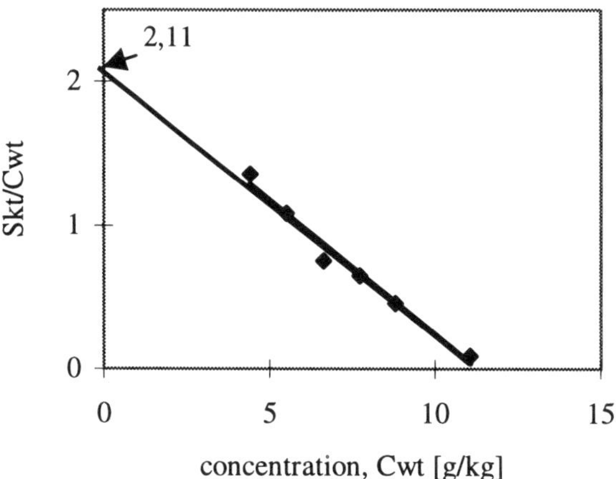

Fig. 8.15: Definition of K_{cal}.

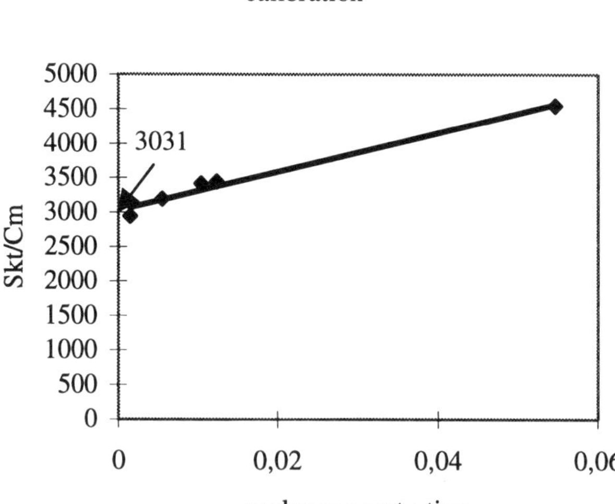

Fig. 8.16: Definition of K.

Table 8.6: Molecular weight for different fractions of Bitumen 200 Elf.

Fraction	Molecular weight
Asphaltene	1440
Hard resin	1220
Paraffin	1090
Soft resin	1150
Oil	607.5

From table 8.6, it is evident that the heaviest fraction of crude oil vacuum residue is the asphaltene fraction. The hard resin fraction has approximately the same molecular weight as asphaltenes. From the molecular determination and ^{13}C-NMR analysis of asphaltenes and hard resins (see preceding section), it is obvious that these two fractions have approximately the same chemical properties. Oil is the lightest fraction of the crude oil residue, and this is why this fraction is not as important as the asphaltene and resin fractions from the chemical perspective of coke formation.

8.3.3 Temperature Influence on Molecular Weight Determination

In the preceding section, the vapor pressure osmometry method for the determination of the molecular weight was described. It is important that the value of the molecular weight from vapor pressure osmometry be weighed carefully since, in general, the measured value of the molecular weight is a function of temperature and the solvent molecular properties. This influence of the temperature and solvent can be explained by a very high tendency of asphaltenes to micelle formation. Figure 8.17 shows one example of micelle formation from two asphaltene molecules.

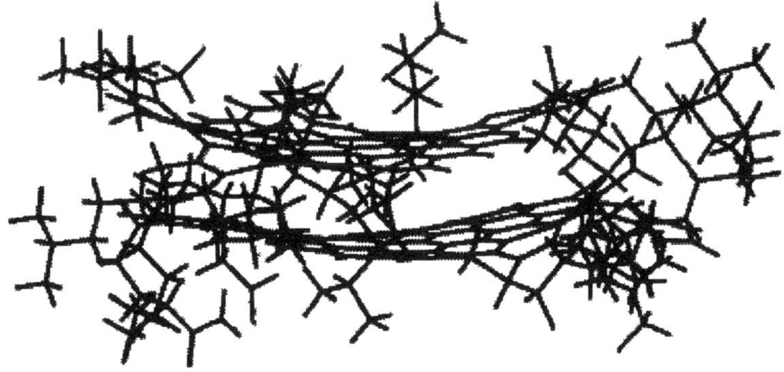

Fig. 8.17: Example of micelle formation.

It is obvious that the solubility of asphaltenes in a given solvent increases as the temperature of measurement is increased. For example, the molecular weight of the asphaltenes from Bitumen 200 Elf is 1440 mol/g defined at 60°C and 10050 mol/g defined at 37 °C. It is clear that at 37°C, asphaltenes were not completely solved in the toluene and the defined molecular weight is probably the molecular weight of the micelles which are formed from 6 to 8 asphaltene molecules. The molecular weight of 1440 is very close to the molecular weight of asphaltenes reported in many references [1, 9].

In order to provide a good explanation of asphaltene chemistry, it is particularly important that all measurements of molecular weights are carried out at the same temperature, because values of molecular weight defined at different temperatures cannot be compared with each other. Doing so can lead to wrong results. All the molecular weight determinations in our investigation were carried out at the temperature of 60°C, since the molecular weight defined at this temperature is closest to values provided by other authors (see above).

9

Processing of Heavy Crude Oils and Crude Oil Residues

9.1 INTRODUCTION

This chapter of the book is devoted to investigations on the processing of heavy crude oil residues. There are many processes already used in the oil industry for this purpose, but usage of the existing or developed processes has not been based on economic justifications all the time, because of the presence of asphaltene in heavy crude oil or residue (see section 8.1).

Methods to treat crude oil vacuum residue with a co-feed that will initiate the cracking of the residues constitute the main topic presented in this chapter. These methods are new and have not yet been applied in industry. However, based on the results of this investigation, many new properties of asphaltenes have been elucidated, which can be employed for deeper treatment of the crude oil.

This chapter as well as chapter 8 is engaged in asphaltene chemistry. In this chapter, however, the reader will see some practical applications of the asphaltene chemistry to heavy crude oils or vacuum residues.

One of the objects of this chapter is to show how many potential possibilities of crude oil treatment are yet to be discovered. Again, it will attempt to answer the question whether it is still possible to develop processes which can strongly increase the production of light fractions from crude oil; that is to increase the deepness of crude oil treatment.

9.2 CHEMISTRY AND REACTION OF ASPHALTENES DURING CO-PROCESSING OF CRUDE OIL RESIDUE AND PLASTICS

9.2.1 Change of Asphaltene Structure During Thermal Processing

Crude oil residues and bitumen are colloidal disperse systems. In these systems, high-molecular solid structure units (asphaltenes) are dispersed in an oily phase (maltenes) (see section 8.2). In industrial thermal cracking processes, these units precipitate as coke. Coke formation is caused by polycondensation reactions of aromatic cores of asphaltenes, which lose the paraffinic periphery. The main objective of a substantial portion of this chapter is to show how deep cracking of bitumen at low temperature can be achieved without coke formation (i.e., without polycondensation of asphaltenes). The main reactions of asphaltenes that lead to coke formation are described. Also described are ways to reduce the negative influence of these reactions on the process.

The behavior of asphaltenes during thermal treatment of pure vacuum residue (VR) is remarkable. At 400°C, the molecular weight of the asphaltenes decreases to a residence time of 30 minutes and then increases up to almost double the value of the molecular weight of native asphaltenes (see Fig. 9.1).

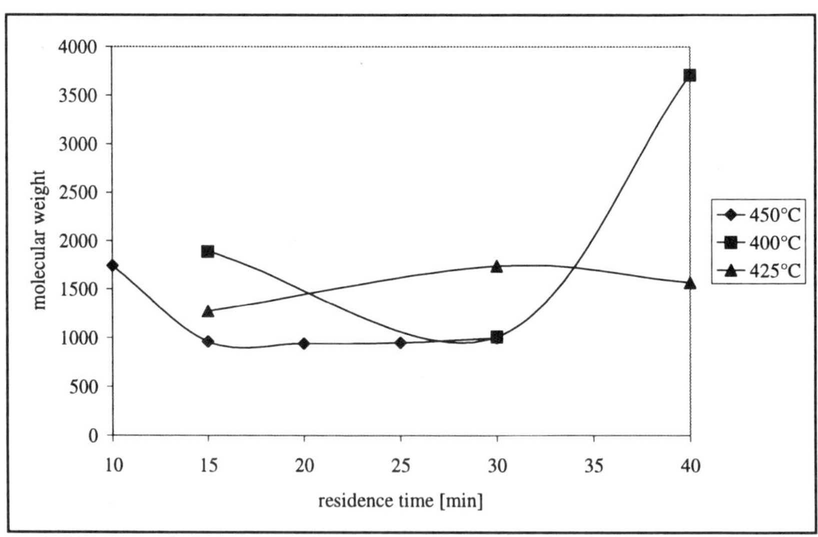

Fig. 9.1: Molecular weight of asphaltenes versus residence time during co-processing.

The reduction in molecular weight as the reaction proceeds is attributed to cracking of the paraffinic periphery. This cracking can be described by the following equation (9.1):

$$A_R\text{-}CH_2\text{-}CH_2\text{-}CH_3 \rightarrow A_R\text{-}CH{=}CH_2 + CH_4 \qquad (9.1)$$

Following this explanation, the decrease by about 30 wt.% in molecular weight has to be directly proportional to the content of paraffinic chains in the asphaltene molecule. ^{13}C-NMR analysis of native asphaltenes for bitumen B 200 Elf showed an amount of 29 wt.% of paraffinic periphery in the asphaltenes. This means that after a residence time of 15 minutes (at 400°C), the aromatic cores of native asphaltenes (see Fig. 9.2) are what are left of the original asphaltene. The strong tendency of aromatic asphaltene cores to polycondensation reactions has been reported in the literature [1]. Polycondensation reactions are reactions that lead to nascence of products heavier than the feedstock and gas. Generally, polycondensation reactions of asphaltenes can be described by the following equation (9.2):

$$n \cdot A_c \rightarrow \text{-}[Ac]\text{-}_n + A \cdot H_2 + B \cdot C_xH_y \qquad (9.2)$$

In contrast to the results at 400°C, no reduction in asphaltene molecular weight was observed for residence times up to 40 minutes for reactions conducted at 425°C (see Fig. 9.2). This means that at higher temperatures, polycondensation reactions proceed faster than decomposition reactions. At any temperature, the determined molecular weight of asphaltenes shows that it reaches equilibrium as the reaction proceeds. This implies that at a longer residence time, the molecular weight of the asphaltene fraction will not increase any further because after achieving equilibrium molecular weight, they become less soluble in the maltenes. This leads to their flocculation from the maltenes fraction (Fig. 9.2) and, finally, to coke formation.

The decrease in molecular weight at longer residence times shows that polycondensed asphaltenes contribute to the molecular structure that can be decomposed at 425°C. The chemical path for the formation of such asphaltenes can be described, for example, by reaction (9.3):

$$R^1 \text{-} A_R = CH_2 + CH_2 = Ar \rightarrow R^1 - A_R = CH - CH = A_R + H_2 \qquad (9.3)$$

R^1 represents a paraffinic side chain. This means that R^1 can be cracked in the same way as was shown in reaction (9.1). At 425°C, the cracking of naphthenic and/or aromatic rings is not possible as this reaction requires a temperature of over 550°C. However, it is especially interesting that hydrogen transfer and hydrogenation of aromatic cores of asphaltenes take place by the mechanism described in reactions (9.4-6) [9].

This means that cracking of the paraffinic periphery (reaction (9.1)) is the only reaction that can cause the reduction of the molecular weight of asphaltenes at this temperature level. In other words, during polycondensation of asphaltenes at 425°C, it is not only the aromatic cores of native asphaltenes that react. Asphaltenes with paraffinic chains also undergo a reaction (reaction (9.3)).

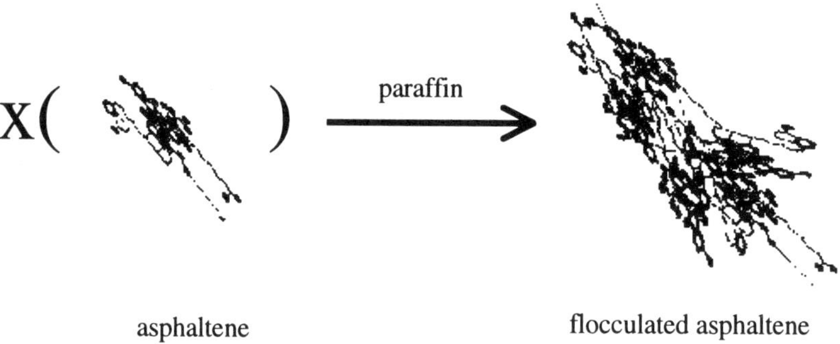

Fig. 9.2: Scheme of asphaltene flocculation.

→ + H· (9.4)

+ H· → (9.5)

+ → + (9.6)

The fact that at the low temperature no cracking of naphthenic rings occurs (which could lead to the nascence of olefins) permits us to say that the quality of the product formed is better if crude oil is thermally processed at a low temperature. This is because the chemical stability of naphthenes and aromatics is higher in comparison with olefins that arise during the cracking of naphthenes at higher temperature.

On account of the tendency of the asphaltene core to polycondensation, the increase in molecular weight at residence time over 15 minutes and a reaction temperature of 400°C can be explained by polycondensation of aromatic cores of asphaltenes (reaction (9.2)). Polycondensation reactions affect coke formation. This is noticed at 40 minutes residence time. By the thermal treatment of Bitumen 200, approximately 1 wt.% coke (based on water free feed) was formed. The fact that coke formation begins at the temperature of 400°C only if asphaltenes attain a very high molecular weight in comparison to higher temperature levels is caused by the formation of a stable steric colloid with the resins. This stable colloid is not cracked at the temperature of 400°C as deeply as at higher temperature levels. The formation of such a colloid is represented in Figure 9.3:

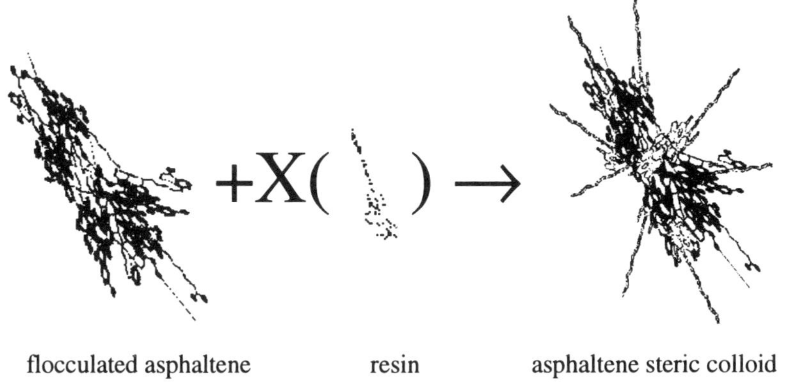

Fig. 9.3: Scheme of formation of asphaltene steric colloid.

At 450°C, asphaltenes achieve the equilibrium molecular weight at a residence time of 15 minutes (see Fig. 9.1). At this temperature, the molecular weight of asphaltenes is reduced by 33 wt.%. This molecular weight degradation implies the complete decomposition of the paraffinic periphery of the asphaltene molecule. At 450°C, the flocculation and polycondensation of asphaltenes to coke begins with the transformation of asphaltenes to the aromatic core of native asphaltenes. This is caused by a faster change of asphaltene solubility in maltenes and a higher reaction rate of polycondensation at this temperature level than at lower temperatures. The formation of a stable colloid solution via reaction (Fig. 9.3) is not possible either, since coagulation analysis has shown that resins are already completely cracked at a residence time of 15 minutes and a temperature of 450°C.

Based on the knowledge about the possible reactions of asphaltenes during thermal processing of pure bitumen, we can suggest the following general scheme of asphaltene reactions during thermal treatment (9.7):

$$\text{Gas} \leftarrow \text{Maltene} \leftrightarrow \text{Resin} \leftrightarrow \text{Asphaltene} \rightarrow \text{Coke} \qquad (9.7)$$

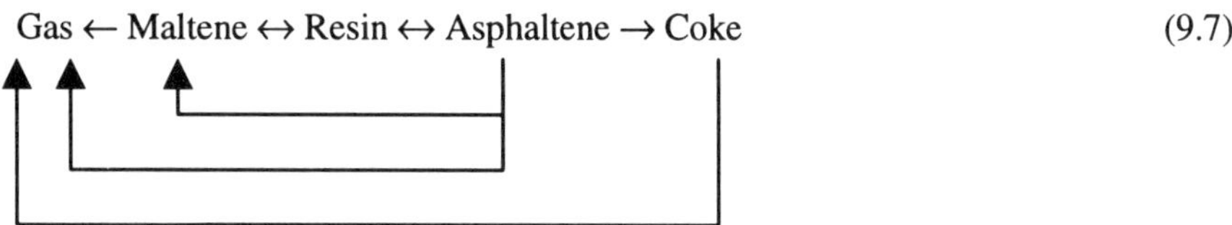

The reaction of coke to gas proceeds only if the coke was formed via the mechanism described by reaction (9.3), i.e. if paraffin side chains are available in the coke structure. These side chains can be cracked via reaction 9.1.

Based on ^{13}C-NMR analysis, ultimate analysis and molecular weight determination (table 9.1), a model of the average molecule was built for native asphaltene (Fig. 9.4a) and asphaltene from visbreaking at 425°C and a residence time of 30 minutes (Fig. 9.4b). These figures show clearly that during the thermal treatment of crude oil residue, only the cracking of paraffinic chains can cause asphaltenes reactivity to yield coke.

In the same way, a model of asphaltenes as a product of the thermal treatment of VR/plastics blends was investigated (Fig. 9.5a-c). The destruction of aromatic cores of asphaltenes can be clearly observed during the co-processing of VR and plastics. This can explain the minimization of coke formation during the co-cracking of VR and plastics.

Table 9.1: Data for development of the average molecule model of asphaltenes .

C-Atom (wt. %)	VR	VR after visbreaking	VR / PP / PS after visbreaking	VR / PE / PP / PS / PVC after visbreaking
$-CH_3$	4.9	4.9	4.6	4.6
α-CH_3	9.5	9.5	8.6	9.7
α-CH_2	0.0	0.0	3.1	3.5
$-(CH_2)_{n>4}-$	9.1	9.1	5.5	7.0
$-CH_2-_{bridges}$	2.8	2.8	4.1	5.4
$\rightarrow$CH	3.9	3.9	5.4	4.2
$C_{protoned\ aromatics}$	44.4	44.4	46.7	43.8
$C_{alcyl.\ aromatics}$ $C_{cond.\ aromatics}$	23.1	23.1	20.5	20.2
$C_{phenol\ aromatics}$	2.2	2.2	1.4	1.5
(wt. %)	ultimate analysis			
C	84.58	86.15	85.70	81.65
H	7.50	4.34	4.94	4.77
N	1.50	1.32	1.49	1.80
$O_{difference}$	6.42	8.19	7.87	11.78
	molecular weight determination			
Molecular weight	1430	1270	350	380

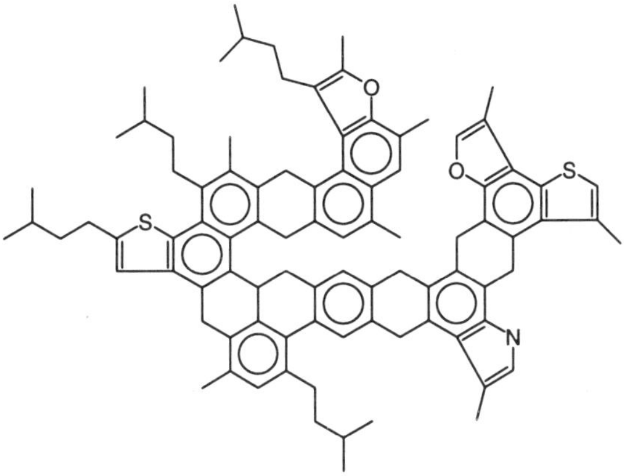

a)

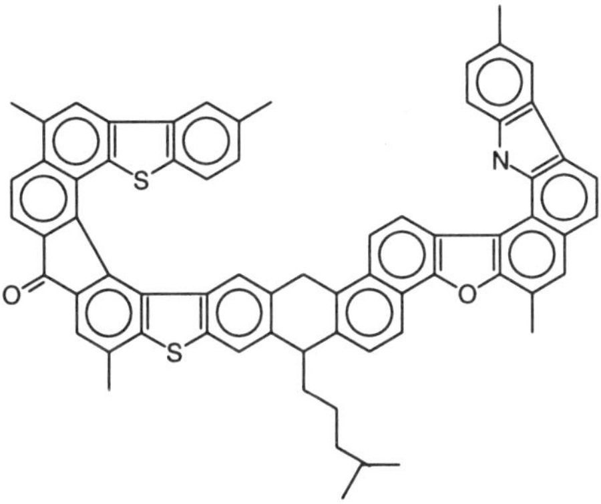

b)

Fig. 9.4: Model of the average molecule of asphaltene.
a) Native asphaltene
b) Asphaltene from visbreaking

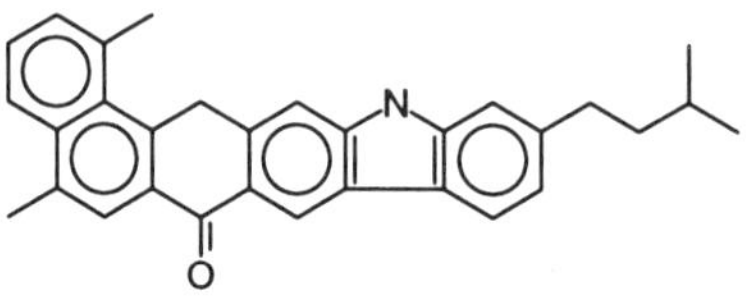

a)

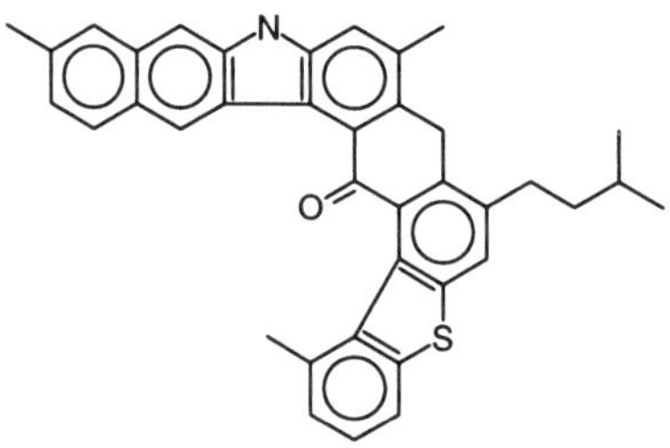

b)

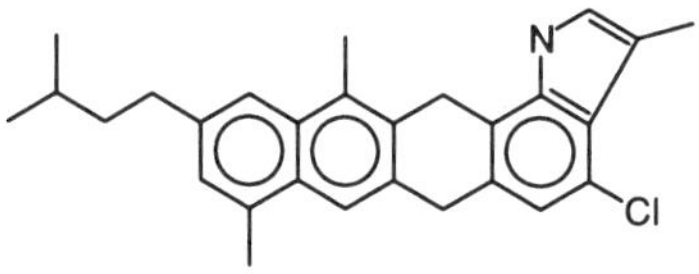

c)

Fig. 9.5: Model of average asphaltene molecule as a product of the thermal treatment of:
a) VR/PP/PS blends at 425°C and residence time 15 minutes
b) VR/PE/PP/PS blends at 425°C and residence time 30 minutes
c) VR/PE/PP/PS/PVC blends at 425°C and residence time 15 minutes

From Figure 9.5c, it is evident that during the joint cracking of bitumen and PVC a reaction takes place between asphaltenes and chlorine. GC analysis of the product gas of the joint cracking of bitumen and PVC has shown that these reactions proceed very extensively, since in the product gas was found no hydrogen chloride which is formed in very large amounts for the thermal cracking of pure PVC. This shows a very high tendency of asphaltenes towards reactions with chlorine.

The main structural elements of the asphaltene molecule can be classified based on the data from Figures 9.4 and 9.5 as follows:

· Paraffin periphery

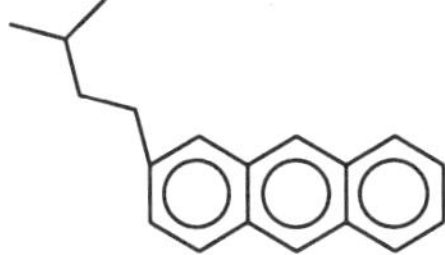

· Bridge-ring structure

or

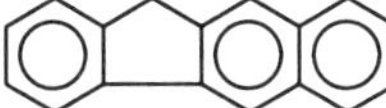

· Aromatic structure

· Heteroatom structure

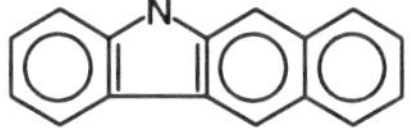

From the investigation into asphaltene chemistry, it is obvious that paraffin side chains can be readily cracked during the thermal treatment of pure bitumen at a relatively low temperature. Aromatic and heteroatom structures are inclined to polycondensation reactions and finally to coke formation. Especially interesting are bridge-ring structures since these structural elements will only be cracked either at high temperature or by the addition of plastics to the feedstock.

9.2.2 Evaluation of Possibilities of Various Asphaltene Reactions Based on Thermodynamics

The reactions of the main structural elements of asphaltene were investigated. First of all, it is necessary to calculate the thermodynamic possibility of the reactions to be investigated. Using equation (9.8) does this evaluation:

$$\Delta G = G_{products} - G_{feed} \tag{9.8}$$

ΔG – difference of the free enthalpy of the reaction
$G_{products}$ – free enthalpy of the reaction product
G_{feed} – free enthalpy of the feed

The values for free enthalpies can be taken from the literature [25]. But for complicated compounds such as asphaltenes, no data are available in the literature. Thus, for the evaluation of the reaction possibility between high molecular weight compounds, the free enthalpy of each of the products and feed have to be calculated. One example for this calculation is given below.

Example

The free enthalpy for the following compound (Fig. 9.6) has to be calculated. The free enthalpy for such complicated compounds can be calculated based on the knowledge of their structural elements. The values for free enthalpy of the structural elements can be found in ref. 25. The enthalpy values for important structures of asphaltene are presented in table 9.2.

In the example (Fig. 9.6), there are the following structural elements:

a) Aromatic bond
   >C-H : seven bonds
   >C-R : three bonds
   >C - : four bonds

b) Acyclic bond
   $-CH_3$: three bonds
   $>CH_2$: one bond

c) Corrections for acyclic substitutes on aromatic core
   1,2-substitutes: two substitutes.

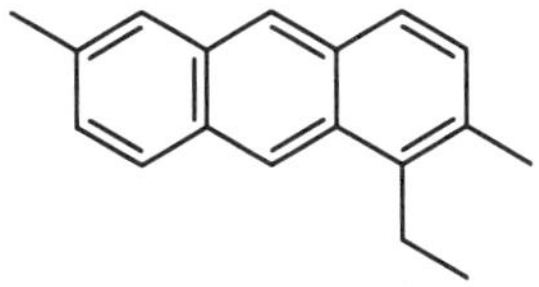

Fig. 9.6: Example of compound used to illustrate the calculation of free enthalpy.

Table 9.2: Values of free enthalpy for structural elements (kJ/mol).

	0°K	298°K	400°K	500°K	600°K	800°K	1000°K	1200°K	1500°K
aromatic bond									
>C-H	16.74	20.26	23.02	25.66	28.38	34.12	40.02	45.92	54.88
>C-R	24.11	36.67	39.10	42.82	46.80	54.75	61.79	69.45	80.50
>C -	22.14	22.73	24.28	27.17	30.47	33.70	38.43	45.84	51.45
naphthenic bond									
$>CH_2$	46.59	55.13	59.32	63.88	68.69	78.49	88.49	98.62	113.82
>CH -	52.95	68.48	74.09	80.16	86.02	97.49	109.09	120.31	136.97
acyclic bond									
$-CH_3$	-34.58	-17.33	-8.37	1.00	10.76	30.98	51.82	72.84	104.48
$>CH_2$	-15.38	8.57	18.75	29.01	39.47	60.63	81.96	103.26	135.34
>CH -	0.75	31.23	43.49	55.51	67.60	91.51	114.91	138.22	171.46
>C<	7.28	47.89	62.79	77.23	91.34	118.71	144.84	170.41	208.09
$H_2C=$	30.39	33.24	35.87	41.73	42.03	48.93	56.26	63.71	75.05
HC≡	113.69	103.81	100.51	97.37	94.19	88.07	82.09	76.27	67.69
Corrections for acyclic substitutes on aromatic core									
1,2-Sub.	---	2.09	2.09	2.09	2.09	2.09	2.09	2.09	2.09
1,3-Sub.	---	-2.93	-2.93	-2.93	-2.93	-2.93	-2.93	-2.93	-2.93
1,2,3-Sub.	4.19	4.19	4.19	4.19	4.19	4.19	4.19	4.19	4.19
heteroatom structures									
>CO	-128.09	-119.72	---	---	-108.42	---	94.60	---	---
-O-	---	-99.63	---	---	---	---	---	---	---
$-NH_2$	---	-26.79	---	---	---	---	---	---	---
-SH	---	12.98	---	---	---	---	---	---	---
- S -	---	32.65	---	---	---	---	---	---	---

In order to calculate the free enthalpy of the compound, the values of the free enthalpies of the various bonds in the compound are added. For the shown example at the temperature of 600°K (326.85°C):

$$G = 7 \cdot 23.38 + 3 \cdot 46.80 + 4 \cdot 30.47 + 3 \cdot 10.76 + 1 \cdot 39.47 + 2 \cdot 2.09 = 501.87 \text{ kJ/ mol}$$

This calculation method has a divergence of approximately ± 2.5 kJ /mol [25].

This example shows that the free enthalpy can be calculated for every compound with various difficulty levels depending on its structure. In the same way, free enthalpy values were calculated for the evaluation of various reaction possibilities for asphaltenes.

From Figures 9.3 and 9.4, we can see that the cracking of the aromatic core of asphaltenes can take place via the destruction of the bridge-ring system. This cracking reaction is the reversible reaction of polycondensation (9.2) or (9.3). It means that a gas will be used during this reaction. This gas can be hydrogen or hydrocarbon. The reaction with hydrogen can be described by equation (9.9):

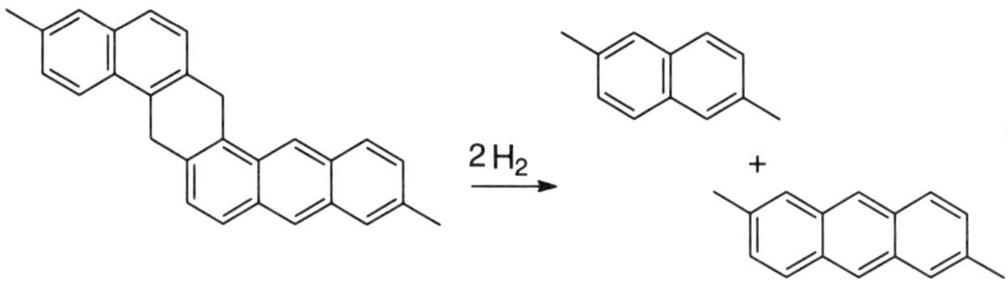

$$\xrightarrow{2\,H_2} \quad + \qquad (9.9)$$

From Figure 9.7, it is evident that the destruction of the bridge-ring structures by hydrogenation of the cracked bond is thermodynamically possible only at temperatures under approximately 25°C. Studies [26] concerning the reaction of asphaltenes and atomic hydrogen at low temperatures have been reported. It was shown in the studies that the aromatic core of asphaltenes can be cracked and its decomposition (of the aromatic core) proceeds especially extensively, if the reaction occurs at a temperature below 0°C.

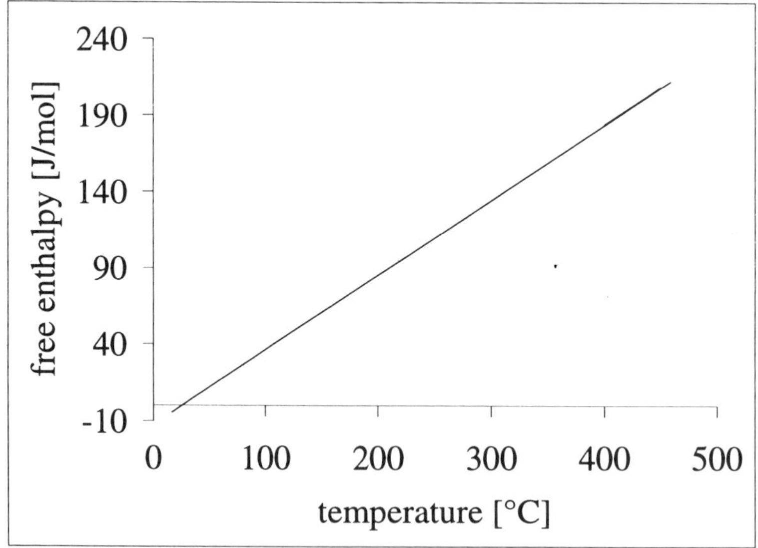

Fig. 9.7: Change of free enthalpy for reaction (e9.9).

This means that hydrogenation of asphaltenes cannot be responsible for asphaltene cracking during the thermal treatment of bitumen and plastics at temperatures over 400°C.

Another type of asphaltene cracking reaction is reaction with hydrocarbons. Equations (9.10) and (9.11) can describe this type of reaction:

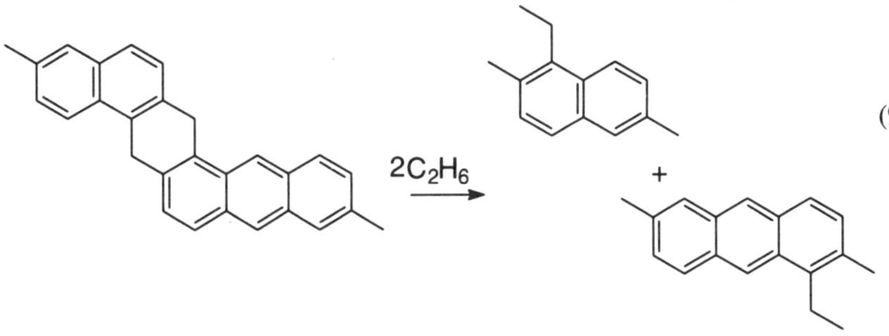

$$\xrightarrow{2C_2H_6} \quad + \qquad (9.10)$$

The use of hydrocarbons heavier than ethane requires a higher reaction temperature (see Fig. 9.8).

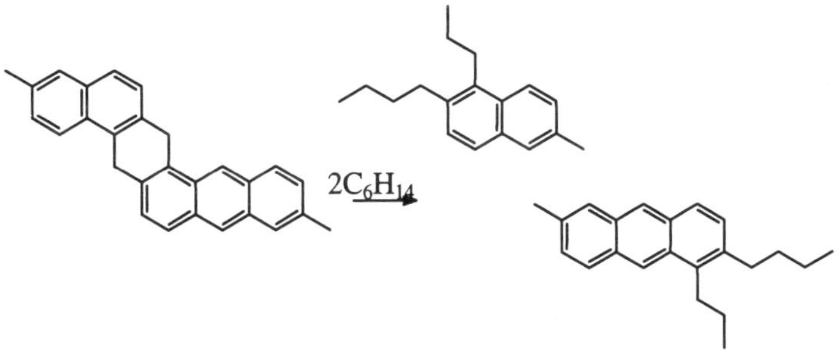

$$\xrightarrow{2C_6H_{14}} \qquad (9.11)$$

From Figure 9.8, we see that only reactions with ethane and/or methyl radicals are thermodynamically possible at temperatures over 300°C. All reactions with heavier hydrocarbons or radicals are possible only at temperatures below 0°C.

Reaction (9.10) is thermodynamically possible. However, for the reaction to proceed to a large extent, it is necessary that the methyl radical is produced on a continuous basis with relatively high velocity. Ethane cracking at a temperature under 500°C is thermodynamically not possible, i.e. reaction (9.10) proceeds only very slowly with ethane. But during the common cracking of bitumen and plastics, the methyl radical can form from cracked products of plastics and then enhance deep asphaltene cracking. These reactions can be described by equation (9.12).

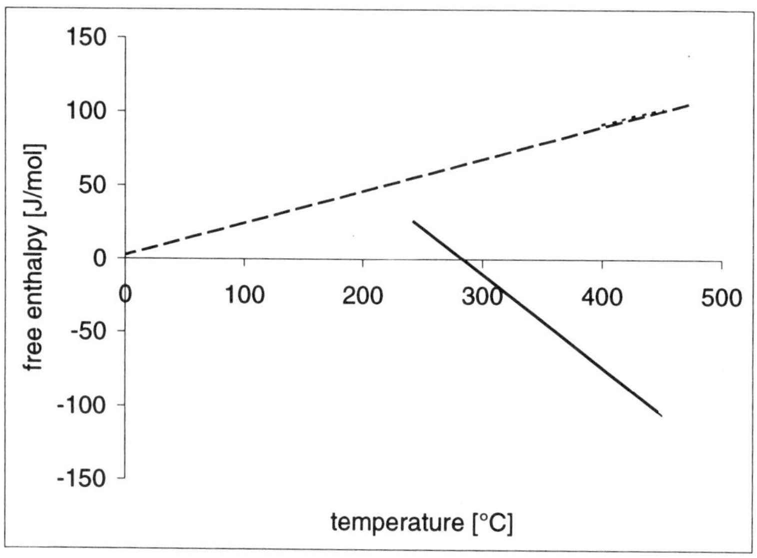

Fig. 9.8: Change of free enthalpy. Solid: for reaction (9.10); doted: for reaction (9.11).

In this reaction example, the cracking products of polypropylene were used as methyl radical donors.

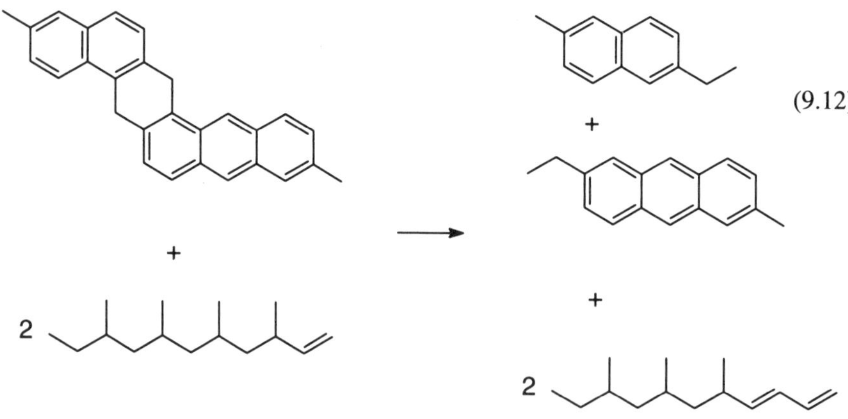

+ 2 → + + 2 (9.12)

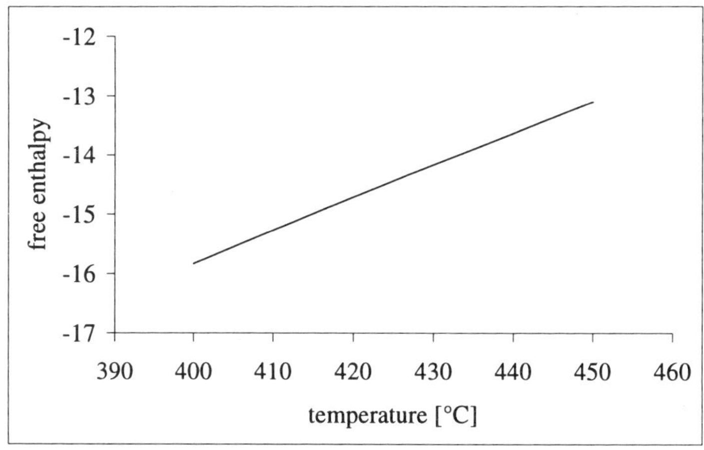

Fig. 9.9: Change of free enthalpy for reaction (9.12).

Figure 9.9 shows that reaction (9.12) is thermodynamically possible. In other words, reactions such as represented by (9.12) are responsible for deep asphaltene cracking during the common cracking of bitumen and plastics.

Experiments with pure asphaltene structural compounds were carried out to show the possibility and extent of asphaltene cracking via the bridged-ring structures. Flouren and dihydroanthracene were used for these experiments, since these compounds have two different bridged-ring structures. Both experiments showed that cracking of bridged-ring structures during the reaction of the model compounds and polypropylene (PP) was possible and proceeded very extensively. One example of the results for this investigation is presented in Figure 9.10.

From Figure 9.10, it is seen that during the thermal treatment of pure model compounds, polycondensation is the only reaction that occurs. In contrast to this, during the co-treatment of the model compound with polypropylene, the flouren is decomposed. This decomposition reaction proceeds via the bridged-ring system, because one-ring aromatic compounds were found in the reaction product in case b) of Figure 9.10.

The aromaticity of the asphaltenes increased during their thermal treatment. This increase proceeded via a dehydrogenation reaction, an example of which is described by reaction (9.13):

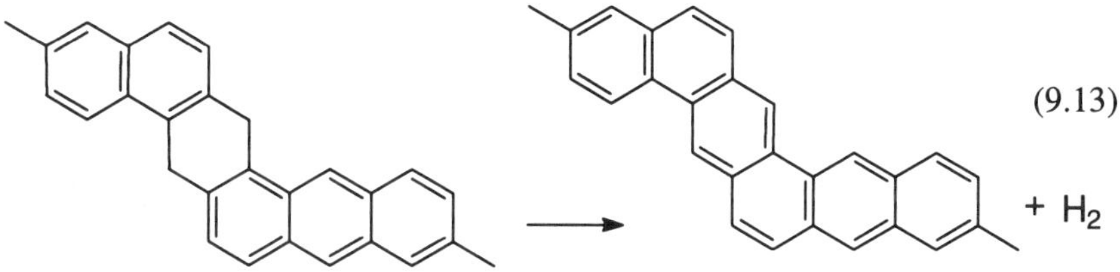

(9.13)

$\longrightarrow$ + H_2

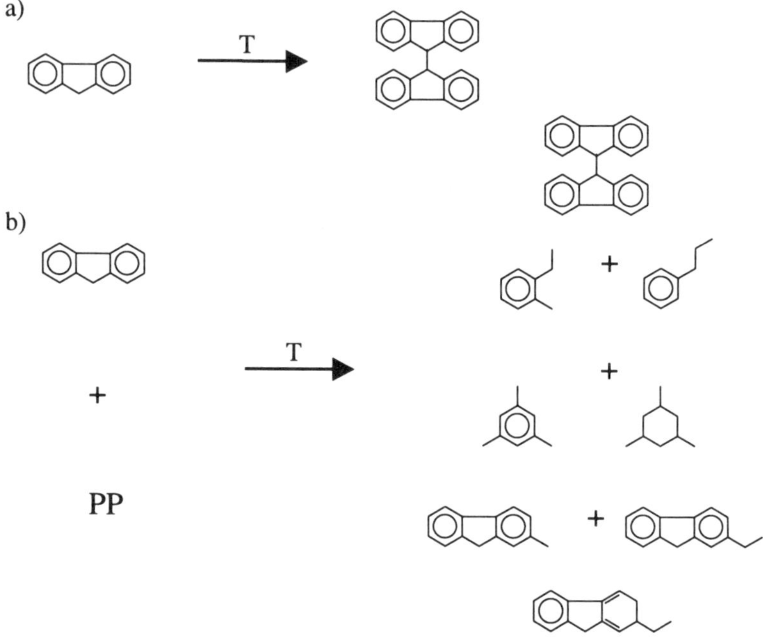

Fig. 9.10: Sample reaction of the thermal treatment of

a) Pure flouren
b) Flouren with polypropylene

The change of free enthalpy for reaction (9.13) is presented in Figure 9.11.

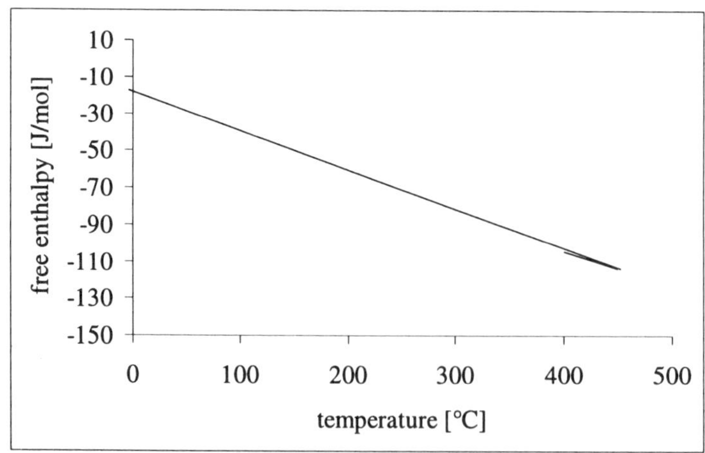

Fig. 9.11: Change of free enthalpy for reaction (9.13).

From Figure 9.11, it is seen that dehydrogenation proceeds with a higher probability at a higher temperature. Since the compounds with higher aromaticity have a higher tendency towards coke formation, thermal treatment of bitumen or vacuum residues has to be carried out at the lowest possible reaction temperature.

9.2.3 Hydrogen Transfer

This section discusses the possibility of deep cracking of bitumen or vacuum residues at a low temperature.

The need for hydrogenation in the case of thermal treatment of VR is due to the chemical instability and high tendency to coke formation of the resulting unsaturated components [9]. During industrial thermal cracking, many alkenes are formed which can be oxidized. The alkodienes have an especially strong predisposition to oxidation reactions. Oxidized components of fuels can be polycondensed. This leads to the formation of resins, which if precipitated can plug the pipes.

Hydrogenation can be achieved by utilization of hydrogen under high pressure. This method is broadly used in modern industry. But nowadays, the use of hydrogenation processes under hydrogen pressure is very expensive. As an alternative, hydrogenation can be carried out with hydrogen donors. All materials with a H/C_{atomic} ratio higher than 2 can be used as hydrogen donors in the thermal treatment of oil residues [28]. This ratio is 2 for plastics such as polyethylene and/or polypropylene, i.e. plastics may be used as hydrogen donors during the thermal treatment of crude oil vacuum residues. The process of utilization of hydrogen donors makes the use of expensive hydrogen unnecessary. Also, it can be carried out at a lower pressure than in any of the modern hydrogenation processes.

The present methods of using a hydrogen acceptor are intended to prove the ability of the substance to be a hydrogen donor.

Shaw et al. [27] indicate that anthracene is more readily hydrogenated than other aromatic systems (e.g. naphthalene). The hydrogenation of anthracene can be described by the following equation (9.14):

$$\text{Anthracene} + H_2 \rightleftharpoons \text{9,10-Dihydroanthracene} \qquad (9.14)$$

After hydrogenation, the product is analyzed by GS-MS. By this method, the ratios for dihydroanthracene/remaining anthracene as well as anthracene derivatives/remaining anthracene were determined as 9 and 10, respectively. These two values show the depth and chemical path for hydrogenation, i.e. from these two values it can be proved which reaction (hydrogenation or alkylation) occurred during hydrogen transfer provoked by the investigated hydrogen donor.

The results of this investigation of hydrogen transfer are presented in table 9.3. The experiment with thermal treatment of pure anthracene (E.1) demonstrated that a significant amount of dihydroanthracene was found in the used anthracene after thermal treatment. Therefore, it is important to use this fact in the discussion of the results. Experiment RE.3 is a reproducibility experiment in comparison with E.3. It is evident that the experimental method has a good reproducibility (relative divergence 4%).

Based on the results for runs E.2 and E.3, it is seen that hydrogenation only has a small significance for hydrogen transfer during co-processing. Laux [3] reported that at a temperature over 300°C, the equilibrium between hydrogenation and dehydrogenation is displaced towards dehydrogenation because hydrogenation is an exothermic reaction.

In contrast to hydrogenation, alkylation reactions proceed very extensively (see table 9.3). In the case of plastic blends with polyethylene (PE) (E.3 anthracene/PE/PP/PS blend), alkylation proceeds more extensively in comparison with mixtures without PE (E.2 anthracene/PP/PS blend). This difference in depth of alkylation can be explained by different reaction velocities of cracking during the co-processing of mixtures with and without PE. At the same temperature, PE is cracked more slowly. This implies that under the same conditions for the use of PE, a larger amount of free radicals will stay in the liquid phase during the reaction. Generally, alkylation has a reaction order of 2 or higher. Therefore, this reaction is strongly accelerated by the transfer of the reaction zone from vapor to the liquid phase.

Generally, it can be said that during co-processing, alkylation reactions are responsible for the hydrogen transfer during the co-processing of bitumen and plastics. Therefore, it is important to promote this reaction. It was shown that alkylation proceeded more extensively if the reaction zone was located in the liquid phase. This is why it is important to keep the hydrogen-rich radicals in the liquid phase. Based on this information, the following recommendations can be given for the reaction conditions during co-processing:

- Use plastics with small reaction velocity of cracking.
- The process can be carried out under high pressure.
- The vapor phase should permanently stay in contact with the liquid phase.

Table 9.3: Results of hydrogen transfer investigation.

Experiment number	E1	E2	E3	RE3
Dihydroanthracene/anthracene [mol/mol]*10^3	13	24	23	22
Anthracene derivatives/anthracene [mol/mol] *10^3	0	43	112	113

Additional experiments with bitumen/plastic blends have shown that at optimum conditions, hydrogen transfer can proceed completely. This means that no unsaturated components (alkene and/or alkadiene) will be found in the fuel fractions (boiling point under 350°C) obtained from the co-processing product.

9.3 CO-PROCESSING WITH CRACKED PRODUCTS

The investigations of co-processing represent a special interest because synergistic effects can take place during the joint treatment of VR and different types of materials (plastics in our investigation). In this case, synergy is understood as a positive divergence in the formation of lighter fractions. An example of the synergy in the case of the co-processing of crude oil vacuum residue is shown in Figure 9.12.

These synergy effects can be explained in two ways: by investigating the process kinetics and by investigating the chemistry of the process. In our investigation of the co-processing of blends of crude oil vacuum residue and plastics we have used both methods. Data from thermo-gravimetric analysis were used for study of the process kinetics. The dynamic method for reaction (9.15) was used for the evaluation of the data.

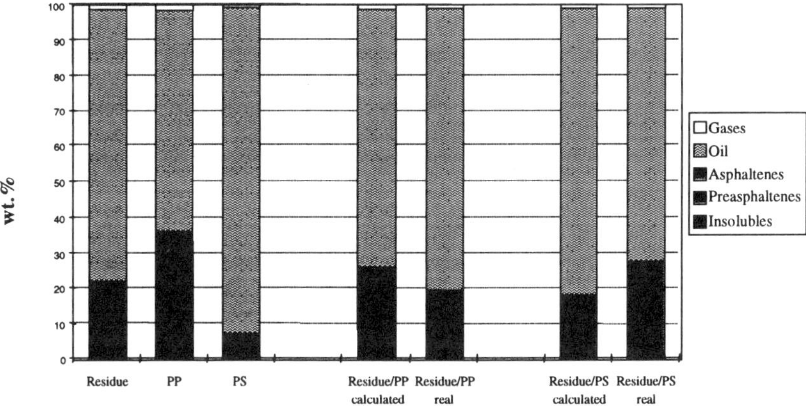

Fig. 9.12: Synergy during the co-processing of blends of VR and PP and VR and PS:
pressure = 20 bar
reaction temperature = 425°C
reaction time = 25 minutes

$$-S + V = 0 \tag{9.15}$$

S - substance
V - volatile product

In the reaction, the rate of disappearance of S may be expressed by equation (9.16):

$$dX/dT = k_0 \exp(-E_a/RT)(1-X)^n/\beta \tag{9.16}$$

X - conversion of S ($0 \leq X \leq 1$)
E_a - energy of activation
k_0 - frequency factor
ß - linear heating rate (20 and 10 deg./min)

The integrated form of the kinetic equation (9.17) or (9.18) is used for the determination of kinetic parameters:

Thus, for reaction order $n \neq 1$:

$$\log_{10}\left(\frac{1-(1-X)^{1-n}}{T^2(1-n)}\right) = \log_{10}\frac{k_0 R}{\beta E_a}\left(1-\frac{2RT}{E_a}\right) - \frac{E_a}{2.3RT} \tag{9.17}$$

and for the reaction order n = 1:

$$\log_{10}\left(\frac{-\log_{10}(1-X)}{T^2}\right) = \log_{10}\frac{k_0 R}{\beta E_a}\left(1-\frac{2RT}{E_a}\right) - \frac{E_a}{2.3RT} \tag{9.18}$$

These equations are solved graphically as in Figure 9.13 in which the left-hand side of the integrated form of the kinetic equations is the ordinate and the term 1/T is the abscissa. The graphical method involves the best approximation of the curve to a straight line [28].

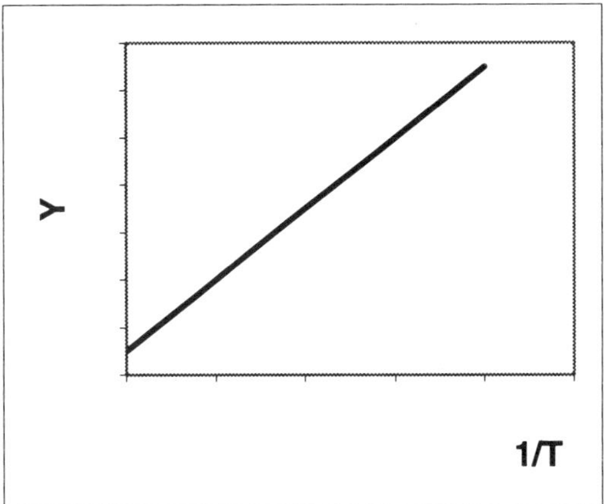

Fig. 9.13: Dynamic evaluation.

A model has been developed for the main reactions in an attempt to obtain a better understanding of the kinetic data obtained for co-processing. The basic principles of this model are:

- Co-processing is a process of thermal cracking of VR and plastic blends.
- Thermal cracking of organic material has a radical chain mechanism.
- The radical chain mechanism includes three phases as presented in Figure 9.14.

The first step (chain start) can be controlled either by a catalyst or change of the reaction conditions (i.e., temperature and/or pressure). On the other hand, however, the second step, chain growth, can be influenced by the input of more radicals to the reaction zone. A material can be used as additive radicals if it can be cracked at a higher reaction velocity than VR with all other conditions remaining the same.

This reaction between plastic radicals and VR molecules has a theoretical reaction order of 2 or higher during the formation of more complicated complexes. This is higher than could be obtained from only one radical and one molecule. This means that the increment of the reaction order during co-processing is evidence of reactions between plastic radicals and VR molecules. The optimum conditions for co-processing are in such ranges of pressure and temperature where the co-cracking has the highest reaction order. On the other hand, recombination and polycondensation reactions have a theoretical reaction order of 2 and higher, too. But these reactions lead to the formation of heavy products that can be condensed into the liquid phase again. This physical process leads to the reduction of the reaction order [29].

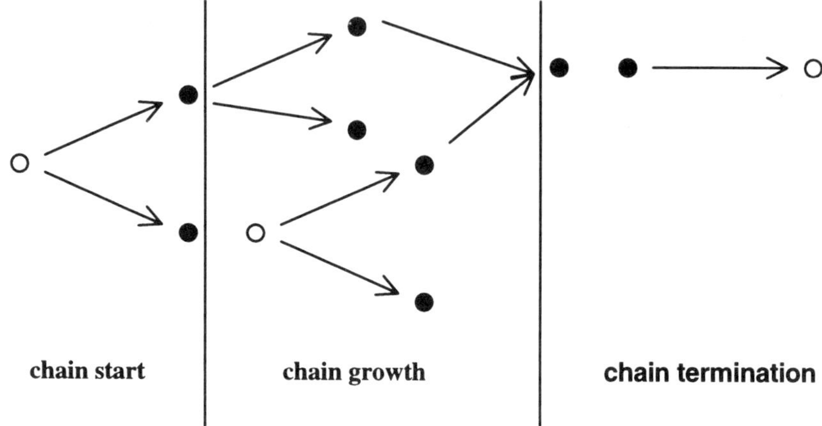

Fig. 9.14: Example of chain reaction.
Black – free radical
White – molecule

The reduction of the reaction order during co-processing shows that under the selected reaction conditions, a large number of radicals stay in the reaction zone leading to recombination and polycondensation reactions, which in turn lead to the formation of heavy products. Therefore, cracking reactions and the activation of co-cracking is decelerated under these conditions.

We have used the method of molecular weight determination and monitoring the molecular weight change during thermal processing to investigate the chemistry of co-processing of blends of crude oil vacuum residue with different types of plastics.

9.3.1 Co-Processing with Cracked Products from Aromatics Containing Plastics (e.g. Polystyrene)

Polystyrene PS 168N was used in our investigation to study the influence of aromatic containing plastics on the co-processing of mixtures of crude oil residue and plastics.

The determined reaction orders for different pressures and various mixtures of vacuum residue and polystyrene are shown in Figure 9.15. This example shows that for vacuum residue/polystyrene mixtures, the optimum reaction condition is in the range between 20-35 bar. In this pressure range, all investigated mixtures showed a maximum reaction order, i.e. reactions with reaction order higher than 1 dominate. Alkylation reactions belong to this type of chemical process (i.e. with high reaction order). Polycondensation also has a high theoretical reaction order, but leads to the formation of heavy products, which can condense back to the liquid product. This leads to smaller reaction orders [29]. During co-processing in the pressure range above 30 bar (see Fig. 9.15), polycondensation reactions dominate.

Our investigation has shown that polystyrene promotes the decomposition reactions of asphaltenes (Figure 9.15). Remarkable is the fact that at a lower temperature (400°C), the decrease in the asphaltenes' molecular weight (65%) is bigger than at the higher temperature (425°C, molecular weight decrease 44%). This can be caused by the poor solubility of asphaltenes in maltenes; at higher temperatures, more aromatic cracking products go to the vapor phase. On the other hand, it can be caused by the tendency of aromatics to polycondensation reactions (see chapter 1). We carried out experiments with vacuum residue/toluene blends in order to prove whether asphaltene cracking is promoted by chemical activation or by better solubility of asphaltenes in highly aromatic maltenes. Since toluene is a thermally stable substance at the reaction temperature of 400°C, it can only influence co-processing by improved solubility of the asphaltenes in the maltenes. The experiments with blends of VR/toluene (Figure 9.16) showed that a better solubility of asphaltenes in maltenes decelerated polycondenstion reactions and coke formation, but this does not activate asphaltenes cracking. This means that the solubility

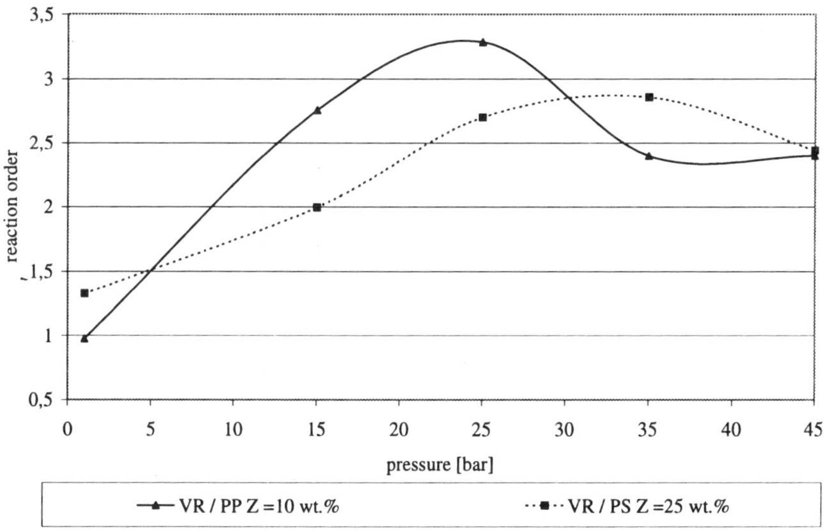

Fig. 9.15: Graphical presentation of the reaction order as a function of pressure.

of asphaltenes in maltene fraction influences polycondensation reactions and shows no influence on the chemistry of asphaltene cracking.

The behavior of asphaltenes during thermal treatment of pure vacuum residue (VR) is in absolute contrast to the co-processing of vacuum residue and polystyrene (Figure 9.16). At 400°C, the molecular weight of the asphaltenes decreases and then increases. This reduction in the molecular weight can be explained by the cracking of the paraffin periphery. This means that the decrease of about 30 wt.% in the molecular weight has to be directly proportional to the content of paraffin chains in the asphaltene molecule. The ^{13}C-NMR analysis of native asphaltenes showed that the portion of paraffin periphery in the asphaltenes was 28.66 wt.%. After a 15 min residence time at 400°C, only the aromatic cores of native asphaltenes are present. The tendency of aromatic asphaltene cores towards polycondensation reactions was reported in ref. 9. On account of this tendency, the increase in molecular weight at a residence time over 15 minutes can be explained by the polycondensation of aromatic cores of asphaltenes. Thus, polycondensation reactions affect coke formation. This is very obvious at 40 min residence time. In this experiment, 1 wt.% coke (based on water free feed) was formed.

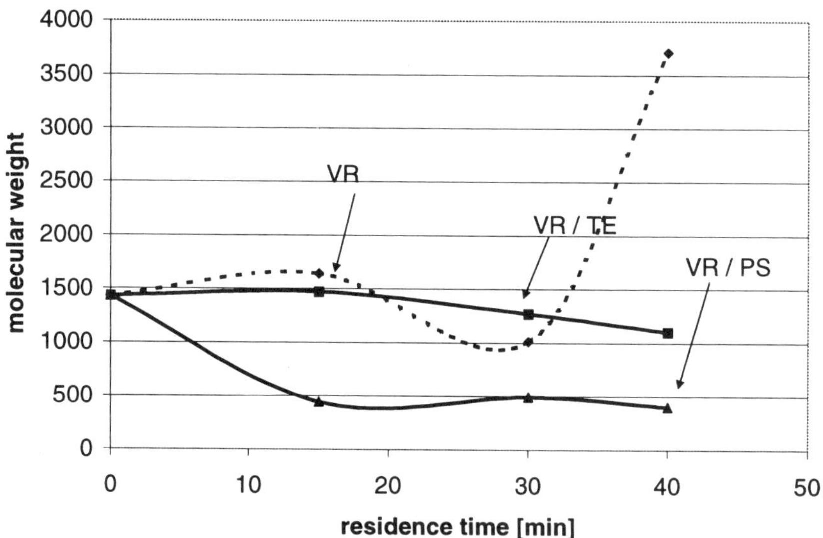

Fig. 9.16: Influence of the asphaltenes' solubility in maltenes on the chemistry of asphaltenes cracking.
Constant parameters:
T = 400 °C
p_{N2cold} = 20 bar
additive concentration = 20 wt.%

9.3.2 Co-Processing with Cracked Products from Plastics Containing Paraffin Groups

Polypropylene PP Novelen 1100H and polyethylene PE Lupolen LDPE were used in our investigation for the study of the influence of the paraffin containing plastics on the co-processing of mixtures of crude oil residue and plastics.

Determination of the kinetics of the co-processing of blends of vacuum residue and paraffin containing plastics shows that the optimum pressure for the cracking of these mixtures is the same as for the mixtures of vacuum residue and polystyrene. Figure 9.17 shows the results of the determination of the activation energies and frequency factors for VR and paraffin containing plastics. It is evident that PP has almost the same activation energy as VR and a higher frequency factor than VR. Data obtained from the kinetic study for co-processing show that the best results for co-processing (with regard to synergy) is obtained only if the plastics and/or plastics blends used have approximately the same activation energy but higher frequency factor than the vacuum residue. This means that PP is the best additive to bitumen in the case of co-processing of vacuum residue-plastics mixtures. In our last investigation on co-processing, we showed that the biggest synergy occurs when mixtures of VR and PP or plastics blends with PP are used in the process (Figure 9.12).

Polyethylene and polypropylene are especially interesting as additives for the activation of the cracking reaction of vacuum residue. The data obtained from thermo-gravimetric analysis of polyethylene and polypropylene show that vacuum residue has its maximum cracking velocity at the temperature 456°C, polypropylene at 450°C and polyethylene at 485°C. This means that the cracking of vacuum residue will only be activated if polypropylene is used. In the case of the thermal treatment of vacuum residue and polyethylene mixtures, the cracking of polyethylene will be activated by free radicals from the vacuum residue (see above).

The investigation into the influence of paraffinic plastics on asphaltene chemistry during thermal cracking showed that pure plastics affect only the equilibrium of alkylation reactions by the increase of the paraffinic radicals in the reaction zone (Figure 9.18). This means that asphaltene decomposition will be slowed down. As such, there will be no decomposition to form aromatic cores without paraffinic periphery. This decelerates polycondensation and coke formation during the thermal treatment of mixtures of vacuum residue and plastics. However, it does not promote the cracking of the asphaltenes.

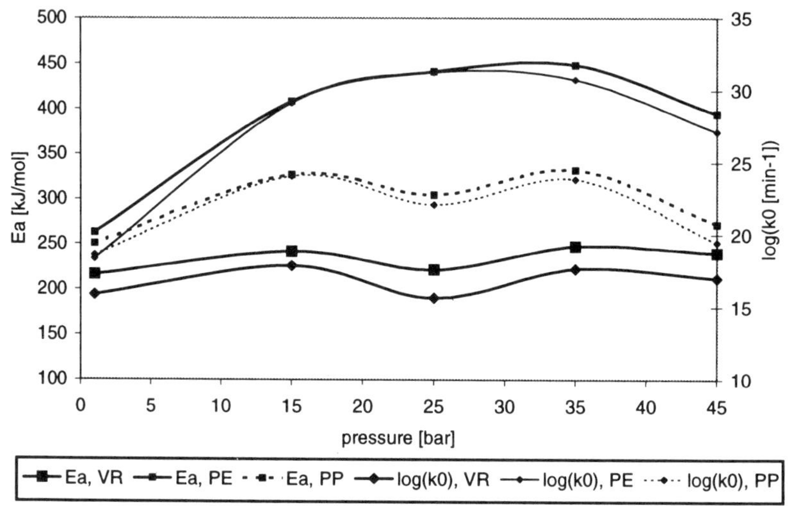

Fig. 9.17: Graphical representation of the kinetic parameters as a function of pressure.

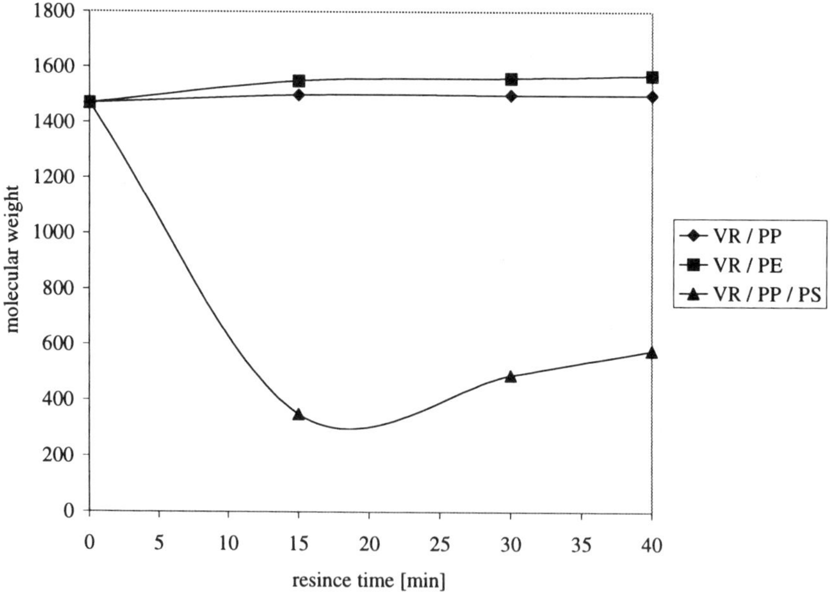

Fig. 9.18: Influence of paraffin containing plastics on the chemistry of asphaltene during thermal cracking.
Constant parameters:
T = 425 °C, p_{N2cold} = 20 bar, additive concentration = 20 wt.%.

9.3.3 Possibilities that Exist for Carrying Out Co-Processing of Heavy Crude Oils and Various Co-Feeds

The possibility of deep asphaltene cracking at a relatively low temperature was shown in Figure 9.16. This possibility was the base for further investigation involving a new type of co-processing called stepped co-processing. The logic in stepped co-processing is that the asphaltenes cracked during the first step need a longer reaction time for polycondensation and formation of fine coke than the native asphaltenes. Thus, vacuum residue can be co-treated with polystyrene during the first step and the product of the first step can be co-treated with polyethylene or polypropylene during the second step. In that way, it is possible to use the property of PS to activate asphaltene cracking at low temperatures and the property of paraffinic plastics to improve the quality of the co-processed products with regards to hydrogen transfer. The first results of investigations on stepped co-processing are shown in table 9.4.

Table 9.4: Comparison between a two-step and a one-step co-processing.

		Two step co-processing	One step co-processing
Conditions			
1. Step	Pressure:	30 bar	35 bar
	Temperature:	400 °C	425 °C
	Reaction time:	15 min	30 min
	Plastic:	Polystyrene	Polyethylene (60 %) Polypropylene (30 %) Polystyrene (10 %)
	Concentration:	15 %	20 %
2. Step	Pressure:	37 bar	
	Temperature:	450 °C	
	Reaction time:	15 min	
	Plastic:	Polyethylene (83.4 %) Polypropylene(16.6 %)	
	Concentration:	20 %	
Results			
Gas	[wt%]	2.63	1.97
Maltenes	[wt%]	81.76	64.02
1. < 180 °C	[wt%]	- 24.56	
2. 180 – 350 °C	[wt%]	- 20.44	
3. 350 – 500 °C	[wt%]	- 15.38	
4. > 500 °C	[wt%]	21.38	
Hard resin	[wt%]	0	18.59
Asphaltene	[wt%]	12.41	4.87
Coke	[wt%]	3.20	10.56

The next possibility for very deep co-processing of crude oil residues and plastics is one that also involves vacuum residue recovery. The main idea of this method is shown in the schematic in Figure 9.19.

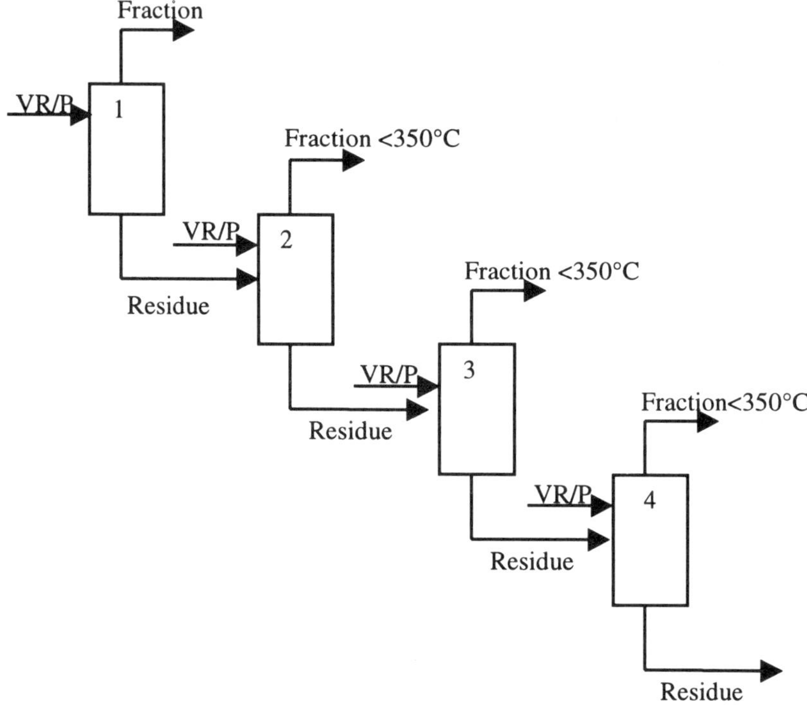

Fig. 9.19: Co-processing of crude oil vacuum residue and plastics with recovery of the vacuum residue, at 425°C, app. 30 bar, 15 min.
VR/P- mixture vacuum residue to PP/PS (5/1), 20 wt.% of plastics
1 – 4 - steps of experiment

The results of investigations of the co-processing with recovery of the vacuum residue that is recycled back to the process are shown in table 9.5.

Table 9.5: Results of investigation into the co-processing with vacuum residue recovery.

Step of experiment / Product	1	2	3	4
Boiling temperature ≤ 150°C [%]	8.40	9.30	9.00	9.120
Boiling temperature 150 – 350°C [%]	15.30	16.50	16.10	16.70
Boiling temperature ≥ 350°C [%]	75.80	73.70	74.40	73.60
Coke [%]	0.00	0.00	0.00	0.50

From the table, it is obvious that the vacuum residue from co-processing can be recovered back to the process. The results show that it is only after the third usage that traces of coke (0.5 wt.%) were found. Such a method can strongly increase the conversion of heavy crude oils and vacuum residues into the desired light fractions.

It is not only plastics that can be used as co-feed for co-processing. Many different organic co-feeds were investigated in our work. For example, investigation into co-processing of crude oil vacuum residue with used fat such as organic co-feed has shown that the optimum conditions for the mixtures of vacuum residue/used fat are similar to those for co-processing of crude oil vacuum residue with plastics. In the case with bio-based co-feed, coke free co-processing was achieved at the temperature of 425°C, pressure 20 bar, and reaction time of 15 minutes.

9.3.4 Behavior of Heavy Metals During Co-Processing

The heavy metals in crude oil residues are agglomerated mostly in asphaltenes in the form of porphyrin compounds. One example of this compound was presented in Figure 8.2.

The investigation into the behavior of the heavy metals has shown that during the co-processing of crude oil residue and plastics more heavy metals are transferred from light fractions to coke and asphaltenes than in the case of thermal treatment of pure crude oil residue. This shows that the fuel fractions from the co-processing are ecologically better than visbreaking fuels. An example of these results is shown in Figure 9.20.

9.3.5 Conclusions of Co-Processing of Crude Oil Residue and Co-Feed

1. Co-processing is marked by reactions between asphaltenes and free radicals, resulting from the decomposition of plastics. For very deep asphaltene cracking, the temperature range for formation of free radicals and those for asphaltene cracking have to be similar.
2. Under optimum conditions, it is possible to achieve deep cracking without coke formation.
3. It is possible to execute step-wise co-processing. In that way, it is possible to achieve a high yield of the desired products without coke formation.
4. Heavy metals are transferred into the heavy fractions or coke during co-processing. It leads to the better ecological properties of co-processing products in comparison to visbreaking products
5. It has been shown that co-processing with different organic co-feeds (for example used fat) is possible.

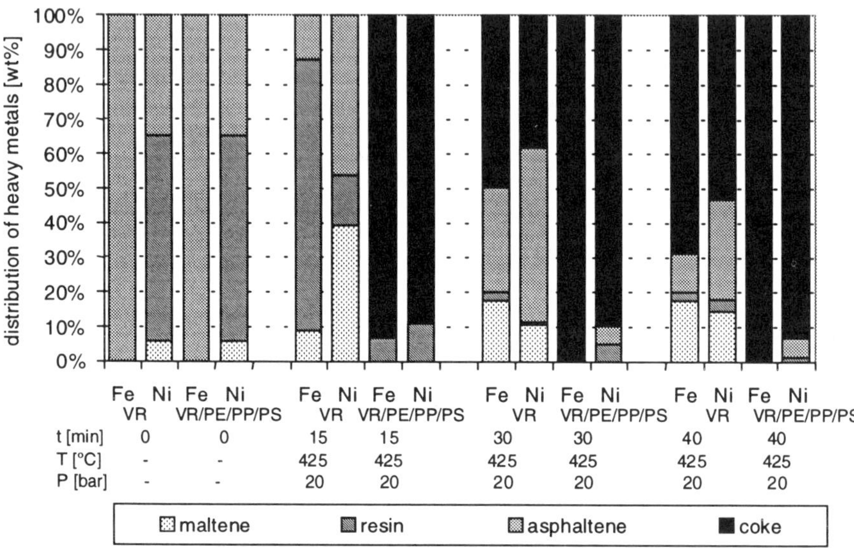

Fig. 9.20: Graphical presentation of heavy metals transfer between vacuum residue fractions during co-processing

9.4 INDUSTRIAL METHODS OF CRUDE OIL RESIDUE TREATMENT

This section deals with current industrial methods of residue treatment.

9.4.1 Fluid Catalytic Cracking

The main task of fluid catalytic cracking is the conversion of a wide range of both virgin and cracked hydrocarbon residues into lower molecular weight and more valuable products.

The main advantages of this process are:

- Selective conversion of different feedstocks to a variety of high quality products
- Production of high pressure steam from waste heat
- Cracking under low pressure conditions
- No hydrogen feed needed

Figure 9.21 shows an example of a fluid catalytic cracking unit. In its simple form, the Fluid catalytic cracking unit consists of three sections:

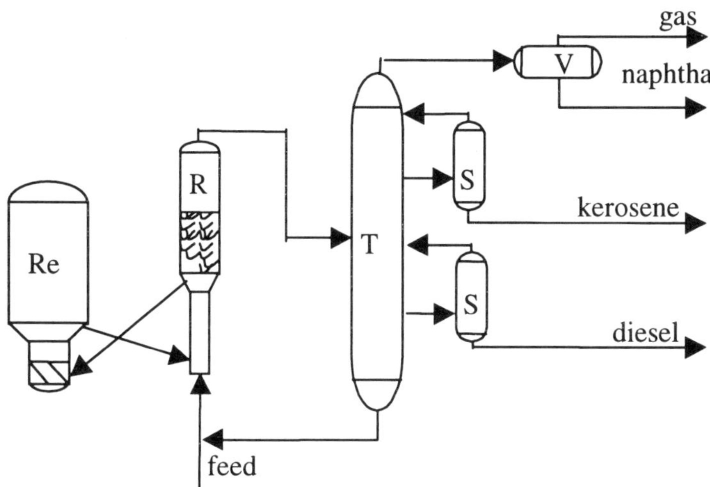

Fig. 9.21: Fluid Catalytic Cracking unit:
Re – regenerator
R – reactor
T – rectification tower
S – stripper
V – separator

- Reactor section
- Fractionation
- Gas concentration

The preheated raw oil feed meets a stream of hot catalyst from the regenerator at the base or lower part of the reactor riser. Heat from the catalyst vaporizes the oil, and the catalyst and oil travel up to the reactor. The cracked hydrocarbons, separated from the catalyst in cyclones, leave the reactor overhead and go to the fractionation column.

The spent catalyst falls down into the stripping section within the reactor. Steam removes most of the hydrocarbon vapor and the catalyst then flows down a standpipe to the regenerator.

The spent catalyst mixes with air and clean catalyst at the base of the regenerator. Here the coke deposited during cracking is burned off to reactivate the catalyst and provide heat for the endothermic cracking reactions. The recirculating loop of clean catalyst provides added heat for initiation of the carbon burn. The catalyst and air flow up the regenerator riser and separate at a T-shaped head. The flue gas is further cleaned of catalyst in cyclones at the top of the regenerator.

Waste heat from the flue gas is used for generation of high-pressure steam in the waste heat boiler. The regenerated catalyst flows down a standpipe to meet the raw oil charge at the bottom or lower part of the reactor riser, and the circulation continues.

The main column is the first step in the separation and recovery of cracked hydrocarbon vapors from the reactor. Heavy naphtha and middle distillates are withdrawn as side cuts, the bottom product is used as fuel oil or recycle stream.

In the gas concentration section the overhead products from the main column are further separated into fuel gas, butane–propane fraction and gasoline.

Typical fluid catalytic cracking processes are operated at a reaction temperature of approximately 500°C and under a pressure of approximately 3 bar in the reaction zone.

9.4.2 Hydrocracking

The main task of hydrocracking is the conversion and desulfurization of a wide variety of hydrocarbon feedstock into more valuable products with lower molecular weight. During hydrocracking, the following typical products are obtained: butane–propane fraction, light olefins, light naphtha, naphtha, jet fuel, diesel, lube oil base stocks, feed for fluid catalytic cracking.

The main advantages of this process are:

- Selective conversion of different feedstocks to a variety of high quality products
- Simultaneous removal of sulfur and nitrogen compounds
- Produced middle distillates are valuable blending compounds for the diesel oil pool

Figure 9.22 shows an example of a hydrocracking unit.

In a single stage hydrocracker, untreated feed and hydrogen-rich gas is charged to the first reactor, where hydrotreating reactions convert sulfur, nitrogen and oxygen compounds. In advance, recycle gas and feed oil are preheated separately. The effluent from the hydrotreating reactor passes directly to the hydrocracking reactor, where 40 to 80% of the reactor charge is converted into products. Reactor temperature control in both reactors is done using quenching by injecting cold recycled gas between the catalyst beds.

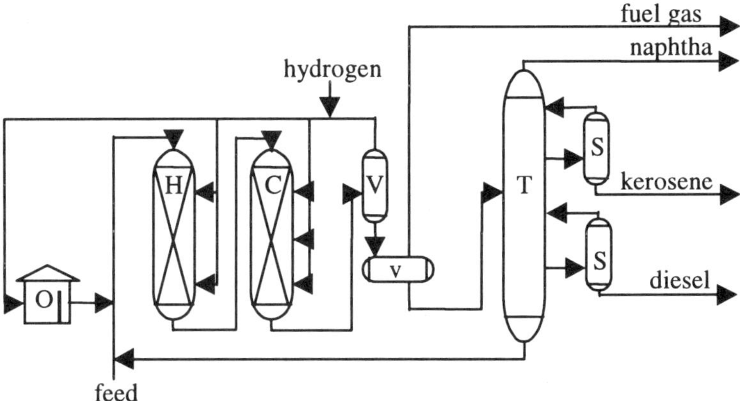

Fig. 9.22: Hydrocracking unit:
O – oven
H – hydrotreating reactor
C – cracking reactor
V – high pressure separator
v – low pressure separator
T – rectification tower
S – stripper

The effluent from the cracking reactor is cooled and condensed. The liquid products are separated from the recycled gas in the high-pressure separator. Recycled gas is returned to the reactors while the liquid product is sent to the low-pressure separator. Here, most of the dissolved gases are removed before product fractionation in the fractionation section. Unconverted oil is usually recycled to the reactors and cracked completely to lower-boiling products. However, if desired, the recycled oil may be withdrawn (one-through operation) to produce feedstock for FCC or ethylene plants.

For large plants, a second stage is added to the basic single-stage hydrocracker. This system permits the unconverted, first-stage oil to be processed in a separate reactor. This configuration can also be used to achieve different product yields and product qualities. Generally, about 40-70% of the total conversion takes place in the second stage. Hydrocracking proceeds in the temperature range 370–430°C and under a pressure of 160–180 bar.

9.4.3 Coking

Coking is used for the conversion of crude oil vacuum residues and cracking residues into coke and a clean liquid product with a high H/C ratio. Typical products of coking are: hydrocarbon gas, naphtha, gas oil, feed oil for downstream processing and coke.

The main advantages of the coking process are:

- High temperature flash distillation of residue oil and simultaneous coking of non-distillable asphaltenes in a unique mixing reactor
- High yields of oil product and lower yields of gas and coke compared to conventional cokers or solvent deasphalters
- Rejection of pollutants like heavy metals, Conradson Carbon Residue, sulfur and nitrogen
- Production of a demetalized product oil with 30-40% vacuum residue content or with 0-5% vacuum residue content
- Autothermal heat supply

Figure 9.23 shows the scheme of the coking reactor.

The coker is based on rapid heat transfer from a fine-grained heat carrier to the feedstock. The heat carrier is produced as coke in the process itself. The main part of the coker is the blender. Here, the solid or liquid feedstock is thoroughly mixed with the hot heat carrier and flash distilled at approximately 500-600°C. Mixing is achieved almost instantaneously by two rotating screws within the reactor which engage and thereby clean each other and the reactor walls. It is a major characteristic of the blender that highly effective radial mixing takes place with very little axial back mixing. The blender thus operates like a plug flow reactor in which all particles have nearly equal residence times. The ultra short vapor residence time in the blender is shorter by an order of magnitude compared to the coke residence time and minimizes further secondary cracking of the vaporized product oil while most of the undesired components remain in the non-distillable asphaltenes and are rejected to the coke.

The coke product is deposited like onionskin on the circulating heat carrier. The gaseous and vapor coker products are withdrawn from the blender, quenched and then condensed in a fractionator. Nearly 100% residue conversion can be achieved if the bottom condensate fraction is recycled to the blender at a suitable process temperature. The heat carrier coke flows from the blender into the surge bin, from where it is fed at a controlled rate through a seal leg to the lift pipe. In the lift pipe the heat carrier is pneumatically elevated with hot air into the collecting bin. The air rate is appropriately adjusted to ensure that the circulating coke is reheated to the required temperature by partial combustion of some coke.

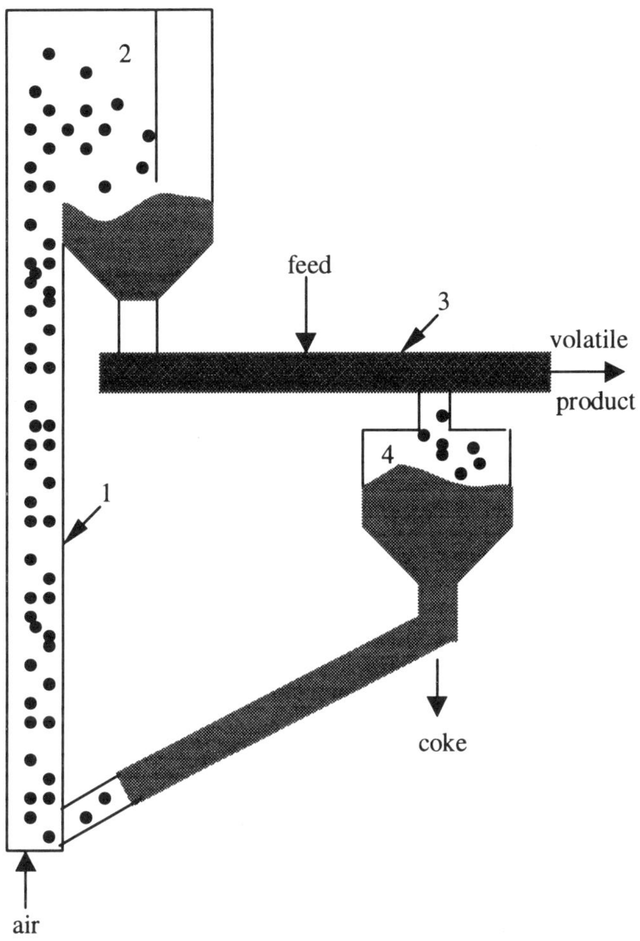

Fig. 9.23: Coker:
1 – lift
2 – collector
3 – blender
4 – reactor

A gas diverter in the collecting bin separates the heat carrier coke from the off gas. The heat carrier collected is recycled at a controlled rate via another seal leg back to the blender.

The seal legs are pipes that are continuously kept filled with fine-grained coke to act as a seal between the coking reactor, i.e. the blender and the combustion reactor.

Coking process proceeds at a temperature ranging from 500 to 600°C.

Bibliography

1 P Seidel. Schweres Erdöl - ein alternativer Rohstoff zur Erzeugung von Treibstoffen. Expert Verlag, Renningen-Malmsheim, 1994

2 C Jentsch. Erdölverarbeitung, in Ullmans Encyclopedie der technischen Chemie. 4. Auflage, Bd. 10, pp. 641-714. Weinheim: Verlag Chemie, 1975

3 H Laux, I Rahimiam. Physikalisch-chemische Grundlagen der Verarbeitung von Erdölrückständen und schweren Erdölen. Hochschulmonographie, TU Claustal, 1994

4 Internet publication http://www.bycosin.se/bcs_asp.htm, 1999

5 I Pfeiffer, R Saal. J. Phys. Chem., 44, 139,1940

6 D Espinat, F du Petrole, J C Ravey. SPE 25187, 1993

7 I A Wiehe. Ind. Eng. Chem. Res., 32, 2447, 1993

8 I A Wiehe. ACS Div. Pet. Chem. Meeting, San Francisco, 1997

9 W N Erih, M G Rasina, M G Rudin. Chimiya i Technologija Nefti i Gaza. Chimiya, Leningrad, 1977

10 J D Elliot. Delayed Coker Design and Operation: Recent Trends and Innovations. Foster Wheeler USA Corporation, 1996

11 J G Speight. The Chemistry and Technology of Petroleum. Marcel Dekker, New York, 1980

12 O V Rogacheva, R N Rimaev, V Z Gubaidullin, D K Khakimov. Investigation of the Surface Activity of the Asphaltenes of Petroleum Residues. Colloid J. USSR, 490, 1980

13 J B Butt, E E Petersen. Activation, Deactivation, and Poisoning of Catalysts. Academic Press Inc., San Diego, 1988

14 E E Petersen. The Fouling of Catalysts: Experimental Observations and Modeling in Catalyst Deactivation. edited by E E Petersen and J B Butt, Marcel Dekker, Inc., New York, 1987

15 S M Davis, F Zaera, G A Somorjai. The reactivity and composition of strongly adsorbed carbonaceous deposits on platinum. Model of the working hydrocarbon conversion catalyst, J. Catal., 77(2), 439-59, 1982

16 J Barbier. Coking of Reforming Catalysts. Stud. Surf. Sci. Catal., 34 (Catal. Deact.), 1-19, 1987

17 D Duprez, M Hadj-Aissa, J Barbier. Effect of steam on the coking of platinum catalysts: I. Inhibiting effect of steam at low partial pressure for the dehydrogenation of cyclopentane and the coking reaction. Appl. Catal., 49, 67-74, 1989

18 D Duprez, M Hadj-Aissa, J Barbier. Effect of steam on the coking and on the regeneration of metal catalysts: A comparative study of alumina-supported platinum, rhenium, iridium and rhodium catalysts. Stud. Surf. Sci. Catal., 68 (Catal. Deact. 1991), 111-118, 1991

19 M C Demicheli, D Duprez, J Barbier, O A Ferretti, E N Ponzi. Deactivation of steam-reforming model catalysts by coke formation: II. Promotion with potassium and effect of water, J. Catal., 145, 437-449, 1994

20 T F Yen. Chemistry of Asphaltenes, ASC Adv. Chem. Ser. No. 195 1981, S. 39

21 E Y Sheu. Physics of asphaltene micelles and microemulsions - theory and experiment. J. Phys.: Condens. Matter 8, A125-A141, 1996

22 Y V Pokonova, A A Geile, V G Spirkin, J B Chertkov, R Z Fahrutdinov, R Z Safieva, V V Tahistov, I Y Batueva. Chimiya Nefti. Chimiya, Leningrad, 1984

23 H J Neumann, I Rahimian, G Zenke. Einfluß der Löslichkeitseigenschaften von Asphaltenen auf die Rückstandsverarbeitung. Erdoel & Kohle, Erdgas, Petrochemie, 42 (1989) 7/8, pp. 278-286

24 J W Pokonova. Chimiya wysokomolekularnich soedinenii nefti. Leningradskowo univerziteta, Leningrad, 1980

25 A F Orlicek, H Poell, H Walenda. Hilfsbuch für Mineralöltechniker. Springer, Wien, 1955

26 D A Loose, L V Tsyro, M V Berezovskaya, L N Andreeva, F G Unger. Treatment of Oils Using Small Amounts of Atomic (Dissociated) Hydrogen. Chemistry and Technology of Fuels and Oils, 486 (1997) 6, pp. 24-26, in Russian

27 R Shaw, D Golden, S Benson. Phys. Chem. 1977, 81, 1716

28 A W Coats, J P Redfern. Kinetic Parameters from Thermogravimetric Data. Nature, 20 (1964), January 4, S. 68-69

29 I Pitault, D Nevicato, M Forissier, J R Bernard. Kinetic model based on a molecular description for catalytic cracking of vacuum gas oil. Chemical Engineering, 49 (1994) 24A, pp. 4249-4262

Appendix A:

Conversion Factors Important for Crude Oil Chemists

INTERNATIONAL SCALE UNITS

Everybody knows the problem arising when using the international scale units: one often finds formulas or product properties in which non-uniform scale units are used. In this short appendix the most important scale units for the petroleum specialist from around the world are collected.

Temperature:
[°C] = 0.566 · [F] – 17.778
[°C] = [K] – 273.15
[°C] = 0.566 · [Rankine] – 273.15
[°C] = 1.25 · [Reamur]

Pressure:
[atm] = 101.325 · [Pa]
[atm] = 1.013 · [bar]
[atm] = 1033.3 · [cm-H_2O]
[atm] = 76 · [cm-Hg]
[atm] = 406.796 · [inch-H_2O(39.2°F)]
[atm] = 407.189 · [inch-H_2O(60°F)]
[atm] = 29.921 · [inch-H_2O(39.2°F)]
[atm] = 30.006 · [inch-H_2O(60°F)]

Volume:
[m^3] = 10^3 · [liter]
[m^3] = 6.29 · [barrel(oil)]
[m^3] = 8.386 · [barrel(US liquid)]
[m^3] = 219.969 · [gallon(UK)]

$[m^3] = 227.021 \cdot$ [gallon(US dry)]
$[m^3] = 264.172 \cdot$ [gallon(US liquid)]

Energy:
$[J] = 2.39 \cdot$ [calorie]
$[J] = [Wh] / 3600$

Density:
$[kg/m^3] = 3.6 \cdot 10^{-5} \cdot [lbm/inch^3]$
$[kg/m^3] = 8.345 \cdot 10^{-3} \cdot$ [lbm/gallon]

Capacity:
[mill ton/year] $\approx 2.066 \cdot 10^{-3} \cdot$ [barrel/day]

Appendix B:

Glossary

In the following glossary, the most important terms used by crude oil chemists have been collected and given brief explanations.

^{13}C-NMR	Nuclear magnetic resonance. Analysis based on measurement of spin of ^{13}C atoms.
Additives	Chemical compounds added to crude oil products in very small amounts in order to improve their properties.
Alkane	Saturated organic compounds of general chemical formula C_nH_{2n+2}, often called paraffin.
Alkene	Unsaturated organic compounds of general chemical formula C_nH_{2n}, often called olefin.
Alkylation	Catalytic process to improve octane number of gasoline fractions. Based on alkylation reactions with formation of isoalkanes with high a octane number.
Aromatics	Organic compounds that include benzene in their structure.
Asphaltene	Heaviest crude oil fraction produced by solution analysis, insoluble in paraffinic solvents and soluble in aromatic solvents.
Atomic absorption spectroscopy	Elemental analysis method based on Kirchhoff law. According to this law, all atoms are able to absorb light quanta with the same wavelength which they are able to emit.
Atomic emission spectroscopy	Elemental analysis method based on Kirchhoff law. According to this law, all atoms are able to absorb light quanta with the

	same wavelength which they are able to emit.
Benzene	Cyclic unsaturated organic compound of general formula C_6H_6.
Bitumen	Crude oil product used in road and building construction. Also see vacuum residue.
Bromine number	Amount of bromine that reacts with sample in the determination of the number of unsaturated bonds.
Carbon residue	Carbonaceous residue formed after evaporation and pyrolysis of a petroleum product.
Catalyst	Chemical compound that accelerates the rate of a chemical reaction. However, the catalyst is not formed or consumed during the reaction.
Catalyst deactivation	Deactivation of catalyst by physically covering the active sites or by physical blocking of the pores in the catalyst by coke.
Catalyst poisoning	Deactivation of the catalyst by chemical reaction the active sites with compounds of the reaction mixture often with sulfur or heavy metals.
Catalytic cracking	Cracking of heavy crude oil fractions using catalysts such as alumino-silicates in order to achieve deep conversion of heavy hydrocarbons of crude oil.
Cetane number	Measure of the ability of diesel fuel to self-ignition under pressure.
Chromatography	Separating method based on differential ability of compounds in sample mixture to adsorb or absorb on an immobilized phase.
Coagulation analysis	Analysis based on different solubility of different compounds or crude oil fractions in different solvents.
Coker	Apparatus used for coking.
Coking	Conversation process used for the conversion of crude oil vacuum residues and cracking residues into coke and a clean liquid product with a high H/C ratio. Typical products of coking are: hydrocarbon gas, naphtha, gas oil, feed oil for downstream processing, and coke.

Colorimetry	Analysis method based on the ability of the substances to absorb a certain wavelengths of visible light.
Complexes formation analysis method	Analysis method based on the formation of complexes of crude oil compounds with other substances.
Conradson Carbon Residue	Carbonaceous residue formed after evaporation and pyrolysis of a petroleum product.
Co-processing	Process having as a feed a mixture of two or more compounds with different derivation.
Cracking	Process used for conversation of heavy hydrocarbons from heavy crude oil fractions into light crude oil fractions.
Crude oil	Often called petroleum. A complex mixture of hydrocarbons that occur in the Earth in liquid, gaseous, or solid forms.
Crude oil refining	conversion of crude oil into useful products as fuels, oils, plastics or chemicals.
Crude oil trap	Natural subsurface reservoir of crude oil. The oil is always accompanied by water and often by natural gas; all are confined in porous rock, usually sedimentary rocks such as sands, sandstones, arkoses, and fissured limestones.
Crystallization analysis	Analysis method based on the differential crystallization temperature of the compounds in the sample. This method is frequently used for the fractionation of lubricating oil into fractions by crude oil chemists.
Delayed coking	Thermal cracking process used in refineries to upgrade and convert crude oil residue into liquid and gas product streams leaving behind a solid concentrated carbon material, coke.
Density	Mass of sample occupying a certain volume.
Diesel	Mixture of flammable liquid hydrocarbons derived from crude oil and products of their treatment and used as fuel for internal-combustion engines. The boiling points of these heavier distillates range from 180 to 350°C.
Diesel engine	Internal-combustion engine in which air is compressed to a

	temperature sufficiently high to ignite fuel injected into the cylinder, where combustion and expansion actuate a piston. Invented by German thermal engineer Rudolf Diesel
Distillation	Process involving the conversion of a liquid into vapor that is subsequently condensed back to liquid form.
Enthalpy	The sum of the internal energies of all the compounds of a thermodynamic system at certain pressure and in a certain volume.
Entropy	The measure of a system's energy that is unavailable for work. Measure for chaos of a thermodynamic system.
Extraction	Separation method based on the different solubilities of the compounds of analyzing sample in solvents.
Fluoremetry	Analysis method often called fluorescence and phosphorescence. Based on the ability of some substances to emit light with a certain wavelength.
Fluorescence	See fluoremetry.
Free energy	See Gibbs energy.
Free enthalpy	See Gibbs energy.
Friction	Force that resists the sliding or rolling of one solid object over another.
Gas chromatography	Analysis method using the principles of chromatography for samples in the gaseous or vapor state.
Gasoline	Mixture of volatile, flammable liquid hydrocarbons derived from crude oil and products of their treatment and used as fuel for internal-combustion engines. It is also used as a solvent for oils and fats in the petrochemical industry. Boiling temperature of gasoline: boiling usually begins at 180°C.
Gasoline engine	Internal-combustion engines that generate power by burning a gasoline with ignition initiated by an electric spark.
Gibbs energy	Full energy of a thermodynamic system including enthalpy and entropy.

Goudron	See vacuum residue.
HPLC	High Pressure Liquid Chromatography. Analysis method for samples in liquid state. Used for Analysis of heavy crude oil fractions.
Hydrocracking	Type of catalytic cracking carried out under high pressure and in a hydrogen atmosphere.
Hydroprocessing	See hydrotreatment.
Hydrotreatment	Catalytic process used to saturate unsaturated compounds in crude oil products or to remove heteroatoms from crude oil fractions. Process carried out under a hydrogen atmosphere.
Infrared spectroscopy	Analysis method based on the ability of substances to absorb certain wavelengths of infrared light.
Iodine number	Amount of iodine that reacts with sample in the determination of the number of unsaturated bonds.
Isomerization	Catalytic process to improve the octane number of gasoline fractions. Based on isomerization reactions of n-alkanes.
Jet fuel	Often called kerosene. A mixture of flammable liquid hydrocarbons derived from crude oil and used as fuel for jet engines. Boiling temperature of jet fuel is usually 200°C – 250°C.
Kerosene	See Jet fuel.
Liquefied gas	Liquefied under pressure. Mixtures of gaseous hydrocarbons such as propene, propane, butene, and butane.
Liquid chromatography	Analysis method used for characterization of crude oil products based on the differential adsorption ability of the product components on an adsorbent.
Lubricant	Amorphous substance, stable colloidal mixture of lubricating oil with solid substances, used for lubrication.
Lubrication	Introduction of any of various substances between sliding surfaces to reduce wear and friction.
Lubrication oil	Oil derived from petroleum or having synthetic nature, used for

	lubrication.
Maltene	Crude oil fraction produced by solution analysis, soluble in paraffinic solvents.
Microcarbon residue	The carbonaceous type residue remaining after sample pyrolysis at 500°C under an inert (nitrogen) atmosphere in a controlled manner for a specific time. ASTM D4530.
Molecular distillation	Analysis method based on the principles of distillation under very low pressures. Usually the pressure for this analysis varies from 0.133 Pa up to 0.013 Pa.
Naphthene	Organic compounds including cyclic structures as, for example, cyclo-heptane.
Octane number	Measure of the ability of gasoline to resist knocking when ignited in a mixture with air in the cylinder of an internal-combustion engine. There are Research Octane Number (RON) and Motor Octane Number (MON). RON is usually 10 points higher than MON for the same gasoline.
Olefin	See Alkene.
OPEC	Organization of Petroleum Exporting Countries.
Paraffin	See Alkane.
Permeation chromatography	Analysis method used for the determination of molecular weight distribution of compounds in a sample mixture.
Petroleum	See Crude oil.
Phosphorescence	See Fluoremetry.
Photometry	Analysis method based on the ability of a substance to absorb certain wavelengths of electromagnetic radiation.
Platforming	Reforming using platinum containing catalyst.
Promoter	Chemical compound that improves the influence of a catalyst on a chemical reaction.
Raman spectros-	Analysis method based on the interactions of monochromatic

copy	infrared radiation with the shell of the atom.
Ramsbottom Carbon Residue	Carbonaceous residue formed after evaporation and pyrolysis of a petroleum product. ASTM D524.
Rectification	Stepwise distillation, used for fractionating crude oil into fractions. There are in existence rectification under atmosphere pressure and under vacuum. The first is used for producing fractions with boiling point up to 350°C. The second is used for producing fractions with boiling point up to 550°C.
Reforming	Catalytic process used to improve octane number of gasoline fractions. Based on cyclization and aromatization reactions.
Residual fuels	Fuel having boiling point over 350°C.
Sedimentation analysis	Analysis method that uses the sedimentation principle for determination of colloidal properties of sample.
Selectivity	Property of catalyst to accelerate only the desired reaction. Property of solvent to solve the compounds with certain properties.
Solution analysis	Analysis based on the differential solubility of different compounds or crude oil fractions in different solvents.
Thermal cracking	Thermal splitting process of heavy crude oil fraction under influence of high temperature. Proceeds according to the free radical chain mechanism.
Thermal diffusion	Analysis method based on separation of compounds in the sample depending on the differential diffusion velocity for different compounds when to the same temperature gradient.
Thin layer chromatography	Qualitative analysis method that uses the principles of chromatography for substances in the dissolved state.
Ultimate analysis	Determination of carbon, hydrogen and nitrogen content in the sample.
Vacuum residue	Bottom product of vacuum rectification. Often called bitumen or goudron.
Vapor pressure	Analysis method used for determination of average molecular

osmometry	weight of the sample.
Visbreaking	Thermal cracking process at moderate conditions used to reduce viscosity of residual fuels.
Viscosity	Measure for cohesion.
Wear	The removal of material from a solid surface as a result of mechanical action exerted by another solid.
X-Ray fluorescence spectroscopy	Analysis method based on the ability of some substances to emit light of a certain wavelength under the influence of X-rays.

Index

Asphalt, 3
Asphaltenes:
  change in structure during processing of, 360-367
  chemistry of, 327-328
    in crude oil, 327
    in crude oil residue, 327
  coke formation, reactions of, 327-331
  definition of, 9, 327, 342
  reactions of:
    during co-processing, 360-367
    during secondary treatment, 327-342

Biodiesel, transesterification, 55

Cetane number, 49-51
Chromatography:
  in crude oil chemistry, 73
  gas, 74-95
    anomalies in, 92-93
    carrier gases for, 93-94
    columns used for, 76-80
    detectors for, 80-90
    injectors for, 90-91
  liquid:
    adsorption, 100-101
    affinity, 100, 102
    columns used for, 99-102
    detectors in, 102-105
    distribution, 100-101
    gel permeation, 100-102
    high performance, 97-107
    ion-pair, 100, 102
  thin layer, 107-112
Crude oil:
  consequences of intensive extraction and processing of, 26-33
  extraction of, 23-26
  methods for characterization and analysis of, 73-163
    adsorptive, 146-147
    chemical analysis, 149-153
    for colloidal properties, 153-157
    evaporative, 146, 147 (*see also* Crude oil distillation and rectification)
    extractive, 147
    n-d-M method, 150-153
    thermal diffusion, 147-148
    for unsaturated compounds, 149-150
  physical properties of:
    density, 157-159
    refractive index, 161-163
    viscosity, 159-161
  products from, 39-57
  prospecting of, 20-23
Crude oil distillation:
  atmospheric rectification, 235-251
    packed towers in, 242-243
    rectification towers in, 236
    reflux, 238-239, 244-247
    tray towers for, 237-240
  heat exchangers in, 256-259
  petroleum and gas preparation for, 221-234
    desalting, 234-235
    petroleum drying, 227-230
    petroleum emulsions, 222-224
    separation of water–oil emulsions, 224-230
    stabilization of petroleum, 230-231
    technological scheme for, 231-234
  vacuum rectification, 251-256
Crude oil residue:
  ash content in, 342
  asphaltenes in,
  co-processing of:
    with aromatics containing plastics, 380-382
    behavior of heavy metals during, 387
    with cracked products, 377-380
    hydrogen transfer in, 375-377
    with paraffin containing plastics, 383-384.

with plastics, 360-367
possibilities with various co-feeds, 385-387
thermodynamics of, 368-375
and heavy crude oil:
asphaltenes from visbreaking, 365
native asphaltene, 327, 365
problems caused by asphaltenes in, 328-329
treatment of, 328
industrial treatment methods of:
coking, 392-394
fluid catalytic cracking, 388-390
hydrocracking, 390-391
methods of analysis of:
chromatography, 350-351
^{13}C-NMR, 351-352
coagulation, 346-349
coke forming tendency, 342-
distillation, 349-350
molecular weight determination, 354-358
solution, 343-346
ultimate, 353-354
treatment possibilities for:
catalytic, 337-340
chemical, 333-342
physical, 331-333
thermal, 333-337, 340-342

Diesel, 49-55
Drilling, 23

Environmental issues, 303-304
for crude oil:
cleaning contaminated soil, 303-306
cleaning contaminated water, 306-309
consequences of intensive extraction, 26-32
gas cleaning, 309-318
in the refining industry, 303-304

Fuels, international standards for, 58

Gasoline, 40-48
average composition of, 45
characteristics of, 43-44

Heavy crude oil (*see* Crude oil residue)
Hydrocarbons:
aromatics, 9
asphaltenene and resins, 9
naphthenes, 7
paraffins, 6

Jet fuel, 48-49

Lubricating oils and lubricants, 59-65
international standards for, 66-72
properties of, 65-66

Octane number, 40-42
Oil formation in world's oceans,16-18
OPEC, 173
background on, 174
members of, 174

Paraffins:
associated, 6-7
iso-, 6
light, 5
molecular, 6
normal-, 6
Petroleum:
bitumen, 3
black, 3
classification of:
chemical, 35-37
by density, 37-38
technological, 39
by viscosity-gravity constant, 38-39
composition of, 5
compounds in:
aromatics, 9
asphaltenes and resins, 9-10
metallic compounds or ash, 13
metals, 13
naphthenes, 7-9
nitrogen compounds, 10-12
oxygen compounds
paraffins, 6-7
sulfur compounds, 12-13
formation of:
modern concept of, 18-26
in world's oceans, 16-18
history and nature of, 3-13

international companies:
British petroleum, 197-200
Castrol, 200-201
ExxonMobil, 201-206
LUKoil, 214-215
Shell, 206-208
TNK, 216-217
Total/Fina/Elf, 208-214
YUKOS, 215-216
origin of, 13-16
producing countries, 173-195
Arabian East, 185-195
Canada, 178
OPEC, 173-175
Russia, 178-185
United States, 175-178
product fractions of, 33-35
white, 3

Petroleum treatment:
alkylation:
catalysts for, 298-300
feed for, 298
function of, 298
reactions in, 298-299
types of, 299-301
blending, 301-302
catalytic cracking:
catalysts for, 282
feed for, 275
functions of, 283
types of, 283-285
catalytic processes, 276-282
coke formation reactions:
catalytic, 337-340
thermal, 340-342
coking, 285-287, 335-337
delayed, 335-337
feed for, 336-337
fluid, 336-337
types of, 335-337
hydroprocessing:
catalyst deactivation in, 291
catalysts for, 289
feed for, 287, 289-291
function of, 287
of hydrocarbons, 287-289
of nitrogen compounds, 287-288
of oxygen compounds, 287-288
of sulfur compounds, 287-288
isomerization:
catalysts for, 296-297
feed for, 297
function of, 296
reactions in, 297
primary processes, 221-260
reforming:
catalysts for, 293
feed for, 294
function of, 292
reactions in, 293
types of, 294
secondary processes:
bimolecular reaction, 264
elementary reaction, 264
monomolecular reaction, 264
reaction kinetics of, 263-275
termolecular reaction, 264
thermodynamics of, 261-274
thermal cracking, 261-275, 333-337
visbreaking, 285, 334-335

Refinery processes, 40

Spectroscopy:
atomic absorption, 139-143
atomic emission, 143-144
colorimetry and photometry, 131-136
in crude oil analysis, 126, 130, 136, 139 143-144, 146
infrared, 112-126
fluorescence and phosphorescence, 136-139
NMR, 351-352
Raman, 127-131
X-ray fluorescence, 144-146